Donald L. Gunter
Gunter Physics, Inc.

Microfluid Mechanics

Nanoscience and Technology Series
Omar Manasreh, Series Editor

MICHAEL H. PETERS • *Molecular Thermodynamics and Transport*
KENNETH GILLEO • *MEMS / MOEM Packaging*
NICOLAE O. LOBONTIU • *Mechanical Design of Microresonators*
ROBERTO PAIELLA • *Intersubband Transitions in Quantum Structures*
JOSEPH H. KOO • *Polymer Nanocomposites*
JENS W. TOMM AND JUAN JIMENEZ • *Quantum-Well High-Power Laser Arrays*

Microfluid Mechanics

Principles and Modeling

William W. Liou

Yichuan Fang

Department of Mechanical and Aeronautical Engineering
Western Michigan University
Kalamazoo, Michigan

McGraw-Hill

New York Chicago San Francisco Lisbon London Madrid
Mexico City Milan New Delhi San Juan Seoul
Singapore Sydney Toronto

The McGraw-Hill Companies

Library of Congress Cataloging-in-Publication Data

Liou, William W.
Microfluid mechanics / William W. Liou, Yichuan Fang.
p. cm.
Includes index.
ISBN 0-07-144322-3
1. Fluid dynamics. I. Fang, Yichuan. II. Title.
TA357.L567 2005
620.1′06—dc22

2005048383

1 2 3 4 5 6 7 8 9 0 DOC/DOC 0 1 0 9 8 7 6 5

ISBN 0-07-144322-3

The sponsoring editor for this book was Kenneth P. McCombs and the production supervisors were Richard Ruzycka and Sherri Souffrance. It was set in Century Schoolbook by International Typesetting and Composition. The art director for the cover was Margaret Webster-Shapiro.

Printed and bound by RR Donnelley.

This book was printed on recycled, acid-free paper containing a minimum of 50% recycled, de-inked fiber.

Contents

Preface

This book is written as a textbook for an analytical-oriented microfluid flow course with graduate students who, preferably, have already taken an advanced fluid mechanics course. The derivations of the equations are presented, whenever possible, in a fairly detailed manner. The intent is that, even without the help of an instructor, students can self-navigate through the materials with confidence and come away with a successful learning experience. Similarly, practicing engineers who are interested in the subject should also be able to pick up the book and follow the flow of the contents without much difficulty. Some background in modern numerical computation tools will help the readers getting the full benefit of the two computer programs, NB2D and DSMC-IP, associated with the book. In fact, it is highly recommended that readers do make use of these Fortran programs.

The book begins with an introduction to the kinetic theory of gas and the Boltzmann equation to build the foundation to the later mathematical modeling approaches. With the dilute gas assumption, the nature of the micro gas flows allows the direct application of the Chapman-Enskog theory, which then brings in the modeling equations at the various orders of the Knudsen number in Chapter 4. The direct simulation Monte Carlo (DSMC) method and the information preservation (IP) method are described as the numerical tools to provide solutions to the Boltzmann equation when the Knudsen number is high. The later chapters cover the hybrid approaches and the important surface mechanisms. Some examples of micro gas flows at high and low speeds are shown. One interesting aspect of micro gas flows that is yet to see extensive examination in the literature is the characteristics of the flow disturbances at microscales. Chapter 11 provides a detailed description of some of our preliminary studies in this area.

Although a part of the content of the book has been used in a one-semester graduate level microfluid dynamics course at Western Michigan University, the book is best used as a textbook for a two-semester course. Chapters 1 to 4 provide the introductory content for the basic mathematical and the physical aspects of micro gas flows. The computer program NB2D may be used, for example, as project exercises. The second part would emphasize the DSMC and the IP solution methods, and their parallelization. The computer program DSMC-IP can be used for term project type of assignments. In the situation where the analytical microfluid course is preceded by another experiment-oriented course on microfabrication or microengineering, and the students have already had somewhat extensive knowledge of micro gas flows, the instructor may wish to concentrate on Chapters 2, 4, and 5 in the lectures and leave the rest as reading assignments. When the book is used in a 16-to-20-hour short-course setting, the instructor may wish to highlight the materials from Chapters 2, 4, 5, and 7. It might be a good idea to provide opportunities to run at least one of the two computer programs onsite. Chapters 1, 3, and 10 can be assigned as overnight reading materials.

The book does not contain extensive updates and details on the current engineering microfluidic devices. We feel that the book's focus is on the fundamental aspects of mirofluid flows and there is a myriad of readily available information on the technologies and the many different microfluidic device applications that have been cleverly designed and painstakingly manufactured by experts in the field. Also, since new devices are being brought to light almost daily, we feel that what is current at the time of this writing may become outdated within a few years.

A portion of the work the authors have accomplished at Western Michigan University has been performed under the support of NASA Langley Research Center. Near the completion of the manuscript, the second author moved to the Georgia Institute of Technology.

We appreciate the help of Dr. James Moss for reviewing a part of the manuscript. Thanks should go to those at McGraw Hill who worked on the book and to those who reviewed it. We should also thank Jin Su and Yang Yang for their work on the codes. The first author (WWL) would like to thank the unconditional support and patience of his wife, Shiou-Huey Lee, during this writing and the love from his children, Alex and Natalie.

William W. Liou
Kalamazoo, Michigan

Yichuan Fang
Atlanta, Georgia

Microfluid Mechanics

Chapter

1

Introduction

Microelectromechanical systems (MEMS) are considered one of the major advances of industrial technologies in the past decades. MEMS technology was derived initially from the integrated circuit (IC) fabrication technologies. Now, microfabrication is a diverse spectrum of processing techniques that involve a wide range of disciplines from chemical sciences to plastic molding. As the name suggests, MEMS covers micron-sized, electrically and/or mechanically driven devices. Compared with the conventional mechanical or electrical systems, these MEMS devices are five to six orders of magnitude smaller in size. In fact, these dimensions are in the same range as the average diameter of human hair (about 50 μm, 50×10^{-6}m, or 50 μm). An MEMS device can be a single piece of hardware that produces outputs directly based on the inputs from external sources. The outputs can be mechanical and fluidic movement, electrical charges, analog signals, and digital signals. Often several microcomponents are integrated, such as the lab-on-a-chip device, which performs the multistage processing of the inputs and produces several different types of outputs, all in one single miniature device. The small sizes of MEMS make them portable and implantable. The manufacturing cost of MEMS is far from prohibitive because of the wide use of the batch-processing technologies that grow out of the well-developed IC industry. MEMS, therefore, offer opportunities to many areas of application, such as biomedical and information technology, that were thought not achievable using conventional devices. Estimates of the potential commercial market size were as high as billions of U.S. dollars by 2010.

Since the early work of Tai et al. (1989) and Mehregany et al. (1990) on the surface-machined micromotors, there has been an explosive growth of the number and the types of potential application of MEMS.

Accompanying this growth is the significant increase of new journals that are dedicated to reporting advances in the field. The universal use of the Internet also helps disseminate new MEMS knowledge quickly and hasten its development effectively. There are also a number of titles written by distinguished researchers in the field. [a sample of them would include Gad-el-Hak (2002), Karniadakis and Beskok (2002), Koch et al. (2000), Nguyen and Wereley (2002), and Li (2004)] MEMS related technologies, ranging from electrokinetics and microfabrication to applications, can be readily found in these and many other forms of publications and media.

As more new applications are proposed and new MEMS devices designed, it was often found that measured quantities could not be interpreted by using conventional correlations developed for macro scale devices. Electric power needed to drive a micromotor was extraordinarily high. The properties of MEMS materials, such as Young's modulus, have been found to differ from that of the bulk material. For MEMS that use fluid as the working media, or a microfluidic device, for instance, the surface mechanisms are more important than mechanisms that scale with the volume. Overcoming surface stiction was found to be important in the early work on micromotors. The surface tension is perhaps among the most challenging issues in microfluidic devices that involve the use of liquid for transporting, sensing, and control purposes. The mass flow rate of microchannels of gas and liquid flows in simple straight microchannels and pipes were found to transition to turbulence at a much lower Reynolds number than their counterparts at the macro scales. Due to the miniature size, there are uncertainties in measuring the various properties of MEMS, such as specimen dimensions, with sufficient accuracy. Nevertheless, it has become increasingly apparent that the physical mechanisms at work in these small-scale devices are different from what can be extrapolated from what is known from experience with macroscaled devices. There is a need to either reexamine or replace the phenomenological modeling tools developed from observations of macro scale devices.

This book covers the fundamentals of microfluid flows. The somewhat limited scope, compared with other titles, allows a detailed examination of the physics of the microfluids from an *ab initio* point of view. Since the first principle theory is far less developed for liquids, the focus in this writing is the microfluid flows of gas. The Boltzmann equation will be introduced first as the mathematical model for micro gas flows. Analytical solutions of the Boltzmann equation can be found for a limited number of cases. The Chapman-Enskog theory assumes that the velocity distribution function of gas is a small perturbation of that in thermodynamic equilibrium. The velocity distribution function is expressed as a series expansion about the Knudsen number. The Chapman-Enskog theory is therefore adequate for micro gas flows where the Knudsen number and the departure from local equilibrium are small. It will be shown that the zeroth-order solution of the Chapman-Enskog theory leads to the Euler equations and the first-order solution results in the Navier-Stokes equations. The linear constitutive relations between the stress and strain, and that between the heat transfer and temperature gradient, which are used in the derivation of the Navier-Stokes equations from the continuum point of view of gas flow, are thus valid only for very small Knudsen number. The second-order solution of the Chapman-Enskog theory produces nonlinear closure models for the stresses and heat fluxes. The resulting equations were referred to as the Burnett equations. Two forms of the Burnett equations will be discussed. For micro gas flows of higher Knudsen number, the Boltzmann equation should be used to model the micro gas flow behavior.

The analytical solutions of these various mathematical model equations for micro gas flows can be found for simple geometry and for limited flow conditions, some of which are discussed in the appropriate chapters. For many of the complex design of microfluidic devices, the flow solutions can only be found by numerically solving the model equations. To this end, two computer programs are provided in the appendix section of the book. The programs are written using the standard FORTRAN language and can be compiled in any platforms. The NB2D code solves the Navier-Stokes equations as well as two forms of the Burnett equation. The all-speed numerical algorithm has been used in the numerical discretization. The density-based numerical method has been shown to be able to handle low-speed as well as high-speed flows, and is appropriate for gas flows commonly seen in microdevices. The DSMC/IP1D code uses the direct simulation Monte Carlo (DSMC) method and the information preservation (IP) method to provide simulations of the gas microflow at the large Knudsen number. The IP method has been shown to be exceptionally efficient in reducing the statistical scatter inherent in the particle-based DSMC-like methods when the flow speed is low. The two computer programs will provide the readers with numerical tools to study the basic properties of micro gas flows in a wide range of flow speeds and in a wide range of Knudsen number. Examples of low- and high-speed micro gas flow simulations are also presented in later chapters. One of the unsolved problems in conventional macro scale fluid dynamics is associated with the flow transition to turbulence. In Chap. 11, the behavior of the flow disturbances in two simulated micro gas flows is described.

The book is geared toward developing an appreciation of the basic physical properties of micro gas flows. The computer software, coupled with the necessary analytical background, enable the reader to develop a detailed understanding of the fundamentals of microfluidic flows and to further validate their findings using computer microflow simulations. The knowledge can then be used in either further studies of the microflows or in the practical design and control of microfludic devices.

References

Gad-el-Hak, M., *The MEMS Handbook*, Boca Raton, FL, CRC Press, 2002.

Karniadakis, G.E. and Beskok, A., *Microflows: Fundamentals and Simulation*, Springer, New York, 2002.

Koch, M., Evens, A., and Brunnshweiler, A., *Microfluidic Technology and Applications*, Research Studies Press, Hertfordshire, England, 2000.

Li, D., Electrokinetics in Microfluidics, *Interface Sciences and Technology*, Vol. 2, Elsevier Academic Press, London, UK, 2004.

Mehregany, M., Nagarkar, P., Senturia, S., and Lang, J., Operation of microfavricated harmonic and ordinary side-drive motor, *IEEE Micro Electro Mechanical System Workshop*, Napa Valley, CA, 1990.

Nguyen, N.-T. and Wereley, S., *Fundamentals and Applications of Microfluidic*, Artech House, Norwood, MA, 2002.

Tai, Y., Fan, L., and Muller, R., IC-Processed micro-motors: Design, technology and testing, *IEEE Micro Electro Mechanical System Workshop*, Salt lake City, UT, 1989.

Chapter

2

Basic Kinetic Theory

2.1 Molecular Model

In kinetic theory, the composition of a gas is considered at the microscopic level. A gas is assumed to be made up of small individual molecules that are constantly in a state of motion. The name molecule can mean a single-atom molecule or a molecule with more than one atom. In each atom, a nucleus is surrounded by orbiting electrons. The internal structure of the molecule may change during an interaction with other molecules, such as collision. Collisions with other molecules occur continuously as the molecules move freely in their state of motion. Collision will also happen when a surface is present in its path. The intermolecular collision causes the magnitude and the direction of the velocity of the molecule to change, often in a discontinuous manner. If there is no collective, or macroscopic, movement, the motion of the molecule is completely random. This freedom of movement is not shared by liquid or solid molecules.

A molecular model for gas would then describe the nature of the molecule, such as the mass, the size, the velocity, and the internal state of each molecule. A measure for the number of molecules per unit volume, or number density n, would also be a parameter. The model also describes the force field acting between the molecules. The force field is normally assumed to be spherically symmetric. This is physically reasonable in light of the random nature of the large number of collisions in most cases. The force field is then a function of the distance between the molecules. Figure 2.1.1 shows a typical form of the force field $F(r)$ between two molecules with distance r. At a large distance, the weak attractive force approaches zero. The attractive force increases as the distance decreases. In close range, the force reverses to become repulsive

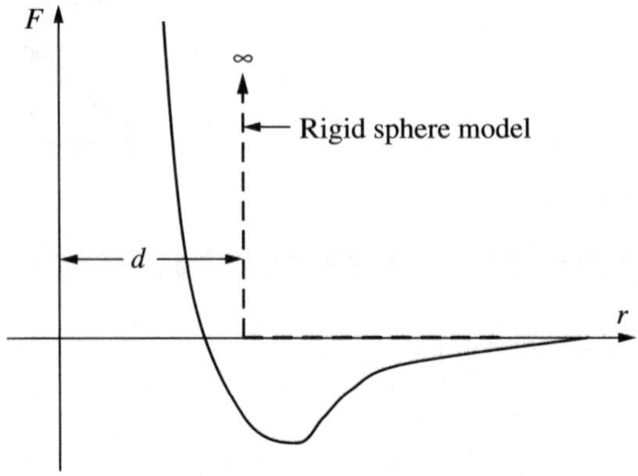

Figure 2.1.1 Sketch of spherically symmetric intermolecular force field.

as the orbiting electrons of the two molecules intermingle. Analyses with nonspherically symmetric force field are complicated. In fact, it is generally found that, the exact form of the force field is less important than other collision parameters. The force field of a simple rigid sphere model is shown in Fig. 2.1.1. The model assumes an infinitive repulsive force when the molecules are in contact and zero otherwise. The contact occurs when the distance between the centers of the molecules are the same as the assumed diameter of the sphere d. Use of the rigid sphere model can lead to accurate results if the diameter d is properly chosen according to some basic properties of the gas. The internal structure of the molecules affects the energy content of the gas. With the nuclei and the electrons in motion, the molecule can have, for instance, rotational and vibrational modes of energy in addition to the energy associated with the molecular translational motion.

These molecular quantities need to be related to macroscopic properties for analyses. This is especially true when there is a general macroscopic movement of the gas. As will be seen in the following section, a macroscopic property is merely the sample averaged value of the corresponding molecular quantity. The motion of the molecules is then not completely random when there is macroscopic motion.

2.2 Micro and Macroscopic Properties

In this section, we will use a simplified model to introduce the relations between the molecular behavior and the macroscopic properties of gases. We consider an equilibrium monatomic gas of single species

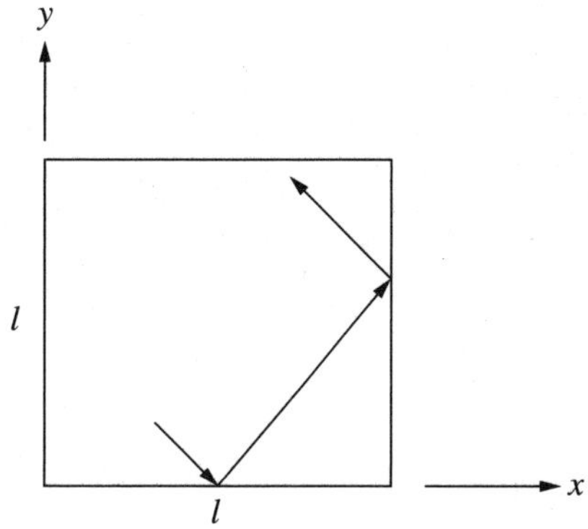

Figure 2.2.1 A planar projection of a molecular path.

inside a cubic box at rest of length l on each side (see Fig. 2.2.1). A gas in equilibrium would exhibit no gradients of macroscopic quantities in space or time. The average velocity of the molecule is therefore zero. Molecular motion is random with velocity vector $\mathbf{c}$. The molecular velocity components are c_x, c_y, and c_z in the Cartesian coordinate system (x, y, z). Again, molecules here can represent an atom, monatomic molecules, diatomic molecules, or polyatomic gas molecules. Assuming specular reflection at the wall, the x momentum change of a molecule during a collision with the right, vertical side of the box is $(2mc_x)$ where m denotes the molecular mass. If we assume that there are no intermolecular collisions, the rate of the momentum change, or the force F_x exerted on the wall by the molecule, is

$$F_x = (2mc_x)\left(\frac{1}{2l/c_x}\right) = \frac{mc_x^2}{l}$$

For a total number N molecules in the box, the total force becomes

$$F_x = \frac{\sum_N mc_x^2}{l}$$

The pressure p_x on the wall then is

$$p_x = \frac{F_x}{A} = \frac{F_x}{l^2} = \frac{\sum_N mc_x^2}{l^3} = \frac{\sum_N mc_x^2}{V}$$

Similarly

$$p_y = \frac{\sum_N mc_y^2}{V} \qquad \text{and} \qquad p_z = \frac{\sum_N mc_z^2}{V}$$

Therefore, pressure p becomes

$$\begin{aligned} p &= \frac{1}{3}(p_x + p_y + p_z) \\ &= \frac{\sum_N m|c|^2}{3V} \\ &= \frac{2E_{\text{tr}}}{3V} \end{aligned} \tag{2.2.1}$$

where E_{tr} is the total energy of translation of the molecules

$$E_{\text{tr}} = \frac{\sum_N m|c|^2}{2}$$

By using the total system mass $M = \sum_N m = Nm$, we get

$$\begin{aligned} p &= \frac{2E_{\text{tr}}/M}{3V/M} \\ &= \frac{2}{3}\rho e_{\text{tr}} \end{aligned} \tag{2.2.2}$$

where ρ is the density and e_{tr} the molecular translational kinetic energy per unit mass, or the specific molecular translational kinetic energy. Therefore, from the kinetic theory point of view, the pressure is proportional to the gas density and the specific translational kinetic energy. The empirical equation of state for a thermally perfect gas can provide pressure from the thermodynamics consideration. That is,

$$p = \rho RT$$

where R represents the gas constant. The two different expressions for pressure will give the same quantity if

$$T = \frac{2e_{\text{tr}}}{3R} \tag{2.2.3}$$

This equation relates the temperature defined in the classical thermodynamics to the specific kinetic energy of molecular translational motion in kinetic theory. Therefore, temperature, a macroscopic gas property, can then be used as a measure of the specific molecular energy.

We can also relate temperature and pressure to the average kinetic energy per molecule. For instance, Eq. (2.2.2) gives

$$\begin{aligned} p &= \frac{2}{3}\rho e_{\text{tr}} \\ &= \frac{2}{3}\frac{\rho}{M}(Me_{\text{tr}}) \\ &= \frac{2}{3}\frac{\rho}{M}(N\hat{e}_{\text{tr}}) \\ &= \frac{2}{3}\frac{N}{V}\hat{e}_{\text{tr}} \\ &= \frac{2}{3}n\hat{e}_{\text{tr}} \end{aligned} \tag{2.2.4}$$

where

$$\hat{e}_{\text{tr}} = \frac{1}{2N}\sum_{N}|\mathbf{c}|^2$$

denotes the average kinetic energy per molecule. Similarly, we can find that

$$T = \frac{2\hat{e}_{\text{tr}}}{3k} \tag{2.2.5}$$

where k is the Boltzmann constant. It is the ratio of the universal gas constant $\hat{R}$ to the Avogadro's number $\hat{N}$. As this temperature includes only the translational kinetic energy, it is sometime referred to as the translational kinetic temperature T_{tr}. For monatomic gases, the molecule can be assumed to possess translational kinetic energy only and the translational kinetic temperature may simply be referred to as the temperature. For diatomic and polyatomic molecules, the rotational and the vibrational modes can also be associated with temperature. A general principle of equipartition of energy states that for every part of the molecular energy that can be expressed as the sum of square terms, an average energy of $kT/2$ per molecule is contributed by each such term. For instance, in the kinetic energy of translation, there are three such terms: c_x^2, c_y^2, and c_z^2. Therefore, Eq. (2.2.5) shows that

$$\hat{e}_{\text{tr}} = 3\left(\frac{kT}{2}\right)$$

Or

$$e_{\text{tr}} = 3\left(\frac{RT}{2}\right)$$

In general, for a gas with ξ number of square terms, the average kinetic energy is then

$$e = \xi\left(\frac{RT}{2}\right)$$

The number of terms that can be expressed as quadratic in some appropriate variables depends on the internal structure of the molecule. For a diatomic gas, such as air, there are two additional rotational energy that can be expressed as square terms, therefore $\xi = 5$ and

$$e = 5\left(\frac{RT}{2}\right)$$

The specific heat at constant volume c_v is $(5/2)R$ and

$$\begin{aligned} c_p &= c_v + R \\ &= (7/2)R \end{aligned}$$

The ratio of the specific heats

$$\gamma = \frac{c_p}{c_v} = \frac{7}{5}$$

which is the value normally assumed for diatomic gases. Therefore, even though the assumptions that the molecular collision can be ignored and the molecular interaction with the wall being specular have greatly simplified the gas behavior, the results are still valid. More rigorous derivation of these results can be found in Bird (1994) and Gombosi (1994). To estimate the molecular speed by using the macroscopic properties, divide both sides of Eq. (2.2.1) by the total mass M, then

$$\begin{aligned} \frac{p}{\rho} &= \frac{1}{3}\frac{\sum_N m|\mathbf{c}|^2}{M} \\ &= \frac{\overline{|\mathbf{c}|^2}}{3} \end{aligned} \tag{2.2.6}$$

where $\overline{|\mathbf{c}|^2}$ represents the mean-square molecular speed.

Note that we have defined a sample average $\bar{q}$ as an average of a molecular quantity q over all the molecules in the sample. That is

$$\bar{q} = \frac{1}{N}\sum_N q \tag{2.2.7}$$

For example, the average molecular velocity is

$$\bar{\mathbf{c}} = \frac{1}{N}\sum_{N}\mathbf{c} \tag{2.2.8}$$

This is a macroscopic or stream velocity. The difference between the molecular velocity and the stream velocity is called thermal or random velocity

$$\mathbf{c}' = \mathbf{c} - \bar{\mathbf{c}} \tag{2.2.9}$$

Note that

$$\overline{\mathbf{c}'} = \bar{\mathbf{c}} - \bar{\mathbf{c}} = 0 \tag{2.2.10}$$

That is, the average thermal velocity is zero.

2.3 Binary Collisions

Molecular collisions are the most fundamental process in gases. Molecular collision frequently occurs. There can be any number of molecules involved in a collision. In Sec. 2.3.1 the kinematics and the dynamics of collision that involves only two molecules, or binary collision, will be introduced. Binary collision is most relevant to most microflow analyses with rarefied gas effects.

2.3.1 Kinematics

We will assume that the size of the molecules is small and the molecules can be regarded as point centers of the intermolecular force. The force is along the direction connecting the centers of the molecules and depends only on the relative positions of the molecules. The force is also assumed conservative. Consider two molecules with masses m_1 and m_2, position vectors $\mathbf{r}_1$ and $\mathbf{r}_2$, in a fixed Cartesian coordinate system (see Fig. 2.3.1). The pre-collision velocities of the molecules are $\mathbf{c}_1$ and $\mathbf{c}_2$, respectively. The trajectories or orbits of the two particles are twisted curves in space.

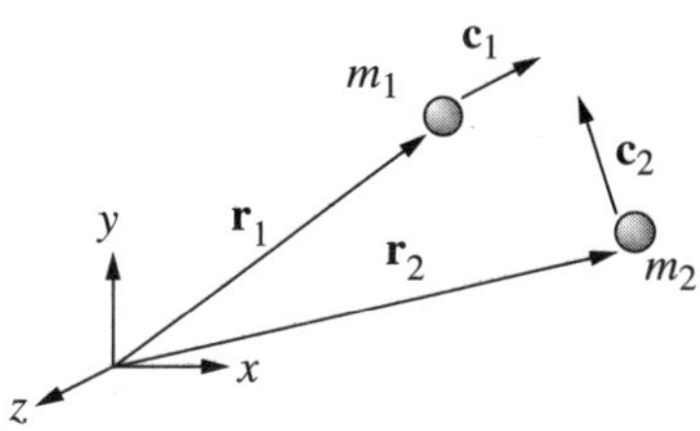

Figure 2.3.1 Binary collision.

The location of the center of mass is

$$\mathbf{r}_m = \frac{m_1}{m_1 + m_2}\mathbf{r}_1 + \frac{m_2}{m_1 + m_2}\mathbf{r}_2 \tag{2.3.1}$$

The velocity of the center of mass is

$$\mathbf{c}_m = \frac{m_1}{m_1 + m_2}\mathbf{c}_1 + \frac{m_2}{m_1 + m_2}\mathbf{c}_2 \tag{2.3.2}$$

In an elastic collision, the linear momentum is conserved. Therefore,

$$\begin{aligned} m_1\mathbf{c}_1 + m_2\mathbf{c}_2 &= m_1\mathbf{c}_1^* + m_2\mathbf{c}_2^* \\ &= (m_1 + m_2)\mathbf{c}_m \end{aligned} \tag{2.3.3}$$

where asterisk (*) denotes post collision quantities. Equation (2.3.3) shows that the velocity of the center of mass does not change in the collision. The conservation of energy gives

$$m_1c_1^2 + m_2c_2^2 = m_1c_1^{*2} + m_2c_2^{*2} \tag{2.3.4}$$

The relative location between the two molecules $\mathbf{r}$ is defined as

$$\mathbf{r} = \mathbf{r}_1 - \mathbf{r}_2$$

The pre- and the post-collision velocities of m_1 relative to those of m_2 are

$$\mathbf{c}_r = \mathbf{c}_1 - \mathbf{c}_2 \qquad \text{and} \qquad \mathbf{c}_r^* = \mathbf{c}_1^* - \mathbf{c}_2^* \tag{2.3.5}$$

From Eqs. (2.3.2) and (2.3.5), one can write the pre-collision velocities in terms of the center of mass velocity and the relative velocity

$$\mathbf{c}_1 = \mathbf{c}_m + \frac{m_2}{m_1 + m_2}\mathbf{c}_r \qquad \text{and} \qquad \mathbf{c}_2 = \mathbf{c}_m - \frac{m_1}{m_1 + m_2}\mathbf{c}_r \tag{2.3.6}$$

Equation (2.3.6) shows that $\mathbf{c}_1 - \mathbf{c}_m // \mathbf{c}_2 - \mathbf{c}_m$. That is, in the center-of-mass reference of frame, the approach velocities of the collision partner are parallel to each other. For a spherically symmetric repulsive intermolecular force $\boldsymbol{F}$

$$\begin{aligned} \boldsymbol{F} &= F\,\mathbf{e}_r \\ &= -\frac{dU(r)}{dr}\mathbf{e}_r \end{aligned} \tag{2.3.7}$$

$U(r)$ is the potential of the intermolecular force and $\mathbf{e}_r$ the unit vector along $\mathbf{r}$. The conservative force field is a function of the distance between the collision partner and is on the same plane formed by $\mathbf{c}_1 - \mathbf{c}_m$ and

$\mathbf{c}_2 - \mathbf{c}_m$. For the collision molecules, the force is equal in magnitude and opposite in direction. That is,

$$\begin{aligned} \boldsymbol{F} &= m_1 \frac{d\mathbf{c}_1}{dt} \\ &= -m_2 \frac{d\mathbf{c}_2}{dt} \end{aligned} \tag{2.3.8}$$

Or

$$m_1 \frac{d\mathbf{c}_1}{dt} + m_2 \frac{d\mathbf{c}_2}{dt} = 0 = \frac{d\mathbf{c}_m}{dt} \tag{2.3.9}$$

Again, Eq. (2.3.9) shows that the velocity of the center of mass does not change during the interaction. We can also find from Eq. (2.3.8) that

$$m_1 m_2 \frac{d\mathbf{c}_1}{dt} - m_1 m_2 \frac{d\mathbf{c}_2}{dt} = (m_1 + m_2)\mathbf{F}$$

Or

$$\begin{aligned} \mathbf{F} &= \frac{m_1 m_2}{m_1 + m_2} \frac{d\mathbf{c}_r}{dt} \\ &= m_r \frac{d\mathbf{c}_r}{dt} \end{aligned} \tag{2.3.10}$$

Therefore, the motion of a particle of mass m_1 with respect to that of mass m_2 is equivalent to that of a particle of reduced mass m_r in a central field of conservative force. The postcollision velocities can also be found from Eqs. (2.3.2) and (2.3.5)

$$\mathbf{c}_1^* = \mathbf{c}_m + \frac{m_2}{m_1 + m_2}\mathbf{c}_r^* \qquad \text{and} \qquad \mathbf{c}_2^* = \mathbf{c}_m - \frac{m_1}{m_1 + m_2}\mathbf{c}_r^* \tag{2.3.11}$$

Equation (2.3.11) shows that $\mathbf{c}_1^* - \mathbf{c}_m // \mathbf{c}_2^* - \mathbf{c}_m$. That is, in the center-of-mass reference of frame, the post-collision velocities of the collision partner are parallel to each other. In fact, since the force field is assumed spherically symmetric, there is no azimuthal acceleration during the interaction and the angular momentum $\mathbf{L}$ is conserved. That is,

$$\begin{aligned} \frac{d\mathbf{L}}{dt} &= \frac{d}{dt}(m_r \mathbf{c}_r \times \mathbf{r}) \\ &= 0 \end{aligned} \tag{2.3.12}$$

Therefore, $\mathbf{c}_r$ and $\mathbf{c}_r^*$ lie on the same plane, commonly referred to as the collision plane. In this accelerating frame of reference attached to m_2, the two-body problem is reduced to a one-body problem with a plane trajectory. Moreover, it can be shown that $|\mathbf{c}_r| = |\mathbf{c}_r^*| = c_r = c_r^*$. First, summing the inner product of Eq. (2.3.3) with $\mathbf{c}_1$ and Eq. (2.3.3) with $\mathbf{c}_2$, we get

$$m_1c_1^2 + m_2c_2^2 = (m_1 + m_2)[\mathbf{c}_m \bullet (\mathbf{c}_1 + \mathbf{c}_2)] - (m_1 + m_2)\mathbf{c}_1 \bullet \mathbf{c}_2$$

By using Eq. (2.3.6) to write $\mathbf{c}_1$ and $\mathbf{c}_2$ in terms of $\mathbf{c}_m$ and $\mathbf{c}_r$, the equation can be written as

$$m_1c_1^2 + m_2c_2^2 = (m_1 + m_2)c_m^2 + \frac{m_1m_2}{m_1 + m_2}c_r^2 \tag{2.3.13}$$

Similarly, we may also obtain

$$m_1c_1^{*2} + m_2c_2^{*2} = (m_1 + m_2)c_m^2 + \frac{m_1m_2}{m_1 + m_2}c_r^{*2} \tag{2.3.14}$$

Since the translation kinetic energy is conserved during the binary collision, Eqs. (2.3.13) and (2.3.14) show that

$$c_r = c_r^* \tag{2.3.15}$$

The velocities $\mathbf{c}_m$ and $\mathbf{c}_r$ can be determined from the pre-collision velocities $\mathbf{c}_1$ and $\mathbf{c}_2$, therefore, the post-collision velocities $\mathbf{c}_1^*$ and $\mathbf{c}_2^*$ can be calculated if the angle between $\mathbf{c}_r$ and $\mathbf{c}_r^*$ is known. This is called the angle of deflection χ. The angle of deflection measures the change of direction of the relative velocity vectors due to a collision and can be obtained by examining the dynamics of the binary collision.

2.3.2 Dynamics and postcollision properties

In the polar coordinate (r, ϕ) with its origin at $\mathbf{r}_2$ on the collision plane, as shown in Fig. 2.3.2, Eq. (2.3.12) becomes

$$m_r r^2 \frac{d\phi}{dt} = m_r c_r b \tag{2.3.16}$$

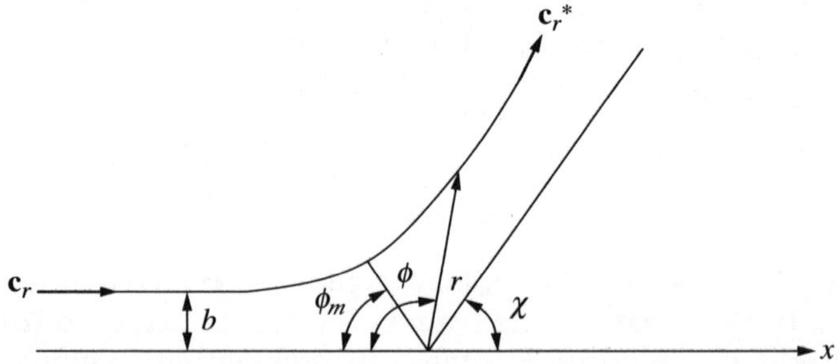

Figure 2.3.2 Collision plane.

where b is the smallest distance between the trajectories of the molecules before the collision and is called the impact parameter. Since the intermolecular force is a short-range force that vanishes at large intermolecular distances, the conservation of the total energy, including the kinetic and the potential energy, becomes

$$\frac{1}{2}m_r\left(\frac{dr}{dt}\right)^2+\frac{m_r c_r^2 b^2}{2r^2}+U(r)=\frac{1}{2}m_r c_r^2 \tag{2.3.17}$$

The trajectory $r(\phi)$ can be deduced by eliminating the time variables from Eqs. (2.3.16) and (2.3.17). That is

$$\frac{dr}{d\phi}=\pm\frac{r}{b}\sqrt{r^2-b^2-\frac{2r^2U(r)}{m_r c_r^2}} \tag{2.3.18}$$

The trajectory is a function of the collision parameter, the ratio of the potential and the kinetic energies. For each r, there are two ϕ values, corresponding to the approach and the departure that are symmetric with respect to (r_m, ϕ_m) described by the zero of the square term. That is

$$r_m^2-b^2=\frac{2r_m^2U(r_m)}{m_r c_r^2}$$

r_m and ϕ_m are referred to as the distance and the angle of the closest approach, respectively. The symmetry in the approach and the departure trajectories also means that for a collision with precollision velocity of $\mathbf{c}_1^*$ and $\mathbf{c}_2^*$, the postcollision velocities will be $\mathbf{c}_1$ and $\mathbf{c}_2$. This is called the inverse collision of the original, direct collision. Also as a result of the symmetry, the angle of deflection becomes,

$$\chi=\pi-2\phi_m \tag{2.3.19}$$

Equation (2.3.18) can then be solved,

$$\phi-\phi_m=\pm\int_{r_m}^{r}\frac{b}{r'}\left(r'^2-b^2-\frac{2r'^2U(r')}{m_r c_r^2}\right)^{-1/2}dr' \tag{2.3.20}$$

Since $\phi\to 0$ when $r\to\infty$, Eq. (2.3.20) gives

$$\phi_m=\int_{r_m}^{\infty}\frac{b}{r'}\left(r'^2-b^2-\frac{2r'^2U(r')}{m_r c_r^2}\right)^{-1/2}dr' \tag{2.3.21}$$

Therefore,

$$\chi=\pi-2\int_{r_m}^{\infty}\frac{b}{r'}\left(r'^2-b^2-\frac{2r'^2U(r')}{m_r c_r^2}\right)^{-1/2}dr' \tag{2.3.22}$$

Equation (2.3.22) shows that given the intermolecular potential and the relative speed of the approach, the deflection angle χ is determined by the value of the impact parameter b.

For hard sphere molecular model, $U = 0$ for $r > d_{12}$, and

$$\phi_m = \int_{d_{12}}^{\infty} \frac{b}{r'} \left(r'^2 - b^2\right)^{-1/2} dr'$$
$$= \frac{\pi}{2} - \cos^{-1}\left(\frac{b}{d_{12}}\right)$$

and Eq. (2.3.19) gives

$$\chi = 2\cos^{-1}\left(\frac{b}{d_{12}}\right) \tag{2.3.23}$$

To calculate the postcollision velocities, $\mathbf{c}_1^*$ and $\mathbf{c}_2^*$, by using Eq. (2.3.11), we still need to calculate $\mathbf{c}_r^*$. Noting that $c_r = c_r^*$, we will use a new coordinate system (x', y', z') with the x' axis aligned with the direction of $\mathbf{c}_r$ and y' on the reference plane, as shown in Fig. 2.3.3.

In the new coordinate system (x', y', z') shown in Fig. 2.3.3, the postcollision relative velocity can be written as

$$\mathbf{c}_r^{*\prime} = (c_r \cos\chi, c_r \sin\chi \cos\varepsilon, c_r \sin\chi \sin\varepsilon) \tag{2.3.24}$$

Its components in the (x, y, z) coordinate system can be obtained by using the direction cosines. That is

$$\mathbf{c}_r^* = \mathbf{X}\mathbf{c}_r^{*\prime} \tag{2.3.25}$$

$\mathbf{X}$ is a second order tensor with components being the direction cosines of the primed coordinate system. That is,

$$X_{ij} = \cos(x_i, x_j') \tag{2.3.26}$$

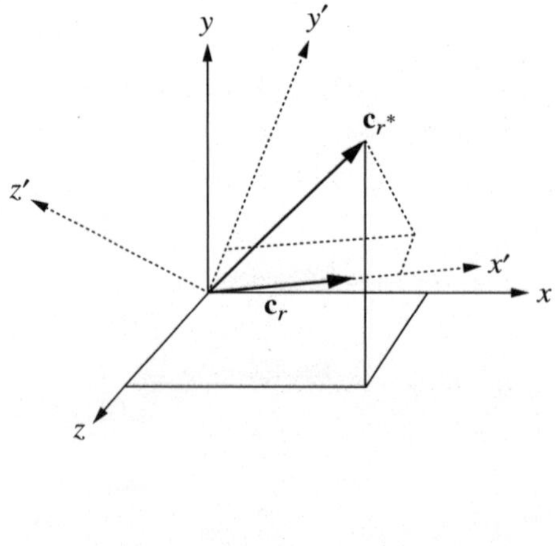

Figure 2.3.3 Collision coordinate system.

Since x' is aligned with the direction of $\mathbf{c}_r$, its direction cosines are

$$X_{11} = \frac{u_r}{c_r} \qquad X_{21} = \frac{v_r}{c_r} \qquad \text{and} \qquad X_{31} = \frac{w_r}{c_r} \tag{2.3.27}$$

where u_r, v_r, and w_r denote the components of $\mathbf{c_r}$ in the x, y, and z directions, respectively. Let y' be normal to the x-axis, so that $X_{12} = 0$. By requiring that the y'-axis be also normal to the x'-axis, we get the other two components of X_{i2}

$$X_{12} = 0 \qquad X_{22} = \frac{w_r}{\sqrt{v_r^2 + w_r^2}} \qquad \text{and} \qquad X_{32} = \frac{-v_r}{\sqrt{v_r^2 + w_r^2}} \tag{2.3.28}$$

The z'-axis is normal to both the x'- and the y'-axes. The following expressions for X_{i3} can be derived. They are

$$X_{13} = \frac{-\sqrt{v_r^2 + w_r^2}}{c_r} \qquad X_{23} = \frac{u_r v_r}{c_r \sqrt{v_r^2 + w_r^2}} \qquad \text{and} \qquad X_{33} = \frac{u_r w_r}{c_r \sqrt{v_r^2 + w_r^2}} \tag{2.3.29}$$

In summary, the postcollision velocities can be found with the following five steps.

Step 1. Calculate the center of mass velocity.

$$\mathbf{c}_m = \frac{m_1}{m_1 + m_2}\mathbf{c}_1 + \frac{m_2}{m_1 + m_2}\mathbf{c}_2 \tag{2.3.2}$$

Step 2. Calculate the precollision relative velocity.

$$\mathbf{c}_r = \frac{m_1 + m_2}{m_2}(\mathbf{c}_1 - \mathbf{c}_m) \tag{2.3.6}$$

Step 3. Calculate the deflection angle.

$$\chi = \pi - 2\int_{r_m}^{\infty} \frac{b}{r'}\left(r'^2 - b^2 - \frac{2r'^2 U(r')}{m_r c_r^2}\right)^{-1/2} dr' \tag{2.3.22}$$

Step 4. Calculate the postcollision relative velocity.

$$\begin{aligned} \mathbf{c}_r^{*\prime} &= (c_r \cos\chi, c_r \sin\chi \cos\varepsilon, c_r \sin\chi \sin\varepsilon) \\ \mathbf{c}_r^* &= \mathrm{X}\, \mathbf{c}_r^{*\prime} \end{aligned} \tag{2.3.24}$$

where the components of $\boldsymbol{X}$ tensor can be found in Eqs. (2.3.25), (2.3.27), (2.3.28), and (2.3.29).

Step 5. Calculate the postcollision velocities $\mathbf{c}_1^*$ and $\mathbf{c}_2^*$.

$$\mathbf{c}_1^* = \mathbf{c}_m + \frac{m_2}{m_1 + m_2}\mathbf{c}_r^* \qquad \text{and} \qquad \mathbf{c}_2^* = \mathbf{c}_m - \frac{m_1}{m_1 + m_2}\mathbf{c}_r^* \tag{2.3.11}$$

Therefore if the nature of the molecular-collision process were known, the motion of the gas could be completely specified. The dynamic trajectory of each particle can be completely determined from their given initial conditions. Since the number of particles in a real gas is large, such processes are cumbersome and are seldom followed in practice.

The concept of collision cross section is important in the evaluation of probabilities of collision in terms of the interaction potential and the collision parameter b (see Fig. 2.3.4). The differential collision cross section is the perpendicular (to $\mathbf{c}_r$) area through which molecules with $\mathbf{c}_r$ are scattered to an infinitesimal velocity space solid angle $d\Omega$ around $\mathbf{c}_r^*$. The solid angle $d\Omega$ can also be written as

$$\sin(\chi) d\varepsilon\, d\chi$$

where ε is the angle between the collision plane and a reference plane. For a spherically symmetric intermolecular potential, the inverse collision exists and

$$\sigma d\Omega = (b d\varepsilon)\, db \tag{2.3.30}$$

where σ is the mapping ratio and is called differential cross section. One can also write,

$$\sigma(\sin(\chi) d\varepsilon d\chi) = -(b d\varepsilon)\, db$$

The negative sign is there to expect that the deflection angle decreases with the increasing collision parameter b. Therefore, the differential cross section becomes

$$\sigma = -\frac{b}{\sin(\chi)}\frac{db}{d\chi} \tag{2.3.31}$$

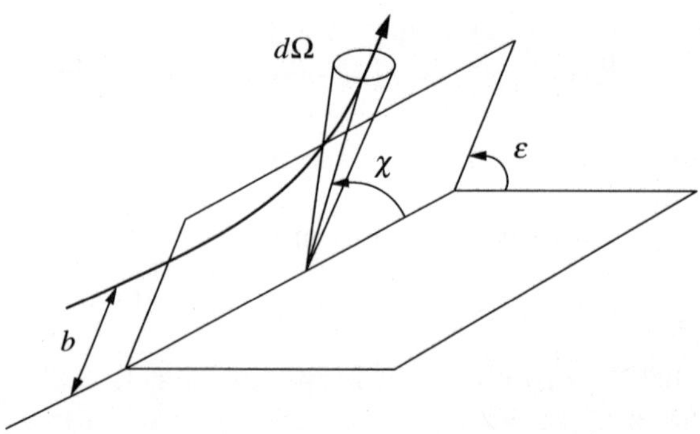

Figure 2.3.4 Collision cross section and solid angle.

The total collision cross section can be obtained by integration over all the possible scattering directions. That is

$$\begin{aligned}\sigma_T &= \int_0^{4\pi} \sigma d\Omega \\ &= \int_0^{\pi}\int_0^{2\pi} \sigma \sin(\chi)\, d\varepsilon d\chi \\ &= 2\pi \int_0^{\pi} \sigma \sin(\chi)\, d\chi\end{aligned} \tag{2.3.32}$$

For hard sphere model for gas of molecular diameter d, $b = \cos\left(\frac{\chi}{2}\right) d_{12}$

$$\begin{aligned}\sigma &= -\frac{b}{\sin(\chi)}\frac{d}{d\chi}\left(d_{12}\cos\left(\frac{\chi}{2}\right)\right) \\ &= \frac{d_{12}^2}{4}\end{aligned} \tag{2.3.33}$$

and

$$\begin{aligned}\sigma_T &= 2\pi \int_0^{\pi} \frac{d_{12}^2}{4} \sin(\chi)\, d\chi \\ &= \pi d_{12}^2 \\ &= \pi d^2\end{aligned} \tag{2.3.34}$$

2.3.3 Molecular force field models

As described earlier, the intermolecular force field consists of attractive and repulsive portions as a function of distance between the two molecules. The best know attractive–repulsive model is the Lennard-Jones model.

Lennard-Jones (L-J) model. The Lennard-Jones 12-6 potential can be written as

$$U = 4\varepsilon\left[\left(\frac{r}{\delta}\right)^{-6} - \left(\frac{r}{\delta}\right)^{-12}\right] \tag{2.3.35}$$

The corresponding force field is then

$$F = 48\frac{\varepsilon}{\delta}\left[\frac{1}{2}\left(\frac{r}{\delta}\right)^{-7} - \left(\frac{r}{\delta}\right)^{-13}\right] \tag{2.3.36}$$

where ε denotes the minimum potential value, and δ a characteristic length scale taken as the distance at which the potential function changes its sign. This model is widely used in the simulations of dense gas and liquid.

Inverse power law model. Most force field models neglect the long-range attractive portion. In this case, the inverse power law model describes the potential and the force fields as follow

$$U = \frac{\kappa}{(\eta - 1)r^{\eta-1}} \qquad F = \frac{\kappa}{r^{\eta}} \tag{2.3.37}$$

where η is the power determining the "hardness" of particles, κ a constant. The deflection angle χ is

$$\chi = \pi - 2\int_0^{W_1} \left[1 - W^2 - \{2/(\eta - 1)\}(W/W_0)^{\eta-1}\right]^{-1/2} dW \tag{2.3.38}$$

where $W_0 = b(m_r c_r^2/\kappa)^{1/\eta-1}$, and the dimensionless W_1 is the positive root of the equation

$$1 - W^2 - \{2/(\eta - 1)\}(W/W_0)^{\eta-1} = 0 \tag{2.3.39}$$

Note that χ is a function of only the dimensionless impact parameter W_0, which makes the inverse power law model easy to implement. The rigid sphere model, often referred to as the hard sphere (HS) model, that has been used in the previous sections can be regarded as a special case of the inverse power law with $\eta = \infty$. The use of $\eta = 5$ results in the Maxwell model. The model can be regarded as the limiting case of a "soft" molecule, contrasting the HS model at the other limit. For a real monatomic molecule, the effective value of η is about 10. The HS model is simple and easy to use. Its primary deficiency is that the resulting scattering law is not realistic.

Variable hard sphere model. The Variable hard sphere (VHS) model was introduced by Bird (1981) to correct the primary deficiency of the HS model. In the VHS model, the molecule is a hard sphere with a diameter d that is a function of c_r. Generally the function can be obtained based on a simple inverse power law α. That is

$$d = d_{\text{ref}}(c_{r,\text{ref}}/c_r)^{\alpha} \tag{2.3.40}$$

where the subscript "ref" denotes the reference values at the reference temperature T_{ref}, and $\alpha = 2/(\eta - 1)$. σ_T can be expressed in terms of reference conditions

$$\sigma_T = \sigma_{\text{ref}}\left(\frac{c_r^2}{c_{r,\text{ref}}^2}\right)^{-a} \tag{2.3.41}$$

For an equilibrium gas, c_r is related to temperature by

$$c_r^2 = \frac{2(2-\alpha)kT}{m_r} \tag{2.3.42}$$

Combining Eqs. (2.3.41) and (2.3.42) yields an expression for σ_T, appropriate for an equilibrium gas, as

$$\sigma_T = \sigma_{\text{ref}} \left(\frac{m_r c_r^2}{2(2-\alpha)kT_{\text{ref}}} \right)^{-a} \tag{2.3.43}$$

The VHS model leads to a power law temperature dependence of the coefficient of viscosity by

$$\mu \propto T^{\omega} \tag{2.3.44}$$

where $\omega = 0.5 + \alpha$, and the deflection angle

$$\chi = 2\cos^{-1}(b/d) \tag{2.3.45}$$

The VHS model is currently the most widely used model in direct simulation Monte Carlo (DSMC) simulations, because of its simplicity and its effective approximation to real intermolecular potential. Typical values of α and d_{ref} are given in Table 2.3.1 (Bird 1994).

The HS and VHS models have the same impact parameter $b = d\cos(\chi/2)$. The effective diameter d is invariant for the HS model, but is an energy-dependent variable for the VHS model. Both models obey the isotropic scattering law and do not correctly predict the diffusion in flows of real gases, especially of gas mixtures.

Variable soft sphere model. The Variable soft sphere (VSS) model introduced by Koura and Matsumoto (1991, 1992) took into account the anisotropic scattering of real gases. The molecular diameter varies in

TABLE 2.3.1 Typical Values of α and d_{ref} for VHS Molecules at 273K

Gas	Gas symbol	α	$d_{\text{ref}} \times 10^{10}$ m
Hydrogen	H_2	0.17	2.92
Helium	He	0.16	2.33
Nitrogen	N_2	0.24	4.17
Oxygen	O_2	0.27	4.07
Argon	Ar	0.31	4.17
Carbon dioxide	CO_2	0.43	5.62

SOURCE: Bird, G.A., *Molecular Gas Dynamics and the Direct Simulation of Gas Flows*, Oxford University Press, New York, 1994.

the same way as the VHS model, but there is a deflection exponent β in relation to b such that

$$b = d\cos^{\beta}(\chi/2) \tag{2.3.46}$$

This cosine exponent β and the diameter d are determined such that the viscosity and diffusion coefficient are consistent with the inverse power law potential. For the VHS model, β is 1. This model is called variable soft sphere because χ_{VSS} is smaller than χ_{VHS}, which can be obtained from $\beta > 1$ and $\cos(\chi_{\mathrm{VHS}}/2) = \cos^{\beta}(\chi_{\mathrm{VSS}}/2)$. The VSS total collision cross section can be expressed as

$$\sigma_{T,\mathrm{VSS}} = \frac{1}{S}\sigma_{T,\mathrm{VHS}} \tag{2.3.47}$$

where S is the softness coefficient given by

$$S = 6\beta/[(\beta+1)(\beta+2)] \tag{2.3.48}$$

The values of the cosine exponent β and softness coefficient S for a large variety of molecular species are compiled in Koura and Matsumoto (1991). In addition, the corresponding viscosity and diffusion cross-section for the VSS model are also given in the reference. It was found that the VSS model is preferable to the VHS model in flows of gas mixtures where molecular diffusion is important.

Generalized hard sphere model. The Generalized hard sphere (GHS) model developed by Hassan and Hash (1993) is an extension of the VHS and VSS models. The scattering distribution is that of the hard or soft sphere, but the variation of the total cross section as a function of the relative translational energy mimics that of the corresponding attractive–repulsive potential. It is implemented through the parameters that describe the intermolecular potentials of the form of Eq. (2.3.22), and can therefore make use of the existing database that has been built up from the measured transport properties of real gases.

Larsen-Borgnakke phenomenological model. There have been molecular models proposed for inelastic collision. The most widely used model for inelastic collisions is the Larsen-Borgnakke phenomenological model (Larsen and Borgnakke 1974, Borgnakke and Larsen 1975). This model allows molecules to have continuous internal energy modes. The exchange of energy among translational, rotational and vibrational modes is accounted for in distribution of post-collision velocity for the molecules involved. The Larsen-Borgnakke model was also extended to include the repulsive force described in the GHS model. The model is applicable to binary collisions in a mixture of polyatomic gases.

2.4 Statistical Gas Properties

An important concept brought in by the statistical consideration of the collision between molecules is the mean free path. The mean free path λ is defined as the average distance a molecule travels between successive collisions. It is defined in a frame of reference that moves with the local stream. We will consider again single species with the average spacing between the molecules, δ, which is much larger than the diameter d of the molecules of hard sphere. That is

$$\delta \gg d \tag{2.4.1}$$

Such gases are referred to as dilute gases. $\delta/d = 7$ is commonly taken as the limit of dilute gas assumption. The average volume a molecule occupies is $1/n$. Therefore, we can also obtain

$$\delta = n^{-1/3} \tag{2.4.2}$$

For the hard sphere gas considered here, a given target molecule will experience a collision with another molecule whenever the distance between the centers of the molecules d_{12} equals to d (see Fig. 2.4.1).

The target molecule then carries a sphere of influence of radius, d. A collision will occur when the center of other field molecules lie on the surface of this sphere. For hard sphere model, according to Eq. (2.3.34), the total collision cross section σ_T becomes the area of the projection of the sphere of influence on to a plane normal to the relative velocity vector of the colliding molecules (see Fig. 2.4.2). The relative velocity vector $\mathbf{c}_r$ is the velocity of the target molecule $\mathbf{c}_1$ and that of the field molecule $\mathbf{c}_2$ or $\mathbf{c}_r = \mathbf{c}_1 - \mathbf{c}_2$. The number of collisions per unit time between the target molecule and the molecules of class $\mathbf{c}_2$ is then the number of molecules in the class $\mathbf{c}_2$ that lies within the cylindrical volume swept by the collision cross section area. That is

$$\Delta n \sigma_T \, c_r$$

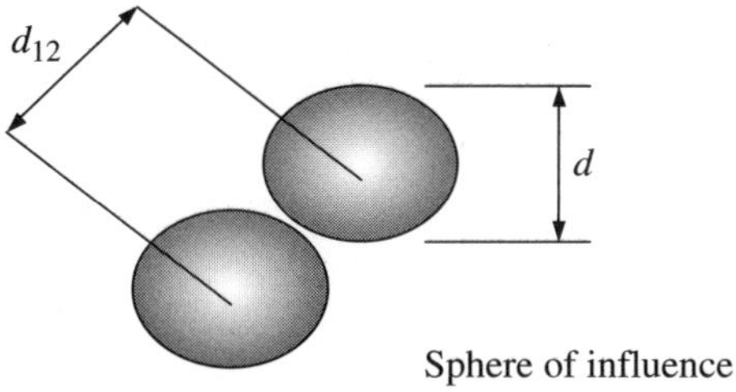

Figure 2.4.1 Sphere of influence for hard sphere model.

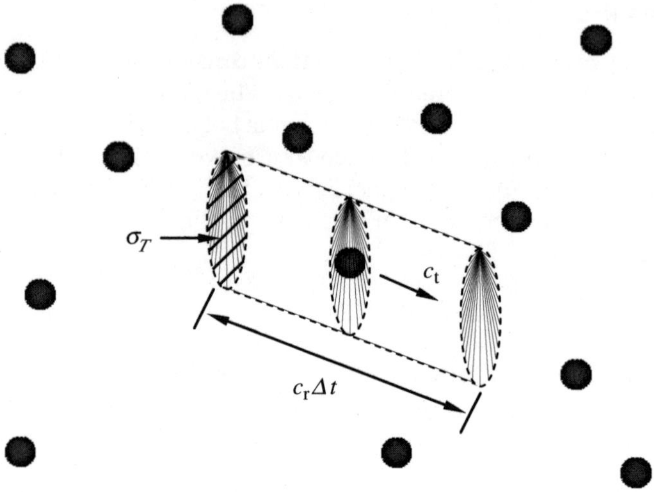

Figure 2.4.2 Collision frequency.

where Δn represents the number density of the molecules of class $\mathbf{c}_2$. For hard sphere molecules, where $\sigma_T = \pi d^2$, the mean collision rate ν is then the sum over all velocity class. Or

$$\begin{aligned}\nu &= \sum \Delta n \sigma_T c_r \\ &= n\pi d^2 \left(\sum \frac{\Delta n}{n} c_r \right)\end{aligned} \tag{2.4.3}$$

Since $\Delta n/n$ represents the fraction of the field particles in the class of $\mathbf{c}_r$, Eq. (2.4.3) can also be written as

$$\nu = n\pi d^2 \overline{c_r} \tag{2.4.4}$$

The mean free path λ is defined as the average distance a molecule travels between successive collisions.

It is defined in a frame of reference that moves with the local stream. Therefore

$$\begin{aligned}\lambda &= \frac{\overline{c'}}{\nu} \\ &= \frac{1}{n\pi d^2} \frac{\overline{c'}}{\overline{c_r}}\end{aligned} \tag{2.4.5}$$

where $\overline{c'} \equiv \overline{|c'|}$ is the mean thermal speed. We can obtain, from Eqs. (2.4.2) and (2.4.5), that

$$\frac{\lambda}{\delta} = \frac{1}{\pi} \frac{\overline{c'}}{\overline{c_r}} \left(\frac{\delta}{d} \right)^2 \tag{2.4.6}$$

The ratio of $\overline{c'}/\overline{c_r}$ is of order unity ($1/\sqrt{2}$ for equilibrium monatomic gas), and $\delta \gg d$. Thus, we obtain, for dilute gas,

$$\lambda \gg \delta \gg d \tag{2.4.7}$$

Since the effective range of the intermolecular force field is of the order of the diameter, this equation suggests that for dilute gas, the gas molecules interact only during the collision process of relatively short duration in time. This is an important result in the development of kinetic theory of gas. It suggests that molecular collision is an instantaneous occurrence. It also suggests that molecular collisions in dilute gas are most likely to involve only two molecules. Such a collision, as was described in the previous section is called a binary collision.

Molecular collisions cause the velocity of the individual molecule to vary. Therefore, the number of molecules in a particular velocity class changes with time. For a gas in equilibrium where there is no gradient in its macroscopic properties, the number of particle in a velocity class remains unchanged. That is, locally, for every molecule that leaves its original velocity class, there will be another molecule entering the velocity class. In a non-equilibrium state where there is nonuniform spatial distribution of certain macroscopic quantity, such as the average velocity or temperature, the molecules in their random thermal motion that move from one region to another find themselves with momentum or energy deficit or excess at the new location. The microscopic molecular motion thus causes the macroscopic properties of the gases to change. The results of these molecular transport processes are reflected upon the macroscopic non-equilibrium phenomena such as viscosity μ and heat conduction k. Molecular collisions are responsible for establishing the equilibrium state where the effects of further collisions cancel each other. To examine the phenomenon of viscosity, we can look at a unidirectional gas flow with nonuniform velocity in only one direction and homogeneous in the other two directions. This is normally referred to as a simple shear flow. Let's say, $\bar{c}_x(y)$ be the only nonzero velocity component in a Cartesian coordinate system (Fig. 2.4.3).

The x momentum mc'_x of a molecule with mass m is transported in y direction at the speed of c'_y. Locally, the sample averaged momentum transport per unit area per unit time becomes

$$-nm\,\overline{c'_x c'_y}$$

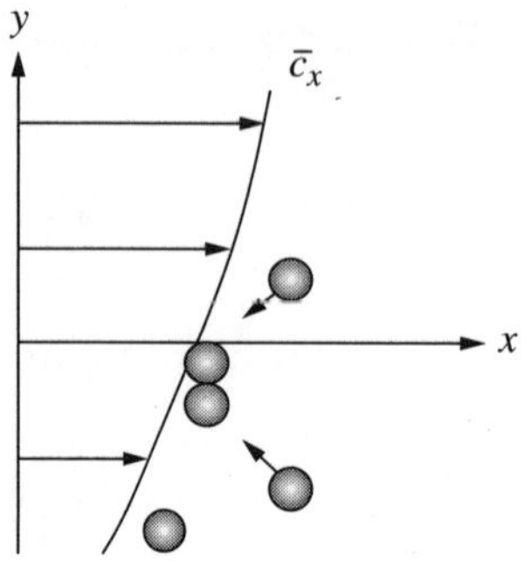

Figure 2.4.3 Transport process.

where n is number density. The minus sign is used since a positive c'_y would carry a momentum deficit at the new location. If one assumes that on an average, a molecule collides with another molecule at the new location with distance λ apart from its original location, say from $y = -\lambda$ to $y = 0$ as indicated in Fig. 2.4.3, and reaches equilibrium with its new environment. The particle of mass m will change its momentum by the amount of

$$m(\bar{c}_x(y=0) - \bar{c}_x(y=-\lambda))$$

Using the Taylor's series expansion

$$\begin{aligned} m(\bar{c}_x(y=0) - \bar{c}_x(y=-\lambda)) &= m\left(\lambda \frac{d\,\bar{c}_x}{dy} + \frac{\lambda^2}{2}\frac{d^{\,2}\bar{c}_x}{dy^2} + \cdots\right) \\ &= m\lambda \frac{d\,\bar{c}_x}{dy} + O(\lambda^2) \end{aligned} \tag{2.4.8}$$

The number of such collisions per unit area per unit time can be estimated by using the average thermal speed and becomes $n\overline{c'}$. Therefore the total momentum change $-nm\overline{c'_x c'_y}$ is

$$-nm\overline{c'_x c'_y} = \beta_\mu n\overline{c'} m\lambda \frac{d\,\bar{c}_x}{dy}$$

where β_μ is a constant of proportionality. For monatomic gases, $\beta_\mu \approx 0.5$. Or

$$-\rho\overline{c'_x c'_y} = \beta_\mu \rho\overline{c'}\lambda \frac{d\,\bar{c}_x}{dy}$$

Or

$$\mu = \beta_\mu \rho \overline{c'}\lambda \tag{2.4.9}$$

Thus, the appropriate velocity scale for viscosity is related to the molecular thermal speed and the length scale is the mean free path.

2.5 Position and Velocity Distribution Functions

The velocity of a molecule changes with time and at any instance of time molecules have different velocities. For a realistic representation of the behavior of gas flow containing a large number of molecules, the distribution of the molecular velocity must be examined by using statistical descriptions. A distribution often used in fluid mechanics is the fluid density. In the physical space, for ΔN number of molecules of mass m in a volume of ΔV located at x_i (shown in Fig. 2.5.1), the local mass density is

$$\begin{aligned}\rho(x_i,t) &= \lim_{\Delta V\to 0}\frac{m\Delta N}{\Delta V}\\ &= m\lim_{\Delta V\to 0}\frac{\Delta N}{\Delta V} \\ &= mn(x_i,t)\end{aligned} \tag{2.5.1}$$

where ΔV should be large compared with the molecular spacing. $n(x_i,t)$ represents the local number of molecules found in a unit volume of physical space as a function of space and time. It is called the local number density or number density. It describes the distribution of the number of molecules in physical space and therefore is a position distribution function. Fluid density is then related to the position distribution function of the number of molecules. For a small element dV $(\equiv dx_1\,dx_2\,dx_3 \equiv dx\,dy\,dz)$ containing dN number of molecules

$$dN = ndV \tag{2.5.2}$$

where the functional dependence has been dropped for brevity. Multiplying both sides by m, we get

$$dM = nmdV = \rho dV$$

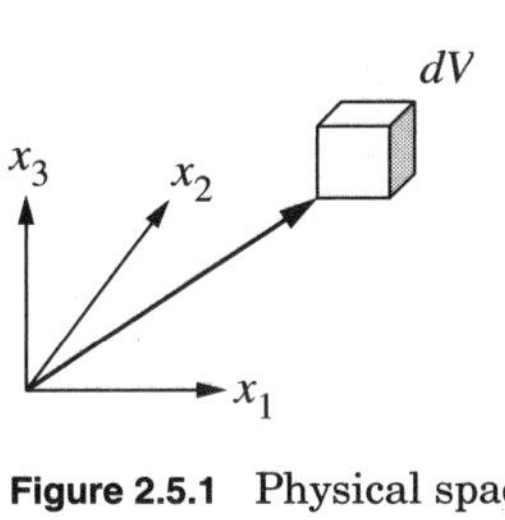

Figure 2.5.1 Physical space element.

The density is also a position distribution function for mass. We can normalize the number density by the total number of molecules in the region of interest N. Thus, Eq. (2.5.2) becomes

$$\begin{aligned} dN &= N\left(\frac{n}{N}\right)dV \\ &= N\varpi dV \end{aligned} \tag{2.5.3}$$

where ϖ represents the normalized number density or a distribution function. Therefore, ϖdV represents the fraction of the molecules that can be found in the volume dV and can be interpreted as the probability that a randomly chosen molecule will reside inside the volume dV. Equation (2.5.3) can be integrated over the entire region of interest,

$$N = \int_V N\varpi dV \tag{2.5.4}$$

Or

$$1 = \int_V \varpi dV \tag{2.5.5}$$

Therefore the function ϖ is also a probability density. The probability of finding a particular molecule inside volume dV is equal to the portion of the number of molecules in the volume dV.

The molecules can also be identified by their three velocity components instead of their three location components in the physical space. In the velocity space, the molecule can be represented uniquely by a point in a Cartesian space with the velocity components as the coordinates. The gas in a region of interest then is represented by a cloud of N points in the velocity space (Fig. 2.5.2).

Following the analysis described above, we can define a velocity distribution function and its normalized form. The normalized velocity distribution can be represented by $f(c_i, x_i, t)$. That is

$$dN = Nf\,d\mathbf{c} \tag{2.5.6}$$

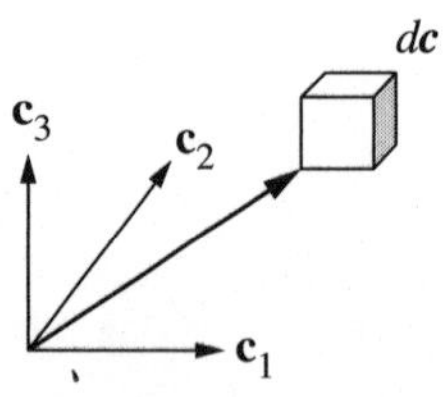

Figure 2.5.2 Velocity space element.

where $d\mathbf{c} \equiv dc_1\,dc_2\,dc_3$ denotes a differential volume in the velocity space. The normalized velocity distribution function satisfies the normalization condition,

$$1 = \int_{V_c} f\,d\mathbf{c} \tag{2.5.7}$$

Therefore $f\,d\mathbf{c}$ represents the number of particles that can be found in a velocity space volume $d\mathbf{c}$, and f the probability of finding a randomly chosen particle resides in a unit volume in the velocity space.

As was mentioned earlier, the average property of the flow can be obtained by sample averaging the microscopic molecular quantity. Let us begin with Eq. (2.2.7) by defining, for any molecular quantity Q,

$$\begin{aligned}\bar{Q} &= \frac{1}{N}\int Q dN' \\ &= \frac{1}{N}\int QNf d\mathbf{c} \\ &= \int Qf d\mathbf{c}\end{aligned} \tag{2.5.8}$$

For example, for $Q = \mathbf{c} = \bar{\mathbf{c}} + \mathbf{c}'$

$$\begin{aligned}\bar{c} &= \frac{1}{N}\int \mathbf{c} dN' \\ &= \frac{1}{N}\int_{-\infty}^{\infty} \mathbf{c}Nf d\mathbf{c} \\ &= \int_{-\infty}^{\infty} \mathbf{c}\,f d\mathbf{c}\end{aligned}$$

where $\bar{\mathbf{c}}$ denotes the average or the mean velocity and $\mathbf{c}'$ the thermal fluctuation velocity. Therefore,

$$\begin{aligned}\overline{\mathbf{c}'} &= \int_{-\infty}^{\infty} (\mathbf{c} - \bar{\mathbf{c}})\,f d\mathbf{c} \\ &= \bar{\mathbf{c}} - \bar{\mathbf{c}} \\ &= 0\end{aligned}$$

The product of $dV d\mathbf{c} = dx_1\,dx_2\,dx_3\,dc_1\,dc_2\,dc_3$ represents a volume in the combination of physical and velocity space. This is called phase space. We can then define a single particle velocity distribution function in phase space $\Im(c_i, x_i, t)$

$$dN = \Im d\mathbf{c}\,dV \tag{2.5.9}$$

Comparing Eq. (2.5.9) with Eq. (2.5.2), Eq. (2.5.9) gives

$$\Im(c_i, x_i, t) = n(x_i, t) f(c_i, x_i, t) \tag{2.5.10}$$

$\Im$ then represents the number of molecules in the phase space element $dV d\mathbf{c}$.

2.6 Boltzmann Equation and Maxwellian Distribution Function

The velocity distribution function provides a statistical description of a gas on a molecular level. For a system of N particle, the dimension of the phase space is $6N$. The complete system is then defined by points in the $6N$ dimensional phase space, which is not practical in seeking its analytical or numerical solutions. The Boltzmann equation (*Boltzmann 1872*) was formulated with a single particle distribution function. The Boltzmann equation can be derived from different points of view. In the following, we will use a procedure similar to that used in the derivation of conservation equations at macroscales.

The Boltzmann equation is the conservative equation to describe dilute gases behavior, considering that molecules move and collide with each other constantly and randomly at the microscopic level. For simplicity and clarity, monatomic gases are assumed and the molecules with velocity distribution $f(\mathbf{c})$ in the range of $\mathbf{c}$ and $\mathbf{c} + d\mathbf{c}$ are considered. With the assumption of dilute gases, Eq. (2.4.7) holds for the distribution of molecules in the phase space. The description of the phase space can be decoupled into physical space and velocity space. That is, the molecule movement in the physical space and the velocity change in the velocity space can be considered independently as shown in Fig. 2.6.1. For an element in the phase space $dV d\mathbf{c}$ shown in Fig. 2.6.1, the local rate of change of the number molecules at an instance of time is

$$\int_{dV} \frac{\partial}{\partial t}(nf) dV' d\mathbf{c}$$

If the shape of the phase element does not change with time, we obtain

$$\frac{\partial}{\partial t}(nf) dV d\mathbf{c} \tag{2.6.1}$$

This change can be caused by influxes of molecules through each side of the phase space element. The net influxes of molecules through the physical space element dV is caused by convection of molecules across

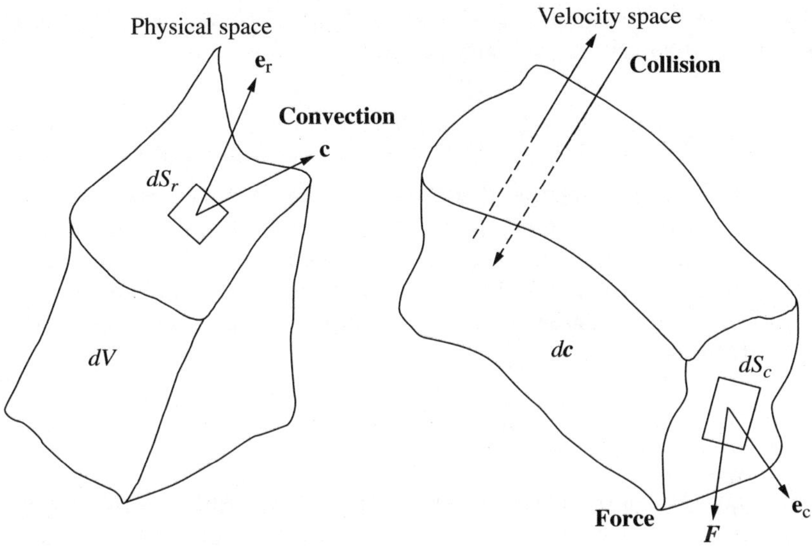

Figure 2.6.1 Phase space element.

the surface S_r of dV. Consider class $\mathbf{c}$ molecules, the number density within dV is $nfd\mathbf{c}$. The net influxes can be written as

$$-\int_{S_r} (nf\ \mathbf{c}) \bullet (\mathbf{e}_r dS_r')d\mathbf{c}$$

where $\mathbf{e}_r$ denotes the unit normal vector of the physical space element dS_r. The surface integral can be written as a volume integral over dV by using the Gauss theorem.

$$-\int_{dV} \nabla \bullet (nf\ \mathbf{c})d\,V'd\mathbf{c}$$

Since the phase space is decoupled into the physical and the velocity space and the velocity of the molecules in class $\mathbf{c}$ does not change in the small physical element dV, the equation becomes

$$-\mathbf{c} \bullet \nabla(nf)dVd\mathbf{c}$$

Since we are considering only class $\mathbf{c}$ molecules, the above equation can be written as

$$-\mathbf{c} \bullet \frac{\partial}{\partial \mathbf{x}}(nf)dVd\mathbf{c} \tag{2.6.2}$$

Similarly, the flux of molecule across sides of $d\mathbf{c}$ can be caused by external force per unit mass $\mathbf{F}$ in a similar fashion to that of velocity $\mathbf{c}$

is to the molecular fluxes across side of dV. Therefore, we can also write the net influx through the velocity space $d\mathbf{c}$ as

$$-\mathbf{F} \bullet \frac{\partial}{\partial \mathbf{c}}(nf)dVd\mathbf{c} \tag{2.6.3}$$

The external force $\mathbf{F}$ is assumed constant in the velocity space and the number density nf is assumed constant in $d\mathbf{c}$ in arriving at Eq. (2.6.3). In addition to these convective effects, the local number of molecules can also be changed by molecular collision in the velocity space. For the number of molecules in class $\mathbf{c}$ as a result of collisions, this mechanism can be written as

$$\left\{\frac{\partial}{\partial t}(nf)\right\}_{\text{collision}} dVd\mathbf{c} \tag{2.6.4}$$

Therefore the equation for the rate of change of the number of molecules in the phase space element $dVd\mathbf{c} = dx_1\, dx_2\, dx_3\, dc_1\, dc_2\, dc_3$ is

$$\frac{\partial}{\partial t}(nf) + \mathbf{c} \bullet \frac{\partial}{\partial \mathbf{x}}(nf) + \mathbf{F} \bullet \frac{\partial}{\partial \mathbf{c}}(nf) = \left\{\frac{\partial}{\partial t}(nf)\right\}_{\text{collision}} \tag{2.6.5}$$

We can further evaluate the collision term by using the binary collision dynamics for dilute gas described in Eq. (2.6.4), considering the collision between a target molecule of class $\mathbf{c}$ with a field particle of class $\mathbf{c}_1$. According to the definition of differential collision cross section, the volume swept by particles of class $\mathbf{c}$ per unit time is $c_r \sigma d\Omega$, and the number of class $\mathbf{c}_1$ particles per volume in the physical space is $nf_1 d\mathbf{c}_1$. Therefore the number of collisions for the class $\mathbf{c}$ particles per unit time is $nf_1\, c_r \sigma\, d\Omega\, d\,\mathbf{c}_1$. The total number of collision for class $\mathbf{c}$ particles per unit time is the product of the total number of class $\mathbf{c}$ target particles $nf\ dV\ d\mathbf{c}$ and the number of collisions. That is

$$n^2 f\, f_1 c_r \sigma\, d\Omega\, d\mathbf{c}_1\, dV\, d\mathbf{c} \tag{2.6.6}$$

This is the time rate at which class $\mathbf{c}$ particles have decreased due to collision. The total depletion rate then is the integration over the entire collision cross section and the class $\mathbf{c}_1$ particles. Or

$$\left(\int_{-\infty}^{\infty}\int_{0}^{4\pi} n^2 f\, f_1 c_r \sigma\, d\Omega\, d\mathbf{c}_1\right) dVd\mathbf{c} \tag{2.6.7}$$

On the other hand, collisions can also bring new particles into class $\mathbf{c}$. The rate at which class $\mathbf{c}$ particles are replenished due to collision can be derived by considering the existing inverse collision and the symmetry between the direct collision between $\mathbf{c}$ and $\mathbf{c}_1$ and the inverse collision between $\mathbf{c}^*$ and $\mathbf{c}_1^*$. The replenish rate can be written as

$$\left(\int_{-\infty}^{\infty}\int_{0}^{4\pi} n^2 f^* f_1^* c_r \sigma \, d\Omega \, d\mathbf{c}_1\right) dV d\mathbf{c} \tag{2.6.8}$$

Therefore the net increase of the number of molecules of class $\mathbf{c}$ becomes

$$\left\{\frac{\partial}{\partial t}(nf)\right\}_{\text{collision}} dV d\mathbf{c} = \int_{-\infty}^{\infty}\int_{0}^{4\pi} n^2(f^* f_1^* - f f_1) c_r \sigma \, d\Omega \, d\mathbf{c}_1 \, dV \, d\mathbf{c} \tag{2.6.9}$$

The equation for the change of the number of molecules of class $\mathbf{c}$ becomes

$$\begin{aligned} &\frac{\partial}{\partial t}(nf) + \mathbf{c} \bullet \frac{\partial}{\partial \mathbf{x}}(nf) + \mathbf{F} \bullet \frac{\partial}{\partial \mathbf{c}}(nf) \\ &= \int_{-\infty}^{\infty}\int_{0}^{4\pi} n^2(f^* f_1^* - f f_1) c_r \sigma \, d\Omega \, d\mathbf{c}_1 \end{aligned} \tag{2.6.10}$$

This is the Boltzmann equation for a simple dilute gas. The Boltzmann equation describes the rate of change, with respect to position and time, of one-particle distribution function. The Boltzmann equation can also be derived by using the Liouville equation, which is a continuity equation for the N-particle distribution function in a $6N$ dimensional phase space. The Liouville equation describes in a more basic manner the mechanics of motion. Discussions of the limitation of the Boltzmann equation can be found in Vincenti and Kruger (1965) and Koga (1970).

The analytical solution of the Boltzmann equation can be found for gases that are in equilibrium. Consider a gas in equilibrium where the number of molecules in every velocity class must be constant at all positions and time. That is

$$\left\{\frac{\partial}{\partial t}(nf)\right\}_{\text{collision}} = 0 \tag{2.6.11}$$

Equation (2.6.9) then becomes

$$f^* f_1^* = f f_1 \tag{2.6.12}$$

Or

$$\ln f^* + \ln f_1^* = \ln f + \ln f_1 \tag{2.6.13}$$

This means that $\ln f$ is a collision invariant. The collision invariants are the momentum and the kinetic energy. Therefore the general solution for $\ln f$ is

$$\ln f = a_1 \left(\frac{1}{2} m(\mathbf{c} \cdot \mathbf{c}) \right) + \mathbf{b} \cdot m\mathbf{c} + a_2 \tag{2.6.14}$$

where a's and $\mathbf{b}$ are constants. It can also be written as

$$\ln f = a_1 \frac{m}{2} (\mathbf{c} + \mathbf{b})^2 + a_2'$$

Or

$$f = A \exp\left[-\frac{\beta m}{2} (\mathbf{c} + \mathbf{b})^2 \right] \tag{2.6.15}$$

A is a constant and β is a positive number for a bounded solution of f. Since the argument of the exponential function appears in a square term, indicating that the probabilities are the same for its value to be positive and negative and the average value should be zero. That is

$$\overline{\mathbf{c} + \mathbf{b}} = 0 \qquad \text{or} \qquad \mathbf{b} = -\bar{\mathbf{c}}$$

This suggests that $\mathbf{c} + \mathbf{b} = \mathbf{c} - \bar{\mathbf{c}} = \mathbf{c}'$. Therefore, Eq. (2.6.15) can be written as

$$f = A \exp\left[-\frac{\beta m}{2} (c'^2) \right] = A \exp\left[-\frac{\beta m}{2} \left(c'^2_1 + c'^2_2 + c'^2_3 \right) \right] \tag{2.6.16}$$

The normalization condition (2.5.7)

$$1 = \int_{V_c} f \, d\mathbf{c}$$

requires that

$$1 = A \int_{-\infty}^{\infty} \exp\left[-\frac{\beta m}{2} c'^2_1 \right] dc'_1 \int_{-\infty}^{\infty} \exp\left[-\frac{\beta m}{2} c'^2_2 \right] dc'_2 \times \int_{-\infty}^{\infty} \exp\left[-\frac{\beta m}{2} c'^2_3 \right] dc'_3 \tag{2.6.17}$$

The integration formulas for this and other similar exponential functions are given in Appendix 2A. It can be shown that Eq. (2.6.17) can be written as

$$1 = A \left[\left(\frac{2\pi}{\beta m} \right)^{1/2} \right]^3 \tag{2.6.18}$$

The constant A can then be related to β according to the following equation,

$$A = \left(\frac{\beta m}{2\pi}\right)^{3/2} \tag{2.6.19}$$

To calculate β, one can make use of Eq. (2.2.5), which gives

$$\overline{|c'|^2} = 3\frac{p}{\rho} = 3RT = \frac{3kT}{m} = A\int_{-\infty}^{\infty}\int_{-\infty}^{\infty}\int_{-\infty}^{\infty} (c'^2_1 + c'^2_2 + c'^2_3)$$
$$\times \exp\left[-\frac{\beta m}{2}(c'^2_1 + c'^2_2 + c'^2_3)\right] dc'_1\, dc'_2\, dc'_3$$

The equation becomes

$$\frac{3kT}{m} = 3A\int_{-\infty}^{\infty} c'^2_1 \exp\left[-\frac{\beta m}{2}(c'^2_1)\right] dc'_1 \int_{-\infty}^{\infty} \exp\left[-\frac{\beta m}{2}(c'^2_2)\right] dc'_2$$
$$\times \int_{-\infty}^{\infty} \exp\left[-\frac{\beta m}{2}(c'^2_3)\right] dc'_3 \tag{2.6.20}$$

By using the integration formulas in Appendix 2A to evaluate the integrals, we get

$$\frac{3kT}{m} = 3\left(\frac{\beta m}{2\pi}\right)^{3/2}\left(\frac{1}{2}\left(\frac{8\pi}{(\beta m)^3}\right)^{1/2}\right)\left(\frac{2\pi}{\beta m}\right)$$

Therefore

$$\beta = \frac{1}{kT} \quad \text{and} \quad A = \left(\frac{m}{2\pi kT}\right)^{3/2} \tag{2.6.21}$$

Therefore the Maxwellian distribution can be obtained by combining Eqs. (2.6.16) and (2.6.21)

$$f_0 = \left(\frac{m}{2\pi kT}\right)^{3/2} \exp\left[-\frac{m}{2kT}c'_i c'_i\right] \tag{2.6.22}$$

This is the famous Maxwellian distribution. Formal mathematical methods can also be used to derive the Maxwellian distribution. It is an important development in relating the microscopic molecular motion to macroscopic, measurable quantities. Further use of the Maxwellian distribution can be found in Chap. 4 where the macroscopic equation of continuity, momentum conservation, and energy conservation are derived for equilibrium and slightly non-equilibrium gases.

The Maxwellian distribution describes the probability of finding a randomly selected particle in a unit volume element in the velocity space for equilibrium gases. Since the distribution is isotropic, the probability of finding a particle in a range between, says c'_1 and $c'_1 + dc'_1$ is the same as those in the other two directions. Therefore, the probability or the fraction of number of particles with a thermal fluctuation component between c'_1 and $c'_1 + dc'_1$ becomes

$$f_0^{c'_1} = \left(\frac{m}{2\pi kT}\right)^{1/2} \exp\left[-\frac{m}{2kT}c'^2_1\right] \tag{2.6.23}$$

Therefore for a single component, the Maxwellian distribution has the form of the Gauss's error-distribution. This curve is shown in Fig. 2.6.2. Note that the positive value of c'_1 and the negative value of c'_1 have equal probability of occurrence. Because the Maxwellian distribution is isotropic and has no directional preference in the velocity space, it then becomes interesting to examine the fraction of particles with certain velocity magnitudes between c' and $c' + dc'$ regardless of the direction. This is equivalent to finding the fraction of particle $f_0 d\mathbf{c}$ that lies in a spherical shell centered at the origin of the velocity space and with thickness dc'. For a spherical coordinate system (c', θ, ϕ) is shown in Fig. 2.6.3.

The unit volume on the shell can be written as

$$c'^2 \sin\phi \, d\phi \, d\theta \, dc'$$

and the particle fraction becomes

$$f_0 c'^2 \sin\phi \, d\phi \, d\theta \, dc' = \left(\frac{m}{2\pi kT}\right)^{3/2} c'^2 \exp\left[-\frac{m}{2kT}c'^2\right] \sin\phi \, d\phi \, d\theta \, dc'$$

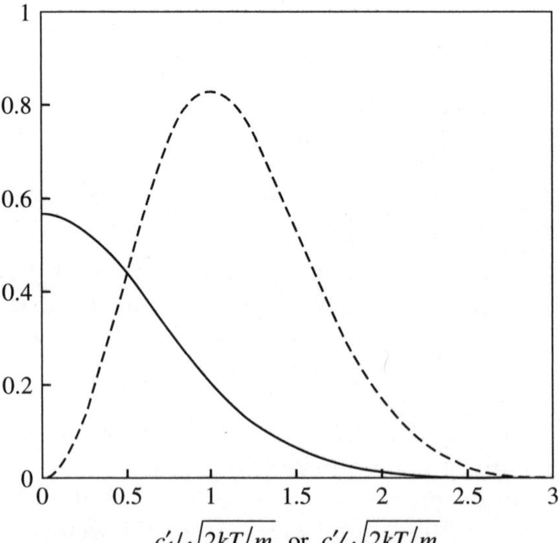

Figure 2.6.2 Distribution functions. Solid line: $\sqrt{2kT/m}\, f_0^{c'_1}$; broken line: $\sqrt{2kT/m}\, f_0^{c'}$.

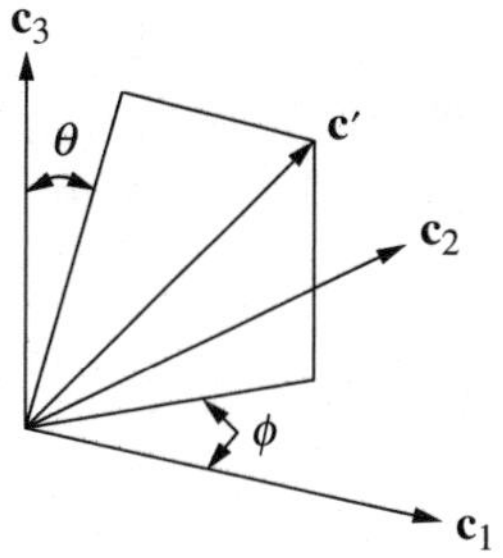

Figure 2.6.3 Spherical coordinates in the velocity space.

By integration over $\theta \in (0, 2\pi)$ and $\phi \epsilon (0, \pi)$ and using $f_0^{c'}$ to represent the speed distribution function, we get

$$f_0^{c'} = 4\pi \left(\frac{m}{2\pi kT}\right)^{3/2} c'^2 \exp\left[-\frac{m}{2kT}c'^2\right] \tag{2.6.24}$$

This distribution is plotted in Fig. 2.6.2. The curve has a global maximum, representing the highest probability. The corresponding c' is the most probable molecular thermal speed c'_{mp} and can be obtained by setting the derivative to zero, which gives

$$c'_{mp} = \left(\frac{2kT}{m}\right)^{1/2} \tag{2.6.25}$$

The average molecular thermal speed $\overline{c'}$ can be obtained by using Eq. (2.4.6)

$$\begin{aligned} \overline{c'} &= \int_0^\infty c' f_0^{c'} dc' \\ &= \frac{2}{\sqrt{\pi}} \left(\frac{2kT}{m}\right)^{1/2} \\ &= \frac{2}{\sqrt{\pi}} c'_{mp} \end{aligned} \tag{2.6.26}$$

The root-mean-square molecular thermal speed $\sqrt{\overline{c'^2}}$ can be obtained in a similar manner, which gives

$$\sqrt{\overline{c'^2}} = \sqrt{\frac{3}{2}} c'_{mp} \tag{2.6.27}$$

Note that the speed of sound is given by $\sqrt{\gamma kT/m}$. For a monatomic gas with $\gamma = 5/3$, the speed of sound is about $0.91c'_{mp}$.

Appendix 2A: Some Useful Integrals

In this and later on in Chap. 4, definite integrals of the distribution function appear in various equation derivation processes. The following gives the integrals that are useful in these cases.

$$\int_0^\infty \exp(-a^2x^2)\,dx = \frac{\sqrt{\pi}}{2a}$$

$$\int_0^\infty x\,\exp(-a^2x^2)\,dx = \frac{1}{2a^2}$$

$$\int_0^\infty x^2\,\exp(-a^2x^2)\,dx = \frac{\sqrt{\pi}}{4a^3}$$

$$\int_0^\infty x^3\,\exp(-a^2x^2)\,dx = \frac{1}{2a^4}$$

$$\int_0^\infty x^4\,\exp(-a^2x^2)\,dx = \frac{3\sqrt{\pi}}{8a^5}$$

References

Bird, G.A., Monte Carlo simulation in an engineering context, *Progr. Astro. Aero.*, Vol. 74, pp. 239–255, 1981.

Bird, G.A., *Molecular Gas Dynamics and the Direct Simulation of Gas Flows*, Oxford University Press, New York, 1994.

Boltzmann, L., *Sber. Akad. Wiss. Wien. Abt. II*, 66, p. 275, 1872.

Borgnakke, C. and Larsen, P.S., Statistical collision model for Monte Carlo simulation of polyatomic gas mixture, *J. Computational Phys.*, Vol. 18, pp. 405–420, 1975.

Gombosi, T.I., *Gaskinetic Theory*, Cambridge University Press, Cambridge, Great Britain, 1994.

Hassan, H.A. and Hash, D.B., A generalized hard-sphere model for Monte Carlo simulations, *Phys. Fluids A*, Vol. 5, pp. 738–744, 1993.

Koga, T., *Introduction to Kinetic Theory Stochastic Process in Gaseous Systems*, Pergamon Press, Oxford, 1970.

Koura, K. and Matsumoto, H., Variable soft sphere molecular model for inverse-power-law or Lennard-Jones potential, *Phys. Fluids A*, Vol. 3, pp. 2459–2465, 1991.

Koura, K. and Matsumoto, H., Variable soft sphere molecular model for air species, *Phys. Fluids A*, Vol. 4, pp. 1083–1085, 1992.

Larsen, P.S. and Borgnakke, C., In *Rarefied Gas Dynamics*, M. Becker and M. Fiebig (eds.), Vol. 1, Paper A7, DFVLR Press, Porz-Wahn, Germany, 1974.

Vincenti, W.G. and Kruger, C.H., *Introduction to Physical Gas Dynamics*, Wiley, New York, 1965.

Chapter

3

Microfluid Flow Properties

3.1 Fundamental Flow Physics

For a flow of gas formed inside or around microscaled devices, the properties of the gas are expected to vary in the flow field. As in the case of gas flows in macroscaled devices, these changes can be caused by the distributed type of external forcing that are, for example, electrical, magnetic, or gravitational in nature. The momentum, heat, and chemical interactions between the gas and the surfaces of the microfluidic devices that are in contact with the gas flows can also have significant influences on the properties of the gas flow. Due to the small molecular weight of gases, the gravitational effects can be small in micro gas flows of single species, similar to what has generally been observed in macro gas flows. The effects of heat transfer through either contact surfaces or dissipation, on the other hand, are expected to be of equal if not greater importance in microflows. However, the difference in the length scale of the devices that the gas flows are associated with brings in additional concerns.

An intrinsic length scale in dilute gases is the mean free path, which measures the average distance the gas molecules travel between collisions. The ratio of the molecular mean free path of gas λ to a flow characteristic length scale L is defined as the Knudsen number Kn.

$$\mathrm{Kn} = \frac{\lambda}{L} \tag{3.1.1}$$

For gas flows found in conventional macroscale devices at atmospheric conditions, the flow characteristic length scale can be orders of magnitude larger than the molecular mean free path and the Knudsen number is essentially zero. For example, the mean free path of air in one standard atmospheric pressure is in the order of 10^{-8} m. The average diameter of the holes in the straight section of a saxophone is about 0.02 m and the opening at the curved end is about 0.1 m. As the air is pushed through the saxophone when it is played, the Knudsen number of the airflow is in the order of 10^{-6} to 10^{-7}. This is the case in more than just wind instruments. In fact, it is normally found in most current engineering systems that operate on gases, where the physical dimensions of the devices can be used as the flow characteristic length scales.

For microscaled devices that also operate with gases in atmospheric conditions, the scaling relation can be quite different. For a microchannel that is 1-μm deep and 1-μm wide, the Knudsen number based on the cross-sectional dimension is then 10^{-2}, which is four to five orders of magnitude larger than that of the airflow in the saxophone, a macroscale device. The Knudsen numbers of micro gas flows are significantly larger than that in the macroscaled devices. Since the mean free path varies inversely with gas density, the Knudsen number can be quite large for microfluidic devices that operate in low-pressure or low-density environment. For microflows where there are localized regions of large gradients of flow property, the device dimension cannot be used as the length scale for the entire flow field and different length scales may be required to characterize the flow in these regions. In this case, the magnitude of the Knudsen number can be even higher when a local length scale of the flow is used, instead of the global length scale. Therefore, for gas flows in microscaled devices, the Knudsen number can be close to zero or large. Since new applications of MEMS technologies are developed perhaps daily, it is difficult to estimate the upper bound.

At zero or extremely small Knudsen number, the physical domain is large compared to the mean free path. For dilute gases, there are then a large number of molecules and, for a time period that is large compared to the mean collision time (about 10^{-10} s for air in standard atmospheric condition), a large number of collisions occur among the molecules. With sufficiently large number of molecules the statistical variation of the number of molecules in the volume can be neglected. Without external forcing, there is then no gradient in the macroscopic flow properties with both time and space due to the large number of molecular collisions. The gas is in an equilibrium state. The fraction of molecules in a velocity class remains unchanged with time, even though the velocity of the individual molecule varies with collisions. The velocity distribution function for equilibrium gases is represented by the Maxwellian distribution function described in Chap. 2. A local equilibrium can be established when the gradient of the macroscopic properties is infinitesimal such that, with sufficiently high collision rate, the velocity distribution of the volume of gas can adjust to the local equilibrium state. The gas flows with such small Knudsen numbers can be regarded as a continuous distribution of matter where the local macroscopic properties can be determined by the sample average values of the appropriate molecular quantities, such as mass, velocity, and energy. Since the statistical fluctuations can be neglected, such averages are defined. The applications of the conservation of mass, momentum, and energy further provide a set of differential equations that regulate the changes of these quantities. By assuming linear relationship between stress and strain, and between heat transfer and temperature, the familiar Navier-Stokes equations can then be obtained.

For flow with high Knudsen number, the number of molecules in a significant volume of gas decreases and there could be insufficient number of molecular collisions to establish an equilibrium state. The velocity distribution function will deviate away from the Maxwellian distribution and is nonisotropic. The properties of the individual molecule then become increasingly prominent in the overall behavior of the gas as the Knudsen number increases. The implication of the larger Knudsen number is that the particulate nature of the gases needs to be included in the study. The continuum approximation used in the small Knudsen number flows becomes invalid. At the extreme end of the Knudsen number spectrum is when its value approaches infinity where the mean free path is so large or the dimension of the device is so small that intermolecular collision is not likely to occur in the device. This is called collisionless or free molecule flows.

Gas flows with similarly wide range of Knudsen number can be found in the field of rarefied gas dynamics. In fact, in rarefied gas dynamics, the Knudsen number is normally used as a measure of the degree of rarefaction. That is, the larger the value of the Knudsen number, the larger the degree of rarefaction. Rarefied gas dynamics has been studied in several different disciplines, such as the aerospace sciences. At high altitude, the air density is low and an adequate account of the rarefied gas effects is important. For example, for high-speed reentry vehicles, accurate predictions of the dynamic and thermodynamic loadings on the vehicle are essential to the design of the thermal protection systems and the safe operation of the vehicle.

If the dynamic similarity associated with the Knudsen number exists between the rarefied gas flow and the gas flow in microdevices, one can apply the findings from the rigorous rarefied gas dynamics studies to the micro gas flows. Figure 3.1.1 shows a classical description of how the gas-flow behavior should be studied from gas kinetic point of view (Bird 1994). The dilute gas assumption is assumed valid for $\delta/d > 7$, which is to the left of the vertical line of $\delta/d = 7$. In the dilute gas region, where the Boltzmann equation is valid, the continuum approach is assumed to hold until its breakdown at $\mathrm{Kn} = 0.1$. For gas with $\mathrm{Kn} > 0.1$, it is then necessary to adopt microscopic approaches, which recognize the molecular nature of gases. Statistical fluctuations become significant in

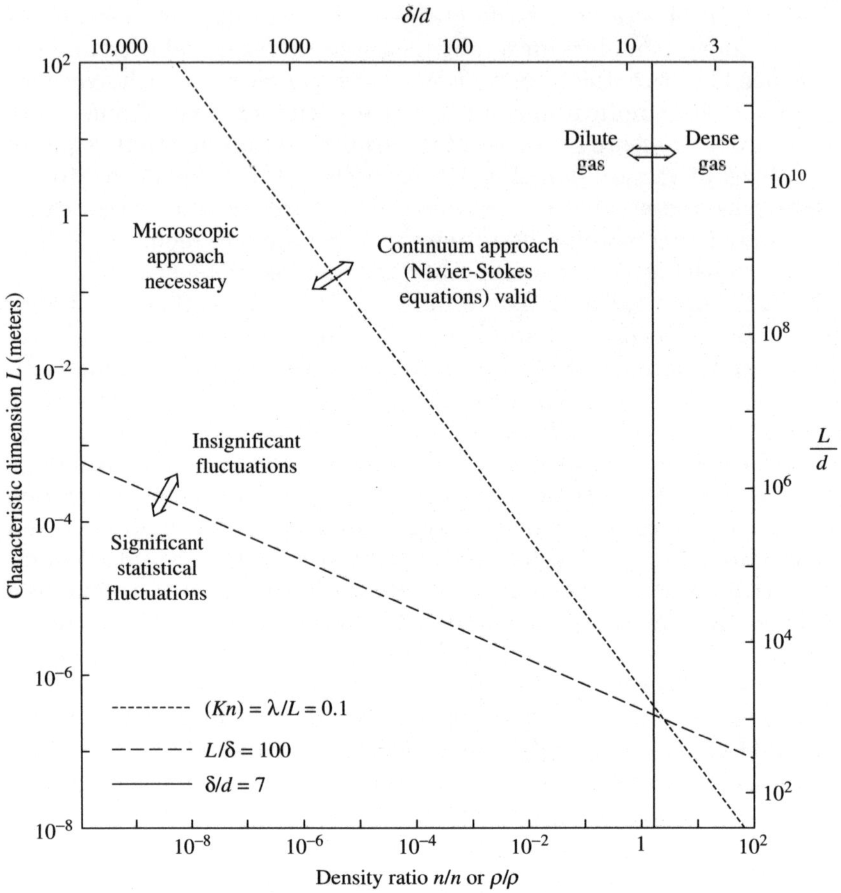

Figure 3.1.1 Limits of gas flow modeling approximations. (*Bird, 1994*)

obtaining macroscopic properties of gas by sample averaging when the number of molecules in a typical flow volume becomes small. The limit given here as L, a local characteristic length scale, is about ten times as large as the side of a cubic element that contain 1000 molecules, which correspond to statistical fluctuations with a standard deviation of approximately three percent. Note that ρ_0 and n_0 represent the mass density and the number density under standard conditions of 1-atm pressure and 0°C temperature. Typically, microflows have a characteristic length scale between 100 and 1 μm (10^{-4} and 10^{-6} m). It can be seen from Fig. 3.1 that, depending on the degree of rarefaction of the gases, the micro gas flows can be modeled as either a continuous medium or a plethora of discrete molecules in the dilute gas regime. The mathematical formulations of these two modeling approaches are different and, as can be expected, so are the processes to find the solutions to the resulting model equations. At the microscopic level, the basic mathematical model is the Boltzmann equation discussed earlier. The Navier-Stokes equations provide the mathematical model for the micro gas flows in the continuum regime. It should be noted that, in the dense gas regime to the right of the vertical line of $\delta/d = 7$, the continuum approach is valid for L in the 100-nm (10^{-9}-m) range. A significant level of statistical fluctuation of the macroscopic properties, however, may occur. The theory of Brownian motion was mentioned by Bird (1994) as one example of such phenomena. This is related to the fluid dynamics at the nanometer (10^{-9}-m) scale.

3.2 Surface Phenomena

The interface boundary conditions of solid surfaces have long been found to affect the overall behavior of gas flows. The no-slip boundary condition on the wall is recognized as a source of vorticity to the flow. In return, the surface experiences a skin friction force that tends to drag it along with the average flow. The condition also demands a conversion of gas energy from kinetic to thermal energies, resulting in problems like aerodynamic heating for high-speed vehicles. For micro gas flows, the Knudsen number, and the departure from equilibrium, is large. The no-slip boundary condition, which results from an equilibrium assumption between the surface and the rebounding molecules, becomes questionable. For instance, the rebounding molecules might not have all assumed the surface velocity, thus creating a slippage over the surface. The magnitude of the gas velocity slip tangential to the surface can be found to depend on the gradients of the velocity and the temperature on the wall. The fact that the macroscopic gas velocity is not the same as the surface speed at high Knudsen number has resulted in significant difference in the micro flow quantities, such as the pressure gradients required to drive a micro flow and the mass flow rates. Due to the small physical scale, the surface heat transfer problem is important to devices that operate at high speeds, such as the read head of the computer hard-disk drive, or that with high external energy sources from, for example, electric fields.

Gas–solid interactions are microscopic in nature. Experiments are often hindered by difficulty in securing the necessary data from various different combinations of surfaces and gases. Mathematical models developed were often correlated using macroscopic data of limited types of surface materials and gases. Recent advances in the use of the *ab initio* simulations techniques, such as the direct simulation Monte Carlo method and the molecular dynamics method, have greatly helped the efforts. Multiscale, multiphysics studies of the surface phenomena are providing very detailed information on the interactions at the gas–solid interface. The advanced three-dimensional molecular beam experimental methods are able to provide data to validate mathematical models. This capability is of particular importance for micro gas sensors and microprocessing, such as thin film vapor deposition. The gas–solid interface interactions will be discussed in detail in Chap. 7.

3.3 Basic Modeling Approaches

Beyond the continuum approach that is valid only at small Knudsen number, the gas flow is governed by the Boltzmann equation discussed in Chap. 2. The equation regulates the molecule distribution f in the phase space. As was shown previously, the macroscopic properties of the micro gas flow can be obtained by sample averaging or taking a moment of the distribution function. That is

$$\overline{Q} = \int Qf d\mathbf{c} \tag{3.3.1}$$

However, the Boltzmann equation is mathematically difficult to solve analytically. Various approximated forms of the Boltzmann equation and its solution methods have been proposed. A method proposed independently by Chapman and Enskog use the Knudsen number as a parameter to seek distribution functions that depart slightly from the equilibrium distribution. The magnitude of the Knudsen number is assumed to be small. In fact, the solutions of the Boltzmann equation are sought in the form of

$$f = f_0(1 + \mathrm{Kn}\phi_1 + \mathrm{Kn}^2\phi_2 + \cdots) \tag{3.3.2}$$

The series expansion method is an important development in gas kinetic theory. At zero Knudsen number, the distribution function is the Maxwellian distribution f_0. As the Knudsen number increases, higher order terms need to be retained. It was found that at equilibrium, the sample average properties of the gases could be described by the Euler equation, with no momentum and energy transport terms. The first-order expansion in Kn showed that the momentum and the heat transport terms are linearly related to the rate of strain and the temperature gradient, producing the same Navier-Stokes equations that can also be derived from the continuum assumption. The Navier-Stokes equations thus constitute appropriate mathematical model equations for gas flows at small Knudsen number. This will be shown in Chap. 4. The Chapman-Enskog expansion essentially validated the linear closure model used in the continuum approach, which was proposed as a phenomenological model. The second-order expansion produces nonlinear as well as high-order derivative terms of the flow properties for the transport of momentum and energy. These belong to the various forms of the Burnett equations.

A numerical simulation called the direct simulation Monte Carlo method has generally been used to simulate gas flow. The method did not solve directly the Boltzmann equation. DSMC, however, has been shown to provide solutions to the Boltzmann equation. Computational molecules are used in a DSMC simulation, each representing a large number of real molecules. The movements of the molecules are decoupled from the collision process. Representative collisions are then performed. The selection of the collision pairs and the distribution of the post collision states are random. Due the statistical nature of the method, a large number of molecules are needed to reduce the statistical scatter. For micro gas flow with a low average flow speed, the thermal fluctuation dominates the average flow and the number of molecules needed for a DSMC simulation becomes impractical. The information preservation (IP) method has been proposed for the simulation of low-speed flows. The statistical error of the IP method is small and a stationary solution of low-speed flow can be obtained more efficiently than that using the traditional DSMC method. In Chap. 5, both DSMC and IP will be described.

References

Bird, G.A., *Molecular Gas Dynamics and the Direct Simulation of Gas Flows*, Oxford University Press, New York, 1994.

Chapter

4

Moment Method: Navier-Stokes and Burnett Equations

4.1 Introduction

The Boltzmann equation can be used to describe the micro gas flow behavior at the microscopic level. The equation describes the rate of change of the number of particles due to convection in the physical and velocity space, and due to molecular collisions. In the equation, nf, the total number of particles of a given velocity class per unit volume, is the only dependent variable. The independent variables are the time, the phase space variables, including the three physical coordinates and the three velocity components. For a one-dimensional problem in the physical space, there are three independent variables. The number of the independent variables becomes five and seven in the two-dimensional and three-dimensional problems, respectively. The mathematical difficulty associated with the dimensions of the problem is further compounded by the integral form of the nonlinear collision term. As a result, it is not yet possible to find analytical solutions to the Boltzmann equation for realistic, complex flow problems such as those in microdevices. Direct numerical solutions would require discretization in the seven independent variables. The mathematical and the numerical efforts will be quite involved.

On the other hand, it is possible to develop equations that describe the macroscopic quantities of microflows by using the microscopic Boltzmann equation. For a certain limited class of microflows at small Knudsen number, it will be shown that the approximate forms of these moment equations bear much resemblance to the equations used in the continuum fluid mechanics, i.e., the Euler and the Navier-Stokes equations. In this chapter we will discuss a reduced set of equations that are valid in the limit of zero or small Knudsen number. For micro gas flows that operate in the continuum and transition regimes, these equations can be used to model the gas flow behavior.

4.2 Moment Equations

The Boltzmann equation can be written as

$$\frac{\partial}{\partial t}(nf) + \mathbf{c} \bullet \frac{\partial}{\partial \mathbf{x}}(nf) + \mathbf{F} \bullet \frac{\partial}{\partial \mathbf{c}}(nf) = \left\{ \frac{\partial}{\partial t}(nf) \right\}_{\text{collision}} \tag{4.2.1}$$

where the collision term can be written for binary collisions,

$$\frac{\partial}{\partial t}(nf) + \mathbf{c} \bullet \frac{\partial}{\partial \mathbf{x}}(nf) + \mathbf{F} \bullet \frac{\partial}{\partial \mathbf{c}}(nf) = \int_{-\infty}^{\infty} \int_{0}^{4\pi} n^2 (f^* f_1^* - f f_1) c_r \sigma d\Omega d\mathbf{c}_1 \tag{4.2.2}$$

The average property of the flow can be obtained by sample averaging the microscopic molecular quantity. Let us begin by defining, for any molecular quantity Q,

$$\begin{aligned} \overline{Q} &= \frac{1}{N} \int Q dN' \\ &= \frac{1}{N} \int QNf d\mathbf{c} \\ &= \int Qf d\mathbf{c} \end{aligned} \tag{4.2.3}$$

For example, for $Q = \mathbf{c} = \bar{\mathbf{c}} + \mathbf{c}'$,

$$\begin{aligned} \bar{\mathbf{c}} &= \frac{1}{N} \int \mathbf{c} dN' \\ &= \frac{1}{N} \int_{-\infty}^{\infty} \mathbf{c} N f d\mathbf{c} \\ &= \int_{-\infty}^{\infty} \mathbf{c} f d\mathbf{c} \end{aligned} \tag{4.2.4}$$

where $\bar{\mathbf{c}}$ denotes the average or the mean velocity and $\mathbf{c}'$ the thermal fluctuation velocity. Therefore

$$\begin{aligned} \overline{\mathbf{c}'} &= \int_{-\infty}^{\infty} (\mathbf{c} - \bar{\mathbf{c}}) f d\mathbf{c} \\ &= \bar{\mathbf{c}} - \bar{\mathbf{c}} \\ &= 0 \end{aligned} \tag{4.2.5}$$

Multiplying the Boltzmann by any molecular property Q, we get

$$Q\frac{\partial}{\partial t}(nf) + Q\mathbf{c}\bullet\frac{\partial}{\partial \mathbf{x}}(nf) + Q\mathbf{F}\bullet\frac{\partial}{\partial \mathbf{c}}(nf)$$
$$= \int_{-\infty}^{\infty}\int_{0}^{4\pi} Qn^2(f^*f_1^* - ff_1)c_r\sigma d\,\Omega d\mathbf{c}_1$$

To use the definition of $\overline{Q}$, we integrate the above equation over the velocity space, which becomes

$$\int_{-\infty}^{\infty} Q\frac{\partial}{\partial t}(nf)d\mathbf{c} + \int_{-\infty}^{\infty} Q\mathbf{c}\bullet\frac{\partial}{\partial \mathbf{x}}(nf)d\mathbf{c} + \int_{-\infty}^{\infty} Q\mathbf{F}\bullet\frac{\partial}{\partial \mathbf{c}}(nf)d\mathbf{c}$$
$$= \int_{-\infty}^{\infty}\int_{-\infty}^{\infty}\int_{0}^{4\pi} Qn^2(f^*f_1^* - ff_1)c_r\sigma d\,\Omega d\mathbf{c}_1 d\mathbf{c}$$

The terms on the left hand side of the equation can be further reduced as is shown in the following equations.

$$\int_{-\infty}^{\infty} Q\frac{\partial}{\partial t}(nf)d\mathbf{c} = \frac{\partial}{\partial t}\int_{-\infty}^{\infty} Qnf d\mathbf{c}$$
$$= \frac{\partial}{\partial t}(n\overline{Q})$$

$$\int_{-\infty}^{\infty} Q\mathbf{c}\bullet\frac{\partial}{\partial \mathbf{x}}(nf)d\mathbf{c} = \frac{\partial}{\partial \mathbf{x}}\left[\int_{-\infty}^{\infty} Q\mathbf{c}(nf)\right] d\mathbf{c}$$
$$= \frac{\partial}{\partial \mathbf{x}}(n\overline{Q\mathbf{c}})$$

$$\int_{-\infty}^{\infty} Q\mathbf{F}\bullet\frac{\partial}{\partial \mathbf{c}}(nf)d\mathbf{c} = \int_{-\infty}^{\infty} \mathbf{F}\bullet\frac{\partial}{\partial \mathbf{c}}(Qnf)d\,\mathrm{c} - \int_{-\infty}^{\infty} n\mathbf{F}\bullet\frac{\partial}{\partial \mathbf{c}}(Q)f d\mathbf{c}$$
$$= -n\mathbf{F}\bullet\overline{\frac{\partial Q}{\partial \mathbf{c}}}$$

Therefore, the moment equations or equations of transfer can now be written as

$$\frac{\partial}{\partial t}(n\overline{Q}) + \frac{\partial}{\partial \mathbf{x}}(n\overline{Q\mathbf{c}}) - n\mathbf{F}\bullet\overline{\frac{\partial Q}{\partial \mathbf{c}}} = \Delta[Q] \tag{4.2.6}$$

where $\Delta[Q]$ represents the collision integral

$$\Delta[Q] = \int_{-\infty}^{\infty}\int_{-\infty}^{\infty}\int_{0}^{4\pi} Qn^2(f^*f_1^* - ff_1)c_r\sigma d\,\Omega d\mathbf{c}_1 d\mathbf{c}$$

For a binary elastic collision, the mass, momentum, and the kinetic energy of the particle remain unchanged. Now let Q be the collision invariants, m, $m\mathbf{c}$, and $\frac{1}{2}m\,|\mathbf{c}|^2$, respectively. Say, $Q = m$

$$n\overline{Q} = nm = \rho$$
$$n\overline{Q\mathbf{c}} = nm\bar{\mathbf{c}} = \rho\bar{\mathbf{c}}$$
$$n\mathbf{F}\bullet\overline{\frac{\partial Q}{\partial \mathbf{c}}} = 0$$

Equation (4.2.6) becomes

$$\frac{\partial}{\partial t}(\rho) + \frac{\partial}{\partial \mathbf{x}}(\rho\bar{\mathbf{c}}) = 0$$

Or in index notation

$$\frac{\partial}{\partial t}(\rho) + \frac{\partial}{\partial x_i}(\rho\bar{c}_i) = 0 \tag{4.2.7}$$

where the summation convention has been used.

For $Q = m\mathbf{c}$, the molecular momentum

$$n\overline{Q} = \rho\bar{\mathbf{c}}$$
$$n\overline{Q\mathbf{c}} = \rho\overline{\mathbf{c}\mathbf{c}}$$
$$n\mathbf{F}\bullet\overline{\frac{\partial(m\mathbf{c})}{\partial \mathbf{c}}} = \rho\mathbf{F}$$

Equation (4.2.6) becomes

$$\frac{\partial}{\partial t}(\rho\bar{\mathbf{c}}) + \frac{\partial}{\partial \mathbf{x}}(\rho\overline{\mathbf{c}\mathbf{c}}) - \rho\mathbf{F} = 0 \tag{4.2.8}$$

where the momentum flux tensor can be written as the sum of the contribution from the average motion and that from the thermal fluctuation,

$$\rho\overline{\mathbf{c}\mathbf{c}} = \rho\overline{(\bar{\mathbf{c}}+\mathbf{c}')(\bar{\mathbf{c}}+\mathbf{c}')}$$
$$= \rho\bar{\mathbf{c}}\bar{\mathbf{c}} + \rho\overline{\mathbf{c}'\mathbf{c}'}$$

The pressure tensor $\rho\overline{\mathbf{c}'\mathbf{c}'}$ or $\rho\overline{c'_i c'_j}$ can be written in terms of the viscous stress tensor τ_{ij}. That is,

$$-\rho\overline{c'_i c'_j} = \tau_{ij} - p\delta_{ij}, \qquad p = \frac{1}{3}\rho\overline{c'_k c'_k} \tag{4.2.9}$$

Substituting Eq. (4.2.9) into Eq. (4.2.8), we obtain

$$\begin{aligned}\frac{\partial}{\partial t}(\rho \bar{c}_i) + \frac{\partial}{\partial x_j}(\rho \bar{c}_i \bar{c}_j) &= \rho F_i - \frac{\partial}{\partial x_j}(\rho \overline{c'_i c'_j}) \\ &= \rho F_i - \frac{\partial p}{\partial x_i} + \frac{\partial \tau_{ij}}{\partial x_j}\end{aligned} \tag{4.2.10}$$

Similarly, for $Q = \frac{1}{2} m \,|\mathbf{c}|^2$, the translational molecular kinetic energy

$$\begin{aligned}n\overline{Q} &= \frac{nm}{2}\overline{|\mathbf{c}|^2} \\ &= \frac{\rho}{2}\int_{-\infty}^{\infty} [(\bar{\mathbf{c}} \bullet \bar{\mathbf{c}}) + 2(\bar{\mathbf{c}} \bullet \mathbf{c}') + (\mathbf{c}' \bullet \mathbf{c}')] f d\mathbf{c} \\ &= \frac{\rho}{2}\left(\bar{\mathbf{c}}^2 + \overline{\mathbf{c}'^2}\right) = \frac{\rho}{2}\left(\bar{c}_k^2 + \overline{c'^2_k}\right)\end{aligned} \tag{4.2.11}$$

where

$$\bar{c}_k^2 = \bar{c}_k \bar{c}_k \qquad \text{and} \qquad \overline{c'^2_k} = \overline{c'_k c'_k}$$

The energy flux becomes

$$\begin{aligned}n\overline{Qc} &= \frac{nm}{2}\overline{|\mathbf{c}|^2\,\mathbf{c}} \\ &= \frac{\rho}{2}\int_{-\infty}^{\infty} [(\bar{\mathbf{c}} \bullet \bar{\mathbf{c}}) + 2(\bar{\mathbf{c}} \bullet \mathbf{c}') + (\mathbf{c}' \bullet \mathbf{c}')](\bar{\mathbf{c}} + \mathbf{c}') f d\mathbf{c} \\ &= \frac{\rho}{2}\left(\bar{c}_k^2 \bar{c}_i + \bar{c}_i \overline{c'^2_k} + 2\bar{c}_j \overline{c'_i c'_j} + \overline{c'_i c'_j c'_j}\right)\end{aligned} \tag{4.2.12}$$

The change of kinetic energy due to work done by external forcing is

$$\begin{aligned}n\mathbf{F} \bullet \overline{\frac{\partial Q}{\partial \mathbf{c}}} &= \frac{\rho}{2}\mathbf{F} \bullet \overline{\frac{\partial (\mathbf{c} \bullet \mathbf{c})}{\partial \mathbf{c}}} \\ &= \rho \mathbf{F} \bullet \bar{\mathbf{c}} \\ &= \rho F_j \bar{c}_j\end{aligned}$$

Therefore the moment equation for $Q = \frac{1}{2} m \,|\mathbf{c}|^2$ becomes

$$\frac{\partial}{\partial t}\left[\frac{\rho}{2}\left(\bar{c}_k^2 + \overline{c'^2_k}\right)\right] + \frac{\partial}{\partial x_i}\left[\frac{\rho}{2}\left(\bar{c}_k^2 \bar{c}_i + \bar{c}_i \overline{c'^2_k} + 2\bar{c}_j \overline{c'_i c'_j} + \overline{c'_i c'_j c'_j}\right)\right] - \rho F_i \bar{c}_i = 0 \tag{4.2.13}$$

The equation can be further reduced by subtracting from it the equation for the average kinetic energy $\frac{\rho}{2}\bar{c}_i^2$, which can be obtained by multiplying Eq. (4.2.10) by $\bar{c}_i$. Eq. (4.2.13) then becomes

$$\frac{\partial}{\partial t}\left(\frac{\rho}{2}\overline{c'^2_k}\right) + \frac{\partial}{\partial x_i}\left(\frac{\rho}{2}\bar{c}_i\overline{c'^2_k}\right) = \tau_{ij}\frac{\partial \bar{c}_i}{\partial x_j} - p\frac{\partial \bar{c}_j}{\partial x_j} - q_{i,i} \tag{4.2.14}$$

where the heat transfer flux vector is defined as

$$q_i = \frac{\rho}{2}\overline{c'_i c'_j c'_j}$$

The moment Eq. (4.2.7), (4.2.10), and (4.2.14) describe the changes of the macroscopic properties of the microflow. They are the density, the average velocity, and the kinetic energy. Their derivations are based on first principles and they are exact conservation laws for these quantities. At this point, the forms of the viscous stress tensor $\rho\overline{c'_i c'_j}$ and the heat transfer flux vector q_i terms are not known. In order to form a closed set of equations, these higher order moments should be related to the lower order moments in the equation. This is then a closure problem. The relations between the viscous stress tensor and the heat flux vector with the lower-order moments are then called constitutive relations. When these closure relations are derived by physical or phenomenological approximations, the resulting closed set of conservation equations can then be used for mathematical modeling of the microfluid flows.

We can consider the gas in local equilibrium where the velocity distribution function is isotropic. The local equilibrium situation requires that the local relaxation time for translational nonequilibrium is negligible compared with the characteristic time of the flow. As described in Chap. 2, the velocity distribution function in equilibrium gas is the Maxwellian distribution,

$$f_0 = \left(\frac{m}{2\pi kT}\right)^{3/2} \exp\left[-\frac{m}{2kT}c'_k c'_k\right]$$

Let's look at the shearing components ($i \neq j$) of the viscous stress tensors first. Say, $i = s$, $j = t$, where s and t denote two different index values.

$$\begin{aligned}
-\rho\overline{c'_s c'_t} &= \tau_{st} \\
&= -\rho\int_{-\infty}^{\infty} c'_s c'_t\, f\, d\mathbf{c} \\
&= -\rho\left(\frac{m}{2\pi kT}\right)^{3/2}\int_{-\infty}^{\infty} c'_s \exp\left[-\frac{m}{2kT}c'_s c'_s\right] dc'_s \int_{-\infty}^{\infty} c'_t \\
&\quad \times \exp\left[-\frac{m}{2kT}c'_t c'_t\right] dc'_t \int_{-\infty}^{\infty} \exp\left[-\frac{m}{2kT}c'_u c'_u\right] dc_u \\
&= 0
\end{aligned}$$

since the first and the second integrals are zero. Note that $s \neq t \neq u$. That is, the three symbols denote different values among 1, 2, and 3. In locally equilibrium gases, the shearing viscous stresses are then zero. For the normal stresses, i.e., $i = j = s$,

$$\begin{aligned}
\tau_{ss} &= -\rho\overline{c'_s c'_s} + p \\
&= -\rho \int_{-\infty}^{\infty} c'_s c'_s f \, d\mathbf{c} + p \\
&= -\rho\left(\frac{m}{2\pi kT}\right)^{3/2} \int_{-\infty}^{\infty} c'^2_s \exp\left[-\frac{m}{2kT}c'_s c'_s\right] dc'_s \\
&\quad \times \int_{-\infty}^{\infty} \exp\left[-\frac{m}{2kT}c'_t c'_t\right] dc'_t \int_{-\infty}^{\infty} \exp\left[-\frac{m}{2kT}c'_u c'_u\right] dc_u + p
\end{aligned}$$

where $s \neq t \neq u$. Or

$$\begin{aligned}
\tau_{ss} &= -\rho\left(\frac{m}{2\pi kT}\right)^{3/2} \left[\frac{\pi^{1/2}}{2(m/2kT)^{3/2}}\right] \left[\frac{\pi^{1/2}}{(m/2kT)^{1/2}}\right] \left[\frac{\pi^{1/2}}{(m/2kT)^{1/2}}\right] + p \\
&= 0
\end{aligned}$$

Therefore, there is no viscous stresses in equilibrium gases. For the heat flux term, say for $i = s$ and $s \neq t \neq u$,

$$\begin{aligned}
q_s &= \frac{\rho}{2} \int_{-\infty}^{\infty} c'_s c'_j c'_j f \, d\mathbf{c} \\
&= \frac{\rho}{2}\left(\frac{m}{2\pi kT}\right)^{3/2} \left\{
\begin{aligned}
&\int_{-\infty}^{\infty} c'^3_s \exp\left[-\frac{m}{2kT}c'_s c'_s\right] dc'_s \\
&\quad \times \int_{-\infty}^{\infty} \exp\left[-\frac{m}{2kT}c'_t c'_t\right] dc'_t \\
&\quad \times \int_{-\infty}^{\infty} \exp\left[-\frac{m}{2kT}c'_u c'_u\right] dc'_u \\
&+2\int_{-\infty}^{\infty} c'_s \exp\left[-\frac{m}{2kT}c'_s c'_s\right] dc'_s \\
&\quad \times \int_{-\infty}^{\infty} c'_t \exp\left[-\frac{m}{2kT}c'_t c'_t\right] dc'_t \\
&\quad \times \int_{-\infty}^{\infty} \exp\left[-\frac{m}{2kT}c'_u c'_u\right] dc'_u
\end{aligned}
\right\} \\
&= 0
\end{aligned}$$

since the leading integrals in each of the two terms inside the bracket are zero. Therefore for locally equilibrium gas flows, the viscous stress tensor and the heat flux vector do not appear in the moment equations.

The moment equations then form a closed set of equations and are equivalent to that of the Euler equations.

4.3 The Chapman-Enskog Expansion

For nonequilibrium gas flows, Chapman and Enskog independently proposed solution methods for the Boltzmann equation. The Chapman-Enskog method uses a power series expansion of the distribution function from the equilibrium state. Solutions of the Boltzmann equation in the various order of the expansion parameter can then be obtained. A summary of the historical development of the Chapman-Enskog method can be found in Chapman and Cowling (1970).

We begin with the nondimensionalized form of the Boltzmann equation

$$\xi\left[\frac{\partial}{\partial t}(nf)+\mathbf{c}\bullet\frac{\partial}{\partial \mathbf{x}}(nf)+\mathbf{F}\bullet\frac{\partial}{\partial \mathbf{c}}(nf)\right] = \int_{-\infty}^{\infty}\int_{0}^{4\pi} n^2(f^*f_1^*-ff_1)c_r\sigma d\,\Omega d\mathbf{c}_1 \tag{4.3.1}$$

where c_r, L, L/c_r, c_r^{-3}, n_r, and ν_r/n_r have been used to nondimensionalize the velocity, length, average flow time, f, n, and the collision volume terms. Also,

$$\xi = \frac{(c_r/\nu_r)}{L}$$

If the thermal speed $\sqrt{c_k'c_k'}$ is used as the reference velocity, ξ becomes the Knudsen number. As was discussed earlier, the Knudsen number can be used as a measure for the level of departure from equilibrium state. For often seen microflows in the continuum and in the transitional regimes, the flows are not highly nonequilibrium and the Knudsen number is small. For such small Knudsen numbers, the form of Eq. (4.3.1) warrants a perturbation approach where the solutions are sought in the various order of the small parameter, in this case, the Knudsen number. Note that, for brevity, we have used in Eq. (4.3.1) the same notations as its dimensional counterpart, with the understanding that the quantities in Eq. (4.3.1) are nondimensional.

For small values of ξ, or small departure from equilibrium, the Chapman-Enskog method assumes a power series expansion of the distribution function f using ξ.

$$f = f_0(1+\xi\phi_1+\xi^2\phi_2+\cdots) = f_0(1+\Phi_1+\Phi_2+\cdots) \tag{4.3.2}$$

Note that in the limit of near equilibrium ($\xi \to 0$), $f \to f_0$ and the Maxwellian distribution function is recovered. As was shown in the previous section, for equilibrium gases, the viscous stress tensor and the heat flux vector assume the value of zero and the moment equations are equivalent to the Euler equations. The first-order approximation of the Chapman-Enskog expansion of the Boltzmann equation will result in a viscous tensor that is related to the rate of strain of the average flow and a heat flux vector that is related to the temperature gradient. Therefore, the first-order Chapman-Enskog expansion of the Boltzmann equation produces an equivalent of the Navier-Stokes equations. The second-order expansion of Chapman-Enskog of the Boltzmann equation will result in, when used with the moment equations, the so call Burnett equations. The viscous stress tensor and the heat flux vector in the Burnett equations contain nonlinear function of the mean flow gradients. Moment equations with higher-order Chapmann-Enskog expansion are called super Burnett equations (Shavaliev 1978). These equations are mathematically complex and are rarely used.

The process of the Chapman-Enskog expansion of the original Boltzmann equation (Chapman and Cowling 1970) is rather involved and will not be shown here. Instead, in the following, we will demonstrate how the Chapman-Enskog expansion proceeds by using a modeled Boltzmann equation. The model equation, known as the Krook equation, uses the Bhatnagar, Gross, and Krook (BGK) approximation model for the collision term in the Boltzmann equation. The resulting Krook equation is physically sound. The Chapman-Enskog solution of the Krook equation is relatively more straightforward than that for the Boltzmann equation. It will be shown that the first-order Chpaman-Enskog expansion results in the Navier-Stokes equations. The conventional Burnett equations will then be introduced. The results of application of the Burnett equations and the Navier-Stokes equations are described in Sec. 4.5, and Chap. 9.

4.3.1 The Krook equation

The evaluation of the nonlinear collision integral term poses a great difficulty to finding a solution of the Boltzmann equation. A simplified model proposed by Bhatnagar, Gross, and Krook (1954) BGK for this term greatly simplifies the equation. BGK model approximates the collision integral in the following manner

$$\left\{ \frac{\partial}{\partial t}(nf) \right\}_{\text{collision}} = n\nu(f_0 - f)$$

and the resulting Boltzmann equation, or the Krook equation, becomes

$$\frac{\partial}{\partial t}(nf) + \mathbf{c} \bullet \frac{\partial}{\partial \mathbf{x}}(nf) + \mathbf{F} \bullet \frac{\partial}{\partial \mathbf{c}}(nf) = n\nu(f_0 - f) \tag{4.3.3}$$

The nondimensionalized form can be written as

$$\xi \left[\frac{\partial}{\partial t}(nf) + \mathbf{c} \bullet \frac{\partial}{\partial \mathbf{x}}(nf) + \mathbf{F} \bullet \frac{\partial}{\partial \mathbf{c}}(nf)\right] = n\nu(f_0 - f)$$

Again, we have continued to use the notations in the nondimensionalized form as the corresponding quantities in the dimensional form with an understanding that they are now representing dimensionless quantities. Using the Chapman-Enskog expansion and retaining the first-order expansion in ξ from both sides of the equation, we obtain

$$\frac{\partial}{\partial t}(nf_0) + c_i \frac{\partial}{\partial x_i}(nf_0) + F_i \frac{\partial}{\partial c_i}(nf_0) = -n\nu\phi_1 f_0$$

The first-order expansion function ϕ_1 can then be written as

$$\phi_1 = -\frac{1}{n\nu f_0}\left[\underbrace{\frac{\partial}{\partial t}(nf_0)}_{A} + \underbrace{c_i \frac{\partial}{\partial x_i}(nf_0)}_{B} + \underbrace{F_i \frac{\partial}{\partial c_i}(nf_0)}_{C}\right] \tag{4.3.4}$$

which can be explicitly, though not straightforwardly, determined. In fact, it can be shown that successive higher order expansion function ϕ_{m+1} can be related to the next lower order function ϕ_m, i.e.,

$$\phi_{m+1} = -\frac{1}{n\nu f_0}\left[\frac{\partial}{\partial t}(nf_m) + c_i \frac{\partial}{\partial x_i}(nf_m) + F_i \frac{\partial}{\partial c_i}(nf_m)\right] \tag{4.3.5}$$

To proceed with the first-order solution to find ϕ_1, we note first that the nondimensional Maxwellian distribution f_0 can be written as

$$f_0 = \left(\frac{1}{2\pi T}\right)^{3/2} \exp\left[-\frac{c'_k c'_k}{2T}\right]$$

where the temperature T has been normalized by $\frac{mc_r^2}{k}$. Since the Maxwellian distribution function depends on $\mathbf{c}$ and T, which are functions of x_i and t, the differential terms in Eq. (4.3.4) need to be expanded using the chain rule. For convenience, the three terms in the square bracket have been designated as terms A, B, and C in Eq. (4.3.4).

The nondimensional form of the moment equations for equilibrium gases will be used in the following derivation and can be written as

$$\frac{\partial}{\partial t}(n) + \frac{\partial}{\partial x_i}(n\bar{c}_i) = 0 \tag{4.3.6}$$

$$\frac{\partial}{\partial t}(\bar{c}_i) + \bar{c}_i\frac{\partial}{\partial x_j}(\bar{c}_j) = F_i - \frac{1}{n}\frac{\partial p}{\partial x_i} + \frac{1}{n}\frac{\partial \tau_{ij}}{\partial x_j} \tag{4.3.7}$$

$$\frac{\partial}{\partial t}(T) + \bar{c}_i\frac{\partial}{\partial x_i}(T) = \frac{2}{3n}\tau_{ij}\frac{\partial \bar{c}_i}{\partial x_j} - \frac{2}{3}\frac{p}{n}\frac{\partial \bar{c}_j}{\partial x_j} - \frac{2}{3n}q_{i,i} \tag{4.3.8}$$

since $\frac{1}{2}\overline{c_i'^2} = \frac{3}{2}\frac{k}{m}T$. The terms in A in Eq. (4.3.4) can be grouped in the following manner:

$$\frac{\partial}{\partial t}(nf_0) = \left(\underbrace{\frac{\partial}{\partial n}}_{A1}\underbrace{\frac{\partial n}{\partial t}}_{A2} + \underbrace{\frac{\partial}{\partial \bar{c}_j}}_{A3}\underbrace{\frac{\partial \bar{c}_j}{\partial t}}_{A4} + \underbrace{\frac{\partial}{\partial T}}_{A5}\underbrace{\frac{\partial T}{\partial t}}_{A6}\right)nf_0$$

and the $A1$ to $A6$ terms can be evaluated by using f_0 and (4.3.6–8).

$$A1: \frac{\partial}{\partial n}(nf_0) = f_0$$

$$A2: \frac{\partial n}{\partial t} = -\bar{c}_k\frac{\partial n}{\partial x_k} - n\frac{\partial \bar{c}_k}{\partial x_k}$$

$$A3: \frac{\partial}{\partial \bar{c}_j}(nf_0) = \frac{\partial}{\partial \bar{c}_j}\left\{n\left(\frac{1}{2\pi T}\right)^{3/2}\exp\left[-\frac{(|c_k - \bar{c}_k|)^2}{2T}\right]\right\}$$

$$= n\left\{\left(\frac{1}{2\pi T}\right)^{3/2}\exp\left[-\frac{(|c_k - \bar{c}_k|)^2}{2T}\right]\frac{c_j'}{T}\right\}$$

$$= nf_0\frac{c_j'}{T}$$

$$A4: \frac{\partial \bar{c}_j}{\partial t} = -\bar{c}_k\frac{\partial \bar{c}_j}{\partial x_k} - \frac{1}{n}\frac{\partial p}{\partial x_j} + \frac{1}{n}\frac{\partial \tau_{jk}}{\partial x_k} + F_j$$

$$A5: \frac{\partial}{\partial T}(nf_0) = \frac{\partial}{\partial T}\left\{n\left(\frac{1}{2\pi T}\right)^{3/2}\exp\left[-\frac{c_k'c_k'}{2T}\right]\right\}$$

$$= n\left(-\frac{3}{2}\right)T^{-1}f_0 + n\left(-\frac{c_k'c_k'}{2}\right)\left(-\frac{1}{T^2}\right)f_0$$

$$= nf_0\left(\frac{c_k'c_k'}{2T^2} - \frac{3}{2T}\right)$$

$$A6: \frac{\partial T}{\partial t} = -\bar{c}_k\frac{\partial}{\partial x_k}(T) + \frac{2}{3n}\tau_{kl}\frac{\partial \bar{c}_k}{\partial x_l} - \frac{2}{3}\frac{p}{n}\frac{\partial \bar{c}_k}{\partial x_k} - \frac{2}{3n}q_{k,k}$$

Similarly, the B term can be written as

$$\bar{c}_i \frac{\partial}{\partial x_i}(nf_0) = \bar{c}_i \left(\underbrace{\frac{\partial}{\partial n}\frac{\partial n}{\partial x_i}}_{A1} + \underbrace{\frac{\partial}{\partial \bar{c}_j}\frac{\partial \bar{c}_j}{\partial x_i}}_{A3} + \underbrace{\frac{\partial}{\partial T}\frac{\partial T}{\partial x_i}}_{A5} \right) nf_0$$

Terms in the square bracket that are the same terms as those that appear in group A are marked as such. By using the relation derived in A, the term C becomes

$$F_i \frac{\partial}{\partial c_i}(nf_0) = -F_i f_0 \frac{c_i'}{T}$$

By substituting the terms A, B, and C into Eq. (4.3.4), we can find the first-order expansion function

$$\phi_1 = -\frac{1}{\nu}\left[c_i'\left(\frac{c_k'c_k'}{2T} - \frac{5}{2}\right)\frac{\partial(\ln T)}{\partial x_i} + \frac{1}{T}\left(c_i'c_j' - \frac{1}{3}c_k'c_k'\delta_{ij}\right)\frac{\partial \bar{c}_j}{\partial x_i}\right] \tag{4.3.9}$$

Or

$$\Phi_1 = \xi\phi_1 = -\frac{\xi}{\nu}\left[c_i'\left(\frac{c_k'c_k'}{2T} - \frac{5}{2}\right)\frac{\partial(\ln T)}{\partial x_i} + \frac{1}{T}\left(c_i'c_j' - \frac{1}{3}c_k'c_k'\delta_{ij}\right)\frac{\partial \bar{c}_j}{\partial x_i}\right] \tag{4.3.10}$$

Therefore the first-order Chapman-Enskog solution of the Krook equation becomes

$$\begin{aligned} f &= f_0(1+\xi\phi_1) \\ &= f_0\left\{1 - \frac{\xi}{\nu}\left[c_i'\left(\frac{c_k'c_k'}{2T} - \frac{5}{2}\right)\frac{\partial(\ln T)}{\partial x_i} + \frac{1}{T}\left(c_i'c_j' - \frac{1}{3}c_k'c_k'\delta_{ij}\right)\frac{\partial \bar{c}_j}{\partial x_i}\right]\right\} \end{aligned} \tag{4.3.11}$$

Note that the shear stress tensor and the heat flux vector are of order ξ and therefore do not contribute to the ϕ_1 function.

4.3.2 The Boltzmann equation

For comparison, a rigorous solution of the first-order Chapman-Enskog expansion of the original form of the Boltzmann equation can be shown to result in a distribution function

$$\begin{aligned} f &= f_0(1+\xi\phi_1) = f_0\left\{1 - \frac{\xi}{n}\left[A(|\hat{\mathbf{c}}'|, T)\hat{c}_i'\sqrt{2T}\frac{\partial(\ln T)}{\partial x_i}\right.\right. \\ &\quad \left.\left. + B(|\hat{\mathbf{c}}'|, T)\left(\hat{c}'_i\hat{c}'_j - \frac{1}{3}\hat{c}'_k\hat{c}'_k\delta_{ij}\right)\frac{\partial \bar{c}_j}{\partial x_i}\right]\right\} \end{aligned} \tag{4.3.12}$$

where $\hat{\mathbf{c}}' = \mathbf{c}'/\sqrt{2T}$. It can be seen that functional form of the first-order velocity distribution for the Boltzmann equation is similar to that for the Krook equation shown above. $A(|\hat{\mathbf{c}}'|, T)$ and $B(|\hat{\mathbf{c}}'|, T)$ are determined by integral equations with closed-form solutions that exist only for Maxwellian molecules. Burnett (1935) introduced a series solution for $A(|\hat{\mathbf{c}}'|, T)$ and $B(|\hat{\mathbf{c}}'|, T)$ by using the Sonine polynomials. The solution process is mathematically involved. Interested readers are referred to Vincenti and Kruger (1965) for the derivation.

Second-order expansions of the Chapman-Enskog method, or

$$f = f_0(1 + \xi\phi_1 + \xi^2\phi_2) \tag{4.3.13}$$

of the Boltzmann equation was carried out by Burnett (1935). Again, the expansion process is complex and beyond the scope of this writing. In the following, we will examine the constitutive relations for $\rho\overline{c_i'c_j'}$ and $q_{i,i}$ that can be derived from Eq. (4.3.11) and obtain a closed set of conservation equations for microfluid flows. The Burnett equations will then be introduced.

4.4 Closure Models

We will frist examine the closure model for the viscous tensor. Nondimensionally, it can be written as

$$\tau_{ij} = -n\overline{c_i'c_j'} + p\delta_{ij}$$

4.4.1 First-order modeling

An expression for the pressure tensor in terms of the sample averaging and the distribution function obtained with the Chapman-Enskog expansion of order ξ can be written as $\tau_{ij}^{(1)}$, the contribution to τ_{ij} from the first-order terms of ξ. Likewise, the terms that are in the order of ξ^2 are represented by $\tau_{ij}^{(2)}$, which will be described later. $\tau_{ij}^{(1)}$ can then be written as

$$\begin{aligned}
\tau_{ij}^{(1)} &= -n\int_{-\infty}^{\infty} c_i'c_j' f d\mathbf{c} + p\delta_{ij} \\
&= -n\int_{-\infty}^{\infty} c_i'c_j' f_0(1+\xi\phi_1) d\mathbf{c} + p\delta_{ij} \\
&= -n\int_{-\infty}^{\infty} c_i'c_j' f_0 d\mathbf{c} + p\delta_{ij} + n\int_{-\infty}^{\infty} c_i'c_j' f_0 \\
&\quad \times \left\{\frac{\xi}{\nu}\left[c_s'\left(\frac{c_k'c_k'}{2T} - \frac{5}{2}\right)\frac{\partial(\ln T)}{\partial x_s} + \frac{1}{T}\left(c_s'c_t' - \frac{1}{3}c_u'c_u'\delta_{st}\right)\frac{\partial \bar{c}_t}{\partial x_s}\right]\right\} d\mathbf{c}
\end{aligned}$$

$$= \frac{n\xi}{\nu}\int_{-\infty}^{\infty} c_i' c_j' f_0 c_s' \left(\frac{c_k' c_k'}{2T} - \frac{5}{2}\right) \frac{\partial(\ln T)}{\partial x_s} d\mathbf{c} \qquad (4.4.1)$$

$$+ \frac{n\xi}{\nu}\int_{-\infty}^{\infty} c_s' c_t' f_0 \frac{1}{T}\left(c_s' c_t' - \frac{1}{3}c_u' c_u' \delta_{st}\right)\frac{\partial \bar{c}_t}{\partial x_s} d\mathbf{c}$$

$$= \frac{n\xi}{\nu T}\int_{-\infty}^{\infty} c_i' c_j' f_0 \left(c_s' c_t' - \frac{1}{3}c_u' c_u' \delta_{st}\right)\frac{\partial \bar{c}_t}{\partial x_s} d\mathbf{c}$$

$$= \frac{n\xi}{\nu T}\frac{2}{15}\overset{o}{S}_{ij}\int_{-\infty}^{\infty} f_0 |\mathbf{c}'|^4 d\mathbf{c}$$

where the strain rate tensor S_{ij} is defined as

$$S_{ij} = \frac{1}{2}\left(\frac{\partial \bar{c}_i}{\partial x_j} + \frac{\partial \bar{c}_j}{\partial x_i}\right)$$

and the corresponding traceless tensor $\overset{o}{S}_{ij}$ is defined as

$$\overset{o}{S}_{ij} = \frac{1}{2}\left(\frac{\partial \bar{c}_i}{\partial x_j} + \frac{\partial \bar{c}_j}{\partial x_i} - \frac{2}{3}\frac{\partial \bar{c}_k}{\partial x_k}\delta_{ij}\right)$$

Note that a tensor equation has been used in deriving Eq. (4.4.1)

$$\int_{-\infty}^{\infty} c_i' c_j' f_0 \left(c_s' c_t' - \frac{1}{3}c_u' c_u' \delta_{ij}\right)\frac{\partial \bar{c}_t}{\partial x_s} d\mathbf{c} = \frac{2}{15}\overset{o}{S}_{ij}\int_{-\infty}^{\infty} f_0 |\mathbf{c}'|^4 d\mathbf{c}$$

Substituting the Maxwellian distribution and writing the equation in the Cartesian coordinate system, we obtain

$$\tau_{ij}^{(1)} = \frac{n\xi}{\nu T}\frac{2}{15}\overset{o}{S}_{ij}$$

$$\left\{\int_{-\infty}^{\infty}\int_{-\infty}^{\infty}\int_{-\infty}^{\infty}\left(\frac{1}{2\pi T}\right)^{3/2}\exp\left[-\frac{c_1'^2 + c_2'^2 + c_3'^2}{2T}\right]\right.$$

$$\left.\times\left[c_1'^4 + c_2'^4 + c_3'^4 + 2(c_1'^2 c_2'^2 + c_2'^2 c_3'^2 + c_3'^2 c_1'^2)\right] dc_1 dc_2 dc_3\right\}$$

The triple integration can be evaluated and the results are

$$\tau_{ij}^{(1)} = \frac{n\xi}{\nu T}\frac{2}{15}\overset{o}{S}_{ij}\left[(3)\left(\frac{6}{8}\right)\pi^{1/2}(2T)^{5/2}\left(\pi^{1/2}(2T)^{1/2}\right)^2\right.$$

$$\left.+(3)(2)(\pi^{1/2}(2T)^{3/2})^2\pi^{1/2}(2T)^{1/2}\right] \qquad (4.4.2)$$

$$= \frac{2n\xi T}{\nu}\overset{o}{S}_{ij}$$

Or if we use the dimensional form of the viscous stress tensor,

$$\tau_{ij}^{(1)} = \frac{2nkT}{\nu} \overset{o}{S}_{ij} \tag{4.4.3}$$

The Chapman-Enskog's first order expansion of the Krook equation thus produces a constitutive relation for viscous tensor that is equivalent to that used in the Navier-Stokes equation of the continuum approach. That is,

$$\tau_{ij}^{(1)} = 2\mu \overset{o}{S}_{ij} \tag{4.4.4}$$

where

$$\mu = \frac{nkT}{\nu}$$

Similarly, it can be shown that the heat flux vector can be written as

$$\begin{aligned} q_i^{(1)} &= \frac{\rho}{2}\overline{c_i' c_j' c_j'} \\ &= -\frac{5}{2}\left(\frac{k}{m}\right)\frac{nkT}{\nu}\frac{\partial T}{\partial x_i} \\ &= -K\frac{\partial T}{\partial x_i} \end{aligned} \tag{4.4.5}$$

where the thermal conductivity K is

$$K = -\frac{5}{2}\left(\frac{k}{m}\right)\frac{nkT}{\nu}$$

Instead, if the first-order Chapman-Enskog expansion for the original Boltzmann equation is used, one would obtain similar linear constitutive relations with

$$\mu = \frac{2kT}{15}\int_{-\infty}^{\infty} B(|\hat{\mathbf{c}}'|, T)|\hat{\mathbf{c}}'|^4 f_0 d\mathbf{c}$$

and

$$K = \frac{2k^2T}{3m}\int_{-\infty}^{\infty} A(|\hat{\mathbf{c}}'|, T)|\hat{\mathbf{c}}'|^4 f_0\, d\mathbf{c}$$

The transport properties μ and K thus depend on $A(|\hat{\mathbf{c}}'|, T)$ and $B(|\hat{\mathbf{c}}'|, T)$. For inverse-power repulsive force gas models where the intermolecular potential is proportional to $r^{-\alpha}$, the transport properties were found to depend on temperature. The transport properties μ and K can be written as

$$\mu = \mu_r \left(\frac{T}{T_r} \right)^{(\alpha+4)/2\alpha} \tag{4.4.6}$$

and

$$K = K_r \left(\frac{T}{T_r} \right)^{(\alpha+4)/2\alpha} \tag{4.4.7}$$

where μ_r, K_r, and T_r denote reference quantities. For Maxwellian molecules with $\alpha = 4$, the transport properties depend linearly on temperature, which was also found in the first-order expansion of the Krook equation shown above.

4.4.2 Second-order modeling

Burnett (1935) used the second-order expansion of the Chapman-Enskog theory to find an approximate solution to the Boltzmann equation. By retaining a higher-order term in the expansion, the solution would then be applicable to microflow problems of higher Knudsen numbers. Compared with the first-order constitutive relations derived in the previous section, the closure models thus obtained contain higher-order derivatives of the average quantities and are nonlinear. The resulting set of moment equations is called the original Burnett equations. Chapman and Cowling (1970) used the substantial derivative of the Euler equation in the original Burnett equations and obtained the "conventional" Burnett equations. The conventional Burnett equations were found (Fiscko and Chapman, 1988) to exhibit instability with fine computational meshes in the order of the mean free path in a normal shock problem. To enhance the stability of the conventional Burnett equations, linear stress and heat flux terms selectively lifted from the super Burnett equations were added by Zhong (1991) and the resulting equations are called the "augmented" Burnett equations. Balakrishnan and Agarwal (1996) developed the "BGK"-Burnett equations by using the second-order Chapman-Enskog expansion of the Krook equation. Linear analyses show that the additional stress and heat flux terms make the equations unconditionally stable at high Knudsen number. It should be noted that Woods (1979) demonstrated that the Burnett equations can also be derived without using the Chapman-Enskog expansion.

The second-order closure relations have been found to describe better the physics of flows in the continuum transitional regime than the first-order closure incorporated in the Navier-Stokes equations. The derivation of the various second-order models are more complex than that for the first-order closure shown in the previous section. In the following, the constitutive relations used in the conventional, the augmented, and the BGK-Burnett equations are shown. Micro Couette flow solutions obtained by using the various forms of the Burnett equations can be found in the literature (Xue et al., 2001; Lockerby and Reese, 2003). In a later chapter, the numerical solutions of conventional and the augmented Burnett equations for the micro Poisuelle flow will be shown. The all-Mach number numerical algorithm used will be described in the last part of this chapter.

For a gas near thermodynamic equilibrium, the constitutive relations for the viscous stress and heat flux terms in the conventional Burnett equations can be written in the following forms:

$$\begin{aligned} \tau_{ij} &= \tau_{ij}^{(1)} + \tau_{ij}^{(2)} + O(Kn^3) \\ q_i &= q_i^{(1)} + q_i^{(2)} + O(Kn^3) \end{aligned} \tag{4.4.8}$$

where

$$\tau_{ij}^{(2)} = -\frac{\mu^2}{p}\left\{\begin{array}{l} \omega_1 \dfrac{\partial \bar{c}_k}{\partial x_k} \overset{o}{S}_{ij} + \omega_2 \left[-\overset{o}{\dfrac{\partial}{\partial x_i}\left(\dfrac{1}{\rho}\dfrac{\partial p}{\partial x_j}\right)} - \overset{o}{\dfrac{\partial \bar{c}_k}{\partial x_i}\dfrac{\partial \bar{c}_j}{\partial x_k}} - 2\overset{o}{\overset{o}{S}_{ik}\dfrac{\partial \bar{c}_k}{\partial x_j}} \right] \\ + \omega_3 R \overset{o}{\dfrac{\partial^2 T}{\partial x_i \partial x_j}} + \omega_4 \dfrac{1}{\rho T} \overset{o}{\dfrac{\partial p}{\partial x_i}\dfrac{\partial T}{\partial x_j}} + \omega_5 \dfrac{R}{T} \overset{o}{\dfrac{\partial T}{\partial x_i}\dfrac{\partial T}{\partial x_j}} + \omega_6 \overset{o}{\overset{o}{S}_{ik}\overset{o}{S}_{kj}} \end{array}\right\} \tag{4.4.9}$$

$$q_i^{(2)} = -\frac{\mu^2}{p}\left\{\begin{array}{l} \vartheta_1 \dfrac{1}{T}\dfrac{\partial \bar{c}_k}{\partial x_k}\dfrac{\partial T}{\partial x_i} + \vartheta_2 \dfrac{1}{T}\left[\dfrac{2}{3}\dfrac{\partial}{\partial x_i}\left(T\dfrac{\partial \bar{c}_k}{\partial x_k}\right) + 2\dfrac{\partial \bar{c}_k}{\partial x_i}\dfrac{\partial T}{\partial x_k}\right] \\ + \vartheta_3 \dfrac{1}{\rho}\dfrac{\partial p}{\partial x_k}\overset{o}{S}_{ki} + \vartheta_4 \dfrac{\partial \overset{o}{S}_{ki}}{\partial x_k} + \vartheta_5 \dfrac{1}{T}\dfrac{\partial T}{\partial x_k}\overset{o}{S}_{ki} \end{array}\right\} \tag{4.4.10}$$

The coefficients $\omega's$ and $\vartheta's$ are determined by gas models. For Maxwellian gas molecules with a repulsive force $\sim r^{-5}$, the coefficients are (Chapman and Cowling 1970) shown in Table 4.1.

TABLE 4.4.1 Maxwellian Gas Model Coefficients

$\omega_1 = \frac{4}{3}\left(\frac{7}{2} - \omega\right)$	$\vartheta_1 = \frac{15}{4}\left(\frac{7}{2} - \omega\right)$
$\omega_2 = 2$	$\vartheta_2 = -45/8$
$\omega_3 = 3$	$\vartheta_3 = -3$
$\omega_4 = 0$	$\vartheta_4 = 3$
$\omega_5 = 3\omega$	$\vartheta_5 = 3\left(\frac{35}{4} + \omega\right)$
$\omega_6 = 8$	$\omega = 1$

The other gas model for which the coefficients have been calculated is the hard-sphere gas. The constitutive relations for the individual components of the viscous stress tensor and the heat flux vector can be obtained by expanding the index notation. The following equations show their three-dimensional forms in the Cartesian coordinate system. Note that a short-hand notation is used where, for example,

$$u_x \equiv \frac{\partial u}{\partial x} \qquad \text{and} \qquad T_{yy} \equiv \frac{\partial^2 T}{\partial y^2} \tag{4.4.11}$$

$$\begin{aligned}
\tau_{xx}^{(2)} = -\frac{\mu^2}{p}\Big[& a_1 u_x^2 + a_2\left(u_y^2 + u_z^2\right) + a_3\left(v_x^2 + w_x^2\right) + a_4\left(v_y^2 + w_z^2\right) \\
&+ a_5\left(v_z^2 + w_y^2\right) + a_6(u_x v_y + u_x w_z) + a_7(u_y v_x + u_z w_x) + a_8 v_y w_z \\
&+ a_9 v_z w_y + a_{10} R T_{xx} + a_{11} R(T_{yy} + T_{zz}) + a_{12}\frac{RT}{\rho}\rho_{xx} \\
&+ a_{13}\frac{RT}{\rho}(\rho_{yy} + \rho_{zz}) + a_{14}\frac{RT}{\rho^2}\rho_x^2 + a_{15}\frac{RT}{\rho^2}\left(\rho_y^2 + \rho_z^2\right) \\
&+ a_{16}\frac{R}{\rho}T_x\rho_x + a_{17}\frac{R}{\rho}(T_y\rho_y + T_z\rho_z) + a_{18}\frac{R}{T}T_x^2 \\
&+ a_{19}\frac{R}{T}\left(T_y^2 + T_z^2\right)\Big]
\end{aligned} \tag{4.4.12}$$

$$\begin{aligned}
\tau_{yy}^{(2)} = -\frac{\mu^2}{p}\Big[& a_1 v_y^2 + a_2\left(v_x^2 + v_z^2\right) + a_3\left(u_y^2 + w_y^2\right) + a_4\left(u_x^2 + w_z^2\right) \\
&+ a_5\left(u_z^2 + w_x^2\right) + a_6(u_x v_y + v_y w_z) + a_7(u_y v_x + v_z w_y) + a_8 u_x w_z \\
&+ a_9 u_z w_x + a_{10} R T_{yy} + a_{11} R(T_{xx} + T_{zz}) + a_{12}\frac{RT}{\rho}\rho_{yy} \\
&+ a_{13}\frac{RT}{\rho}(\rho_{xx} + \rho_{zz}) + a_{14}\frac{RT}{\rho^2}\rho_y^2 + a_{15}\frac{RT}{\rho^2}\left(\rho_x^2 + \rho_z^2\right) \\
&+ a_{16}\frac{R}{\rho}T_y\rho_y + a_{17}\frac{R}{\rho}(T_x\rho_x + T_z\rho_z) + a_{18}\frac{R}{T}T_y^2 \\
&+ a_{19}\frac{R}{T}\left(T_x^2 + T_z^2\right)\Big]
\end{aligned} \tag{4.4.13}$$

$$\begin{aligned}
\tau_{zz}^{(2)} = -\frac{\mu^2}{p}\Big[& a_1 w_z^2 + a_2\left(w_x^2 + w_y^2\right) + a_3\left(u_z^2 + v_z^2\right) + a_4\left(u_x^2 + v_y^2\right) \\
&+ a_5\left(u_y^2 + v_x^2\right) + a_6(u_x w_z + v_y w_z) + a_7(u_z w_x + v_z w_y) + a_8 u_x v_y \\
&+ a_9 u_y v_x + a_{10} R T_{zz} + a_{11} R(T_{xx} + T_{yy}) + a_{12}\frac{RT}{\rho}\rho_{zz}
\end{aligned}$$

$$+a_{13}\frac{RT}{\rho}(\rho_{xx}+\rho_{yy})+a_{14}\frac{RT}{\rho^2}\rho_z^2+a_{15}\frac{RT}{\rho^2}\left(\rho_x^2+\rho_y^2\right)$$

$$+a_{16}\frac{R}{\rho}T_z\rho_z+a_{17}\frac{R}{\rho}(T_x\rho_x+T_y\rho_y)+a_{18}\frac{R}{T}T_z^2$$

$$\left.+a_{19}\frac{R}{T}\left(T_x^2+T_y^2\right)\right] \tag{4.4.14}$$

$$\tau_{xy}^{(2)}=\tau_{yx}^{(2)}$$

$$=-\frac{\mu^2}{p}\Big[b_1(u_xu_y+v_xv_y)+b_2(w_xw_y+v_zw_x+u_zw_y)+b_3(u_xv_x+u_yv_y)$$

$$+b_4u_zv_z+b_5(v_xw_z+u_yw_z)+b_6RT_{xy}+b_7\frac{RT}{\rho}\rho_{xy}$$

$$\left.+b_8\frac{R}{T}T_xT_y+b_9\frac{RT}{\rho^2}\rho_x\rho_y+b_{10}\frac{R}{\rho}(\rho_xT_y+\rho_yT_x)\right] \tag{4.4.15}$$

$$\tau_{xz}^{(2)}=\tau_{zx}^{(2)}$$

$$=-\frac{\mu^2}{p}\Big[b_1(u_xu_z+w_xw_z)+b_2(v_xv_z+v_xw_y+u_yv_z)+b_3(u_xw_x+u_zw_z)$$

$$+b_4u_yw_y+b_5(v_yw_x+u_zv_y)+b_6RT_{xz}+b_7\frac{RT}{\rho}\rho_{xz}$$

$$\left.+b_8\frac{R}{T}T_xT_z+b_9\frac{RT}{\rho^2}\rho_x\rho_z+b_{10}\frac{R}{\rho}(\rho_xT_z+\rho_zT_x)\right] \tag{4.4.16}$$

$$\tau_{yz}^{(2)}=\tau_{zy}^{(2)}$$

$$=-\frac{\mu^2}{p}\Big[b_1(v_yv_z+w_yw_z)+b_2(u_yu_z+u_zv_x+u_yw_x)+b_3(v_yw_y+v_zw_z)$$

$$+b_4v_xw_x+b_5(u_xv_z+u_xw_y)+b_6RT_{yz}+b_7\frac{RT}{\rho}\rho_{yz}$$

$$\left.+b_8\frac{R}{T}T_yT_z+b_9\frac{RT}{\rho^2}\rho_y\rho_z+b_{10}\frac{R}{\rho}(\rho_yT_z+\rho_zT_y)\right] \tag{4.4.17}$$

$$q_x^{(2)}=-\frac{\mu^2}{\rho}\Big\{c_1u_{xx}+c_2(u_{yy}+u_{zz})+c_3(v_{xy}+w_{xz})+c_4\frac{1}{T}u_xT_x$$

$$+c_5\frac{1}{T}(v_yT_x+w_zT_x)+c_6\frac{1}{T}(u_yT_y+u_zT_z)$$

$$+c_7\frac{1}{T}(v_xT_y+w_xT_z)+c_8\frac{1}{\rho}u_x\rho_x+c_9\frac{1}{\rho}(v_y\rho_x+w_z\rho_x)$$

$$\left.+c_{10}\frac{1}{\rho}[(u_y\rho_y+u_z\rho_z)+(v_x\rho_y+w_x\rho_z)]\right\} \tag{4.4.18}$$

$$q_y^{(2)} = -\frac{\mu^2}{\rho}\left\{c_1 v_{yy} + c_2(v_{xx} + v_{zz}) + c_3(u_{xy} + w_{yz}) + c_4\frac{1}{T}v_y T_y\right.$$
$$+ c_5\frac{1}{T}(u_x T_y + w_z T_y) + c_6\frac{1}{T}(v_x T_x + v_z T_z)$$
$$+ c_7\frac{1}{T}(u_y T_x + w_y T_z) + c_8\frac{1}{\rho}v_y \rho_y + c_9\frac{1}{\rho}(u_x \rho_y + w_z \rho_y)$$
$$\left. + c_{10}\frac{1}{\rho}[(v_x \rho_x + v_z \rho_z) + (u_y \rho_x + w_y \rho_z)]\right\} \tag{4.4.19}$$

$$q_z^{(2)} = -\frac{\mu^2}{\rho}\left\{c_1 w_{zz} + c_2(w_{xx} + w_{yy}) + c_3(u_{xz} + v_{yz}) + c_4\frac{1}{T}w_z T_z\right.$$
$$+ c_5\frac{1}{T}(u_x T_z + v_y T_z) + c_6\frac{1}{T}(w_x T_x + w_y T_y)$$
$$+ c_7\frac{1}{T}(u_z T_x + v_z T_y) + c_8\frac{1}{\rho}w_z \rho_z + c_9\frac{1}{\rho}(u_x \rho_z + v_y \rho_z)$$
$$\left. + c_{10}\frac{1}{\rho}[(w_x \rho_x + w_y \rho_y) + (u_z \rho_x + v_z \rho_y)]\right\} \tag{4.4.20}$$

The coefficients $a's$, $b's$ and $c's$ are related to the $\omega's$ and $\vartheta's$, and are given in App. 4A.

For the augumented Burnett equations, the viscous stress tensor and the heat flux vector are denoted by $\tau_{ij}^{(a)}$ and $q_i^{(a)}$, respectively.

$$\tau_{ij}^{(a)} = -\frac{\mu^3}{p^2}\left\{\frac{3}{2}\omega_7 RT\ \frac{\partial}{\partial x_j}\left(\frac{\partial^2 \overset{o}{\bar{c}_i}}{\partial x_k \partial x_k}\right)\right\} \tag{4.4.21}$$

$$q_i^{(a)} = -\frac{\mu^3}{p^2}\left\{\vartheta_7 R\frac{\partial}{\partial x_i}\left(\frac{\partial^2 T}{\partial x_k \partial x_k}\right) + \vartheta_6\frac{RT}{\rho}\frac{\partial}{\partial x_i}\left(\frac{\partial^2 \rho}{\partial x_k \partial x_k}\right)\right\} \tag{4.4.22}$$

where $\omega_7 = 2/9$
$\vartheta_6 = -5/8$
$\vartheta_7 = 11/16$

The constitutive relations for the augmented Burnett equations then become

$$\tau_{ij} = \tau_{ij}^{(1)} + \tau_{ij}^{(2)} + \tau_{ij}^{(a)} + O(Kn^3) \tag{4.4.23}$$
$$q_i = q_i^{(1)} + q_i^{(2)} + q_i^{(a)} + O(Kn^3) \tag{4.4.24}$$

4.5 A Numerical Solver for the Burnett Equations

The three-dimensional, steady Burnett equations written in a generalized nonorthogonal coordinate system (ξ, η, ζ) can be expressed in a strong conservation form as

$$\frac{\partial}{\partial \xi}\left(\tilde{E} - \tilde{E}_v^{(1)} - \tilde{E}_v^{(2)} - \tilde{E}_v^{(a)}\right) + \frac{\partial}{\partial \eta}\left(\tilde{F} - \tilde{F}_v^{(1)} - \tilde{F}_v^{(2)} - \tilde{F}_v^{(a)}\right) + \frac{\partial}{\partial \zeta}\left(\tilde{G} - \tilde{G}_v^{(1)} - \tilde{G}_v^{(2)} - \tilde{G}_v^{(a)}\right) = 0 \tag{4.5.1}$$

where the vector $\tilde{E}$, $\tilde{F}$, $\tilde{G}$, $\tilde{E}_v^{(1)}$, $\tilde{F}_v^{(1)}$, $\tilde{G}_v^{(1)}$, $\tilde{E}_v^{(2)}$, $\tilde{F}_v^{(2)}$, $\tilde{G}_v^{(2)}$, $\tilde{E}_v^{(a)}$, $\tilde{F}_v^{(a)}$, and $\tilde{G}_v^{(a)}$ are defined as

$$\begin{aligned}
\tilde{E} &= (\xi_x E + \xi_y F + \xi_z G)/J \\
\tilde{F} &= (\eta_x E + \eta_y F + \eta_z G)/J \\
\tilde{G} &= (\zeta_x E + \zeta_y F + \zeta_z G)/J \\
\tilde{E}_v^{(k)} &= \left(\xi_x E_v^{(k)} + \xi_y F_v^{(k)} + \xi_z G_v^{(k)}\right)/J \\
\tilde{F}_v^{(k)} &= \left(\eta_x E_v^{(k)} + \eta_y F_v^{(k)} + \eta_z G_v^{(k)}\right)/J \\
\tilde{G}_v^{(k)} &= \left(\zeta_x E_v^{(k)} + \zeta_y F_v^{(k)} + \zeta_z G_v^{(k)}\right)/J
\end{aligned} \tag{4.5.2}$$

In the expressions above, $k \equiv 1,\ 2$, and a, which represent, respectively, the first-order, the second-order, and the augment Burnett terms. ξ, η, and ζ denote the generalized spatial coordinates, and x, y, and z are the physical Cartesian coordinates. J represents the transformation Jacobian. E, F, and G are the convection flux vectors.

$$E = \begin{bmatrix} \rho u \\ \rho u^2 + p \\ \rho u v \\ \rho u w \\ (\rho E_t + p)u \end{bmatrix} \qquad F = \begin{bmatrix} \rho v \\ \rho u v \\ \rho v^2 + p \\ \rho w v \\ (\rho E_t + p)v \end{bmatrix} \qquad G = \begin{bmatrix} \rho w \\ \rho u w \\ \rho v w \\ \rho w^2 + p \\ (\rho E_t + p)w \end{bmatrix} \tag{4.5.3}$$

$E_v^{(k)}$, $F_v^{(k)}$, and $G_v^{(k)}$ represent the viscous flux vectors.

$$E_v^{(k)} = \begin{bmatrix} 0 \\ \tau_{xx}^{(k)} \\ \tau_{xy}^{(k)} \\ \tau_{xz}^{(k)} \\ \phi_x^{(k)} \end{bmatrix} \qquad F_v^{(k)} = \begin{bmatrix} 0 \\ \tau_{yx}^{(k)} \\ \tau_{yy}^{(k)} \\ \tau_{yz}^{(k)} \\ \phi_y^{(k)} \end{bmatrix} \qquad G_v^{(k)} = \begin{bmatrix} 0 \\ \tau_{zx}^{(k)} \\ \tau_{zy}^{(k)} \\ \tau_{zz}^{(k)} \\ \phi_z^{(k)} \end{bmatrix} \tag{4.5.4}$$

where

$$\begin{aligned}
\phi_x^{(k)} &= u\tau_{xx}^{(k)} + v\tau_{xy}^{(k)} + w\tau_{xz}^{(k)} + q_x^{(k)} \\
\phi_y^{(k)} &= u\tau_{xy}^{(k)} + v\tau_{yy}^{(k)} + w\tau_{zy}^{(k)} + q_y^{(k)} \\
\phi_z^{(k)} &= u\tau_{xz}^{(k)} + v\tau_{yz}^{(k)} + w\tau_{zz}^{(k)} + q_z^{(k)}
\end{aligned}$$

ρ, p, u, v, and w are the density, pressure, and Cartesian velocity components, respectively. The total internal energy is defined as

$$E_t = e + (u^2 + v^2 + w^2)/2 \tag{4.5.5}$$

with e being the thermodynamic internal energy. The details of the transformation, the metrics, and the Jacobian are given in App. 4B and 4C.

4.5.1 Numerical method

Microflows are characterized by high Knudsen numbers and low Mach numbers. As the value of the Knudsen number in a microflow is typically higher than 0.001, the rarefied gas effect can be important. The compressible form of the Burnett equations should be used. Many of the computational methods developed for the compressible Navier-Stokes equations are ineffective for low-Mach-number flows. This is due to the round-off error caused by the singular pressure gradient term in the momentum equations and the stiffness caused by the wide disparities in eigenvalues. To circumvent these two problems for the current microflow calculations, an all-Mach-number formulation of Shuen et al. (1993) is adopted in the two-dimensional Burnett equations code NB2D. A brief description of the numerical algorithm is given below.

In the all-Mach-number formulation, the system's eigenvalues are rescaled and the pressure variable is decomposed into a constant reference pressure p_0 and a gauge pressure p_g. The resulting Burnett equations can be written as

$$\Gamma\frac{\partial}{\partial\tau^*}\hat{Q}+\frac{\partial}{\partial\xi}\left(\tilde{E}-\tilde{E}_v^{(1)}-\tilde{E}_v^{(2)}-\tilde{E}_v^{(a)}\right)+\frac{\partial}{\partial\eta}\left(\tilde{F}-\tilde{F}_v^{(1)}-\tilde{F}_v^{(2)}-\tilde{F}_v^{(a)}\right)$$
$$+\frac{\partial}{\partial\zeta}\left(\tilde{G}-\tilde{G}_v^{(1)}-\tilde{G}_v^{(2)}-\tilde{G}_v^{(a)}\right)=0 \tag{4.5.6}$$

where the primitive variable vector $\hat{Q}$ and the preconditioning matrix Γ are

$$\hat{Q}=(p_g,\ u,\ v,\ w,\ h)^T/J$$

$$\Gamma=\begin{pmatrix} 1/\beta & 0 & 0 & 0 & 0\\ u/\beta & \rho & 0 & 0 & 0\\ v/\beta & 0 & \rho & 0 & 0\\ w/\beta & 0 & 0 & \rho & 0\\ H/\beta-1 & \rho u & \rho v & \rho w & \rho \end{pmatrix}$$

τ^* denotes a pseudo time and β a parameter for rescaling the eigenvalues of the new system of equations. $h(=e+p/\rho)$ is the specific enthalpy and H the total specific enthalpy. Equation (4.5.6) is the form of the Burnett equations to be solved numerically. Since only steady-state problems are of interest, numerical solutions of Eq. (4.5.6) are obtained when the numerical iteration from an initial condition in pseudo time has converged.

In the linearization of Eq. (4.5.6), the convective terms and the Navier-Stokes-order terms of the viscous stresses and heat fluxes are treated implicitly. The Burnett-order terms are treated explicitly. After the linearization and applying a first-order forward finite differencing in time, Eq. (4.5.6) can be expressed in the following form

$$\left[\Gamma+\Delta\tau^*\left(\frac{\partial A}{\partial\xi}-\frac{\partial}{\xi}R_{\xi\xi}\frac{\partial}{\xi}\right)+\Delta\tau^*\left(\frac{\partial B}{\partial\eta}-\frac{\partial}{\eta}R_{\eta\eta}\frac{\partial}{\eta}\right)\right.$$
$$\left.+\Delta\tau^*\left(\frac{\partial C}{\partial\zeta}-\frac{\partial}{\zeta}R_{\zeta\zeta}\frac{\partial}{\zeta}\right)+\Delta\tau^*D_2\right]^p\Delta\hat{Q}=-\Delta\tau^*\mathrm{RHS}^p \tag{4.5.7}$$

where

$$\mathrm{RHS}^p=\left[\frac{\partial(\tilde{E}-\tilde{E}_v^{(1)})^p}{\partial\xi}+\frac{\partial(\tilde{F}-\tilde{F}_v^{(1)})^p}{\partial\eta}+\frac{\partial(\tilde{G}-\tilde{G}_v^{(1)})^p}{\partial\zeta}\right]$$
$$-\left[\frac{\partial\left(\tilde{E}_v^{(2)}+\tilde{E}_v^{(\mathrm{a})}\right)^p}{\partial\xi}+\frac{\partial\left(\tilde{F}_v^{(2)}+\tilde{F}_v^{(\mathrm{a})}\right)^p}{\partial\eta}+\frac{\partial\left(\tilde{G}_v^{(2)}+\tilde{G}_v^{(\mathrm{a})}\right)^p}{\partial\zeta}\right]+D_{\mathrm{ex}}$$

and

$$\Delta\hat{Q} = \hat{Q}^{p+1} - \hat{Q}^{p}$$

The superscript p denotes a counter for the numerical iteration in the pseudo time with time step $\Delta\tau^*$. A, B, and C are the convective term Jacobians. $R_{\xi\xi}$, $R_{\eta\eta}$, and $R_{\zeta\zeta}$ are the viscous Jacobians of the Navier-Stokes-order terms. The expressions for these Jacobians can be found in Shuen et al. (1993). Second-order accurate central differencing is used to discretize the spatial derivative terms in Eq. (4.5.7) for both the explicit and the implicit operators. To enhance the numerical stability of the centrally discretized equations, second- and fourth-order artificial dissipation terms can be used. The artificial dissipation terms are treated explicitly and are represented by D_{ex}. There are additional second-order derivatives in the Burnett-order terms of the viscous stresses and the heat fluxes. In the generalized nonorthogonal coordinates, this requires transformations of the second-order derivatives when calculating the right-hand-side terms of Eq. (4.5.7). The transformation equations used in the NB2D code for the second-order derivatives are given in App. 4B.

As it is in the conventional CFD for the Navier-Stokes equations, the boundaries of the microflow impose additional constraints to the Burnett equations and determine the numerical solutions. For a solid boundary, the no-slip velocity boundary conditions are normally used in the Navier-Stokes solutions, where the fluid particles are essentially assumed "stick" to the wall. For microflows, the no-slip boundary conditions, as is described later in Chap. 7, become increasingly questionable with the increase of Knudsen number of the flow. The details of the boundary conditions for the velocity and temperature of microflows will be discussed in Chap. 7. These velocity-slip and temperature-jump boundary conditions implemented explicitly in the NB2D code are in the following form.

$$u^{p+1} - u_w = \left[\frac{2-\sigma_v}{\sigma_v}\lambda\left(\frac{\partial u}{\partial y}\right)_w\right]^p \tag{4.5.8}$$

$$T^{\,p+1} - T_w = \left[\frac{2-\sigma_v}{\sigma_v}\frac{2\gamma}{\gamma+1}\frac{\lambda}{\mathrm{Pr}}\left(\frac{\partial T}{\partial y}\right)_w\right]^p \tag{4.5.9}$$

The boundary conditions [Eqs. (4.5.8) and (4.5.9)], the wall pressure condition, and the discretized Burnett Eq. (4.5.7) form a close set of linear equations that can be written as

$$\mathbf{M}\Delta\hat{Q} = \mathbf{b} \tag{4.5.10}$$

The lower–upper symmetric Gauss-Seidel (LU-SGS) scheme (Yoon and Jameson, 1988) was applied to solve the resulting system of linear equations. The coefficient matrix **M** is first decomposed using the LU decomposition. Or

$$(L+D)D^{-1}(U+D)\Delta\hat{Q} = \mathbf{b}$$

where L, U, and D represent the nonzero, off-diagonal lower, off-diagonal upper, and the diagonal parts of **M**, respectively. The solution vector at the new pseudo time level $\hat{Q}^{p+1}$ is solved using the following steps.

$$\begin{aligned} &\text{Step 1}: (L+D)\Delta\hat{Q}^{*} = \mathbf{b} \\ &\text{Step 2}: (U+D)\Delta\hat{Q} = D(\Delta\hat{Q}^{*}) \\ &\text{Step 3}: \hat{Q}^{p+1} = \hat{Q}^{p} + \Delta\hat{Q} \end{aligned} \tag{4.5.11}$$

4.5.2 An example of NB2D results

The NB2D code results are presented for microflows in a two-dimensional channel (Fig. 4.5.1) of 20 μm in length and 4 μm in height. For the velocity inflow option, the inflow velocities u_{in} and v_{in}, and temperature T_{in} are given as the inflow condition. At the downstream boundary, the outlet pressure p_{out} is assigned. The isothermal wall temperature is T_{w}. The operating conditions are:

$$u_{\text{in}} = 133\ \text{m/s},\quad v_{\text{in}} = 0,\quad T_{\text{in}} = 300\ \text{K},\ T_{w} = 300\ \text{K}, h = 4\ \mu\text{m}$$

The gas constants used are $\gamma = 1.4$, Pr $= 0.72$, $R = 287.04\ \text{m}^2/(\text{s}^2\text{K})$, respectively. The Reynolds number is 9.75. The mean free path of the free stream is 1.782×10^{-7}m, which gives a Knudsen number of

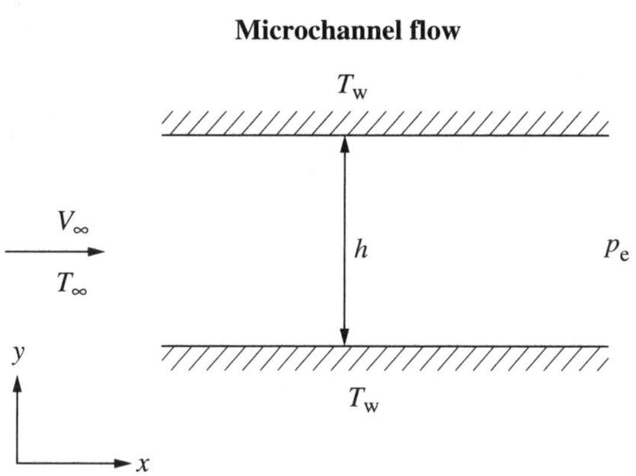

Figure 4.5.1 Simulated microchannel.

0.0445. The first-order slip conditions were used in the calculations. Figure 4.5.2 shows a comparison of the velocity profiles obtained from the Navier-Stokes equations, the conventional Burnett equations, and the augmented Burnett equation, at the same downstream location. The nonuniform grid is 85×35 and the spacing between the wall and the first grid point away from the wall is 1.18×10^{-7}m. The slip boundary conditions have shifted the velocity profiles by nearly an equal amount. There is a good agreement between the solutions. Figure 4.5.3 shows a comparison of the temperature distributions at the same streamwise station. There is a jump of the temperature between that of the wall and the gas temperature. There is a difference of about one degree between the temperature distributions obtained by using the no-slip boundary conditions and the slip boundary conditions. The three sets of equations are also predicting essentially the same temperature distribution, particularly in the center portion of the channel. Figure 4.5.4 shows the streamwise development of the wall slip velocity. The wall slip velocity increases from 5 to about 18 m/s at the end of the domain with the Navier-Stokes equations. This is about the same for the conventional Burnett and the Augment Burnett equations.

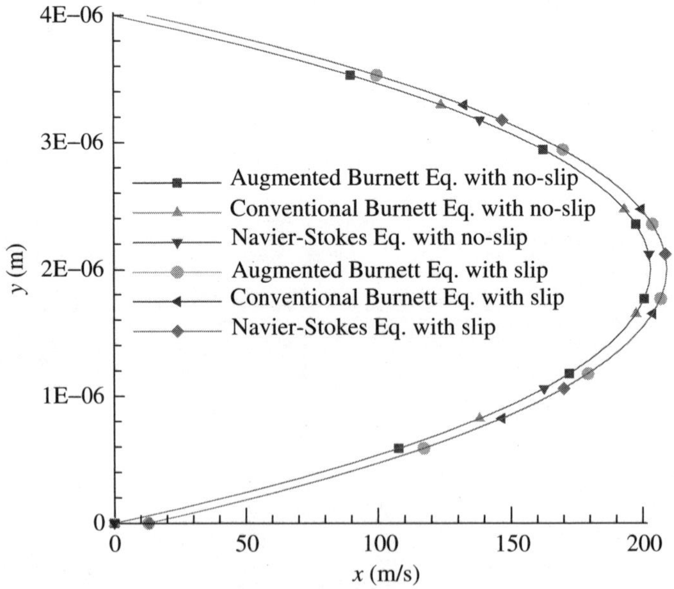

Figure 4.5.2 Streamwise velocity distributions.

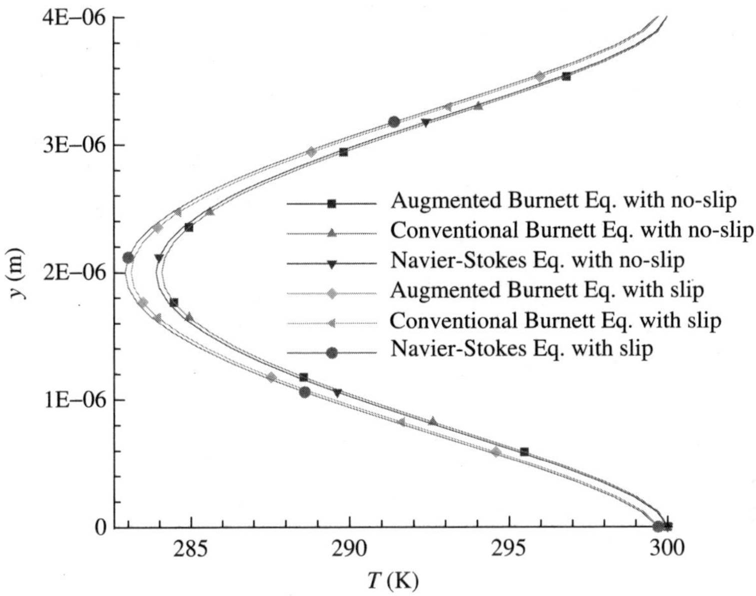

Figure 4.5.3 Temperature distributions.

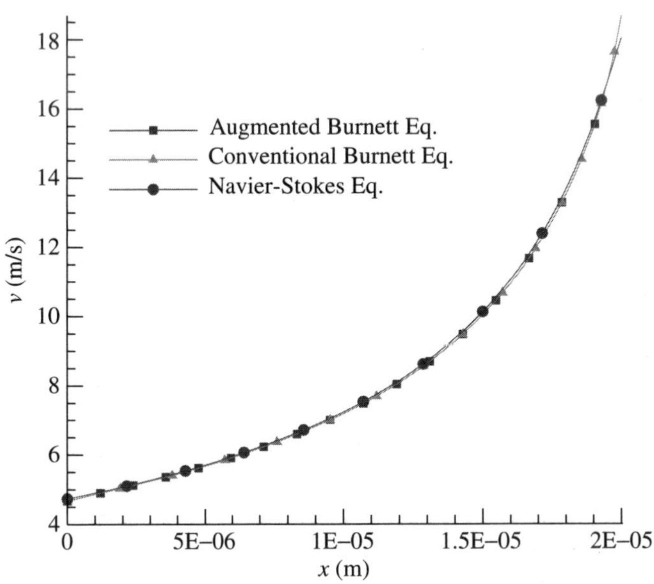

Figure 4.5.4 Streamwise development of the wall slip velocities.

Appendix 4A: Coefficients for Maxwellian Molecules

The coefficients in the conventional Burnett-order viscous stress and heat flux terms for Maxwellian molecules are:

$$a_1 = \frac{2}{3}\omega_1 - \frac{14}{9}\omega_2 + \frac{2}{9}\omega_6 \qquad a_2 = \frac{1}{3}\omega_2 + \frac{1}{12}\omega_6$$

$$a_3 = -\frac{2}{3}\omega_2 + \frac{1}{12}\omega_6 \qquad a_4 = -\frac{1}{3}\omega_1 + \frac{7}{9}\omega_2 - \frac{1}{9}\omega_6$$

$$a_5 = \frac{1}{3}\omega_2 - \frac{1}{6}\omega_6 \qquad a_6 = \frac{1}{3}\omega_1 + \frac{2}{9}\omega_2 - \frac{2}{9}\omega_6$$

$$a_7 = -\frac{2}{3}\omega_2 + \frac{1}{6}\omega_6 \qquad a_8 = -\frac{2}{3}\omega_1 - \frac{4}{9}\omega_2 + \frac{4}{9}\omega_6$$

$$a_9 = \frac{4}{3}\omega_2 - \frac{1}{3}\omega_6 \qquad a_{10} = -\frac{2}{3}\omega_2 + \frac{2}{3}\omega_3$$

$$a_{11} = \frac{1}{3}\omega_2 - \frac{1}{3}\omega_3 \qquad a_{12} = -\frac{2}{3}\omega_2$$

$$a_{13} = \frac{1}{3}\omega_2 \qquad a_{14} = \frac{2}{3}\omega_2$$

$$a_{15} = -\frac{1}{3}\omega_2 \qquad a_{16} = -\frac{2}{3}\omega_2 + \frac{2}{3}\omega_4$$

$$a_{17} = \frac{1}{3}\omega_2 - \frac{1}{3}\omega_4 \qquad a_{18} = \frac{2}{3}\omega_4 + \frac{2}{3}\omega_5$$

$$a_{19} = -\frac{1}{3}\omega_4 - \frac{1}{3}\omega_5$$

$$b_1 = \frac{1}{2}\omega_1 - \frac{5}{3}\omega_2 + \frac{1}{6}\omega_6 \qquad b_2 = -\omega_2 + \frac{1}{4}\omega_6$$

$$b_3 = \frac{1}{2}\omega_1 - \frac{2}{3}\omega_2 + \frac{1}{6}\omega_6 \qquad b_4 = \frac{1}{4}\omega_6$$

$$b_5 = \frac{1}{2}\omega_1 + \frac{1}{3}\omega_2 - \frac{1}{3}\omega_6 \qquad b_6 = -\omega_2 + \omega_3$$

$$b_7 = -\omega_2 \qquad b_8 = \omega_4 + \omega_5$$

$$b_9 = \omega_2 \qquad b_{10} = -\frac{1}{2}\omega_2 + \frac{1}{2}\omega_4$$

$$c_1 = \frac{2}{3}\vartheta_2 + \frac{2}{3}\vartheta_4 \qquad c_2 = \frac{1}{2}\vartheta_4$$

$$c_3 = \frac{2}{3}\vartheta_2 + \frac{1}{6}\vartheta_4 \qquad c_4 = \vartheta_1 + \frac{8}{3}\vartheta_2 + \frac{2}{3}\vartheta_3 + \frac{2}{3}\vartheta_5$$

$$c_5 = \vartheta_1 + \frac{2}{3}\vartheta_2 - \frac{1}{3}\vartheta_3 - \frac{1}{3}\vartheta_5 \qquad c_6 = \frac{1}{2}\vartheta_3 + \frac{1}{2}\vartheta_5$$

$$c_7 = 2\vartheta_2 + \frac{1}{2}\vartheta_3 + \frac{1}{2}\vartheta_5 \qquad c_8 = \frac{2}{3}\vartheta_3$$

$$c_9 = -\frac{1}{3}\vartheta_3 \qquad c_{10} = \frac{1}{2}\vartheta_3$$

Appendix 4B: First-Order and Second-Order Metrics of Transformation

In the computation of the Burnett equations, a finite difference method and a general body-fitted nonorthogonal grid are used. The governing equations (4.2.7), (4.2.10), and (4.2.14) and the constitutive equations (4.4.23) and (4.4.24) written in Cartesian coordinate system (x, y, z) are transformed into a general nonorthogonal computational coordinate system (ξ, η, ζ). The equations of the grid transformation are

$$\begin{cases} x = x(\xi,\ \eta,\ \zeta) \\ y = y(\xi,\ \eta,\ \zeta) \\ z = z(\xi,\ \eta,\ \zeta) \end{cases} \tag{4B.1}$$

Applying the chain rule to Eq. (4B.1) yields the first-order transformation

$$\begin{pmatrix} \dfrac{\partial}{\partial x} \\ \dfrac{\partial}{\partial y} \\ \dfrac{\partial}{\partial z} \end{pmatrix} = \begin{bmatrix} \xi_x & \eta_x & \zeta_x \\ \xi_y & \eta_y & \zeta_y \\ \xi_z & \eta_z & \zeta_z \end{bmatrix} \begin{pmatrix} \dfrac{\partial}{\partial \xi} \\ \dfrac{\partial}{\partial \eta} \\ \dfrac{\partial}{\partial \zeta} \end{pmatrix} \tag{4B.2}$$

where ξ_x, η_x, ζ_x, ξ_y, η_y, ζ_y, ξ_z, η_z, and ζ_z represent the first-order metrics that will be derived in the following.

Taking the first-order derivative of Eq. (4B.1) with respect to x, y, and z, respectively, leads to

$$\begin{bmatrix} x_\xi & x_\eta & x_\zeta \\ y_\xi & y_\eta & y_\zeta \\ z_\xi & z_\eta & z_\zeta \end{bmatrix} \begin{pmatrix} \xi_x \\ \eta_x \\ \zeta_x \end{pmatrix} = \begin{pmatrix} 1 \\ 0 \\ 0 \end{pmatrix} \tag{4B.3a}$$

$$\begin{bmatrix} x_\xi & x_\eta & x_\zeta \\ y_\xi & y_\eta & y_\zeta \\ z_\xi & z_\eta & z_\zeta \end{bmatrix} \begin{pmatrix} \xi_y \\ \eta_y \\ \zeta_y \end{pmatrix} = \begin{pmatrix} 0 \\ 1 \\ 0 \end{pmatrix} \tag{4B.3b}$$

$$\begin{bmatrix} x_\xi & x_\eta & x_\zeta \\ y_\xi & y_\eta & y_\zeta \\ z_\xi & z_\eta & z_\zeta \end{bmatrix} \begin{pmatrix} \xi_z \\ \eta_z \\ \zeta_z \end{pmatrix} = \begin{pmatrix} 0 \\ 0 \\ 1 \end{pmatrix} \tag{4B.3c}$$

By solving the equations set of (4B.3) and denoting the transformation Jacobian J as

$$J^{-1} = \frac{\partial(x,\ y,\ z)}{\partial(\xi,\ \eta,\ \zeta)} = x_\xi(y_\eta z_\zeta - y_\zeta z_\eta) - x_\eta(y_\xi z_\zeta - y_\zeta z_\xi) + x_\zeta(y_\xi z_\eta - y_\eta z_\xi) \tag{4B.4}$$

the first order transformation metrics can be obtained,

$$\xi_x = J\,(y_\eta z_\zeta - y_\zeta z_\eta) \tag{4B.5a}$$

$$\eta_x = J\,(y_\zeta z_\xi - y_\xi z_\zeta) \tag{4B.5b}$$

$$\zeta_x = J\,(y_\xi z_\eta - y_\eta z_\xi) \tag{4B.5c}$$

$$\xi_y = J\,(x_\zeta z_\eta - x_\eta z_\zeta) \tag{4B.5d}$$

$$\eta_y = J\,(x_\xi z_\zeta - x_\zeta z_\xi) \tag{4B.5e}$$

$$\zeta_y = J\,(x_\eta z_\xi - x_\xi z_\eta) \tag{4B.5f}$$

$$\xi_z = J\,(x_\eta y_\xi - x_\xi y_\eta) \tag{4B.5g}$$

$$\eta_z = J\,(x_\zeta y_\xi - x_\xi y_\zeta) \tag{4B.5h}$$

$$\zeta_z = J\,(x_\xi y_\eta - x_\eta y_\xi) \tag{4B.5i}$$

There are also second-order derivatives in the constitutive equations (4.4.23) and (4.4.24) for the viscous stress and heat flux terms. They can be obtained through the chain rule based on the first order transformation.

$$\begin{aligned}\frac{\partial^2}{\partial x^2} &= \xi_x^2\frac{\partial^2}{\partial\xi^2} + \eta_x^2\frac{\partial^2}{\partial\eta^2} + \zeta_x^2\frac{\partial^2}{\partial\zeta^2} + 2\xi_x\eta_x\frac{\partial^2}{\partial\xi\partial\eta} + 2\xi_x\zeta_x\frac{\partial^2}{\partial\xi\partial\zeta}\\ &\quad + 2\eta_x\zeta_x\frac{\partial^2}{\partial\eta\partial\zeta} + \xi_{xx}\frac{\partial}{\partial\xi} + \eta_{xx}\frac{\partial}{\partial\eta} + \zeta_{xx}\frac{\partial}{\partial\zeta}\end{aligned} \tag{4B.6}$$

$$\begin{aligned}\frac{\partial^2}{\partial y^2} &= \xi_y^2\frac{\partial^2}{\partial\xi^2} + \eta_y^2\frac{\partial^2}{\partial\eta^2} + \zeta_y^2\frac{\partial^2}{\partial\zeta^2} + 2\xi_y\eta_y\frac{\partial^2}{\partial\xi\partial\eta} + 2\xi_y\zeta_y\frac{\partial^2}{\partial\xi\partial\zeta}\\ &\quad + 2\eta_y\zeta_y\frac{\partial^2}{\partial\eta\partial\zeta} + \xi_{yy}\frac{\partial}{\partial\xi} + \eta_{yy}\frac{\partial}{\partial\eta} + \zeta_{yy}\frac{\partial}{\partial\zeta}\end{aligned} \tag{4B.7}$$

$$\begin{aligned}\frac{\partial^2}{\partial z^2} &= \xi_z^2\frac{\partial^2}{\partial\xi^2} + \eta_z^2\frac{\partial^2}{\partial\eta^2} + \zeta_z^2\frac{\partial^2}{\partial\zeta^2} + 2\xi_z\eta_z\frac{\partial^2}{\partial\xi\partial\eta} + 2\xi_z\zeta_z\frac{\partial^2}{\partial\xi\partial\zeta}\\ &\quad + 2\eta_z\zeta_z\frac{\partial^2}{\partial\eta\partial\zeta} + \xi_{zz}\frac{\partial}{\partial\xi} + \eta_{zz}\frac{\partial}{\partial\eta} + \zeta_{zz}\frac{\partial}{\partial\zeta}\end{aligned} \tag{4B.8}$$

$$\frac{\partial^2}{\partial x \partial y} = \xi_x \xi_y \frac{\partial^2}{\partial \xi^2} + \eta_x \eta_y \frac{\partial^2}{\partial \eta^2} + \zeta_x \zeta_y \frac{\partial^2}{\partial \zeta^2} + (\xi_x \eta_y + \xi_y \eta_x) \frac{\partial^2}{\partial \xi \partial \eta}$$
$$+ (\xi_x \zeta_y + \xi_y \zeta_x) \frac{\partial^2}{\partial \xi \partial \zeta} + (\eta_x \zeta_y + \eta_y \zeta_x) \frac{\partial^2}{\partial \eta \partial \zeta} + \xi_{xy} \frac{\partial}{\partial \xi}$$
$$+ \eta_{xy} \frac{\partial}{\partial \eta} + \zeta_{xy} \frac{\partial}{\partial \zeta} \tag{4B.9}$$

$$\frac{\partial^2}{\partial x \partial z} = \xi_x \xi_z \frac{\partial^2}{\partial \xi^2} + \eta_x \eta_z \frac{\partial^2}{\partial \eta^2} + \zeta_x \zeta_z \frac{\partial^2}{\partial \zeta^2} + (\xi_x \eta_z + \xi_z \eta_x) \frac{\partial^2}{\partial \xi \partial \eta}$$
$$+ (\xi_x \zeta_z + \xi_z \zeta_x) \frac{\partial^2}{\partial \xi \partial \zeta} + (\eta_x \zeta_z + \eta_z \zeta_x) \frac{\partial^2}{\partial \eta \partial \zeta} + \xi_{xz} \frac{\partial}{\partial \xi}$$
$$+ \eta_{xz} \frac{\partial}{\partial \eta} + \zeta_{xz} \frac{\partial}{\partial \zeta} \tag{4B.10}$$

$$\frac{\partial^2}{\partial y \partial z} = \xi_y \xi_z \frac{\partial^2}{\partial \xi^2} + \eta_y \eta_z \frac{\partial^2}{\partial \eta^2} + \zeta_y \zeta_z \frac{\partial^2}{\partial \zeta^2} + (\xi_y \eta_z + \xi_z \eta_y) \frac{\partial^2}{\partial \xi \partial \eta}$$
$$+ (\xi_y \zeta_z + \xi_z \zeta_y) \frac{\partial^2}{\partial \xi \partial \zeta} + (\eta_y \zeta_z + \eta_y \zeta_x) \frac{\partial^2}{\partial \eta \partial \zeta} + \xi_{yz} \frac{\partial}{\partial \xi}$$
$$+ \eta_{yz} \frac{\partial}{\partial \eta} + \zeta_{yz} \frac{\partial}{\partial \zeta} \tag{4B.11}$$

The second-order metrics can be obtained by taking further derivatives of Eq. (4B.3a–3c) with respect to x, y, and z. For example, taking derivative of Eq. (4B.3a) with respect to x yields

$$\begin{bmatrix} x_{\xi x} & x_{\eta x} & x_{\zeta x} \\ y_{\xi x} & y_{\eta x} & y_{\zeta x} \\ z_{\xi x} & z_{\eta x} & z_{\zeta x} \end{bmatrix} \begin{pmatrix} \xi_x \\ \eta_x \\ \zeta_x \end{pmatrix} + \begin{bmatrix} x_\xi & x_\eta & x_\zeta \\ y_\xi & y_\eta & y_\zeta \\ z_\xi & z_\eta & z_\zeta \end{bmatrix} \begin{pmatrix} \xi_{xx} \\ \eta_{xx} \\ \zeta_{xx} \end{pmatrix} = \begin{pmatrix} 0 \\ 0 \\ 0 \end{pmatrix} \tag{4B.12}$$

with

$$(\,)_{\xi x} = (\,)_{\xi\xi} \xi_x + (\,)_{\xi\eta} \eta_x + (\,)_{\xi\zeta} \zeta_x \tag{4B.13a}$$
$$(\,)_{\eta x} = (\,)_{\xi\eta} \xi_x + (\,)_{\xi\eta} \eta_x + (\,)_{\eta\zeta} \zeta_x \tag{4B.13b}$$
$$(\,)_{\zeta x} = (\,)_{\xi\zeta} \xi_x + (\,)_{\eta\zeta} \eta_x + (\,)_{\zeta\zeta} \zeta_x \tag{4B.13c}$$

and the final results for the second-order mesh metrics with respect to x are

$$\xi_{xx} = -(a_1 \xi_x + b_1 \xi_y + c_1 \xi_z) \tag{4B.14}$$
$$\eta_{xx} = -(a_1 \eta_x + b_1 \eta_y + c_1 \eta_z) \tag{4B.15}$$
$$\zeta_{xx} = -(a_1 \zeta_x + b_1 \zeta_y + c_1 \zeta_z) \tag{4B.16}$$

where

$$a_1 = x_{\xi\xi}\xi_x^2 + x_{\eta\eta}\eta_x^2 + x_{\zeta\zeta}\zeta_x^2 + 2x_{\xi\eta}\xi_x\eta_x + 2x_{\xi\zeta}\xi_x\zeta_x + 2x_{\eta\zeta}\eta_x\zeta_x$$
$$b_1 = y_{\xi\xi}\xi_x^2 + y_{\eta\eta}\eta_x^2 + y_{\zeta\zeta}\zeta_x^2 + 2y_{\xi\eta}\xi_x\eta_x + 2y_{\xi\zeta}\xi_x\zeta_x + 2y_{\eta\zeta}\eta_x\zeta_x$$
$$c_1 = z_{\xi\xi}\xi_x^2 + z_{\eta\eta}\eta_x^2 + z_{\zeta\zeta}\zeta_x^2 + 2z_{\xi\eta}\xi_x\eta_x + 2z_{\xi\zeta}\xi_x\zeta_x + 2z_{\eta\zeta}\eta_x\zeta_x$$

Similarly, the other second-order metrics with respect to y are

$$\xi_{yy} = -(a_2\xi_x + b_2\xi_y + c_2\xi_z) \tag{4B.17}$$
$$\eta_{yy} = -(a_2\eta_x + b_2\eta_y + c_2\eta_z) \tag{4B.18}$$
$$\zeta_{yy} = -(a_2\zeta_x + b_2\zeta_y + c_2\zeta_z) \tag{4B.19}$$

where

$$a_2 = x_{\xi\xi}\xi_y^2 + x_{\eta\eta}\eta_y^2 + x_{\zeta\zeta}\zeta_y^2 + 2x_{\xi\eta}\xi_y\eta_y + 2x_{\xi\zeta}\xi_y\zeta_y + 2x_{\eta\zeta}\eta_y\zeta_y$$
$$b_2 = y_{\xi\xi}\xi_y^2 + y_{\eta\eta}\eta_y^2 + y_{\zeta\zeta}\zeta_y^2 + 2y_{\xi\eta}\xi_y\eta_y + 2y_{\xi\zeta}\xi_y\zeta_y + 2y_{\eta\zeta}\eta_y\zeta_y$$
$$c_2 = z_{\xi\xi}\xi_y^2 + z_{\eta\eta}\eta_y^2 + z_{\zeta\zeta}\zeta_y^2 + 2z_{\xi\eta}\xi_y\eta_y + 2z_{\xi\zeta}\xi_y\zeta_y + 2z_{\eta\zeta}\eta_y\zeta_y$$

$$\xi_{zz} = -(a_3\xi_x + b_3\xi_y + c_3\xi_z) \tag{4B.20}$$
$$\eta_{zz} = -(a_3\eta_x + b_3\eta_y + c_3\eta_z) \tag{4B.21}$$
$$\zeta_{zz} = -(a_3\zeta_x + b_3\zeta_y + c_3\zeta_z) \tag{4B.22}$$

where

$$a_3 = x_{\xi\xi}\xi_z^2 + x_{\eta\eta}\eta_z^2 + x_{\zeta\zeta}\zeta_z^2 + 2x_{\xi\eta}\xi_z\eta_z + 2x_{\xi\zeta}\xi_z\zeta_z + 2x_{\eta\zeta}\eta_z\zeta_z$$
$$b_3 = y_{\xi\xi}\xi_z^2 + y_{\eta\eta}\eta_z^2 + y_{\zeta\zeta}\zeta_z^2 + 2y_{\xi\eta}\xi_z\eta_z + 2y_{\xi\zeta}\xi_z\zeta_z + 2y_{\eta\zeta}\eta_z\zeta_z$$
$$c_3 = z_{\xi\xi}\xi_z^2 + z_{\eta\eta}\eta_z^2 + z_{\zeta\zeta}\zeta_z^2 + 2z_{\xi\eta}\xi_z\eta_z + 2z_{\xi\zeta}\xi_z\zeta_z + 2z_{\eta\zeta}\eta_z\zeta_z$$

$$\xi_{xy} = -(a_4\xi_x + b_4\xi_y + c_4\xi_z) \tag{4B.23}$$
$$\eta_{xy} = -(a_4\eta_x + b_4\eta_y + c_4\eta_z) \tag{4B.24}$$
$$\zeta_{xy} = -(a_4\zeta_x + b_4\zeta_y + c_4\zeta_z) \tag{4B.25}$$

where

$$\begin{aligned} a_4 = {} & x_{\xi\xi}\xi_x\xi_y + x_{\eta\eta}\eta_x\eta_y + x_{\zeta\zeta}\zeta_x\zeta_y + x_{\xi\eta}(\xi_x\eta_y + \xi_y\eta_x) + x_{\xi\zeta}(\xi_x\zeta_y + \xi_y\zeta_x) \\ & + x_{\eta\zeta}(\eta_x\zeta_y + \eta_y\zeta_x) \\ b_4 = {} & y_{\xi\xi}\xi_x\xi_y + y_{\eta\eta}\eta_x\eta_y + y_{\zeta\zeta}\zeta_x\zeta_y + y_{\xi\eta}(\xi_x\eta_y + \xi_y\eta_x) + y_{\xi\zeta}(\xi_x\zeta_y + \xi_y\zeta_x) \\ & + y_{\eta\zeta}(\eta_x\zeta_y + \eta_y\zeta_x) \end{aligned}$$

$$c_4 = z_{\xi\xi}\xi_x\xi_y + z_{\eta\eta}\eta_x\eta_y + z_{\zeta\zeta}\zeta_x\zeta_y + z_{\xi\eta}(\xi_x\eta_y + \xi_y\eta_x) + z_{\xi\zeta}(\xi_x\zeta_y + \xi_y\zeta_x) + z_{\eta\zeta}(\eta_x\zeta_y + \eta_y\zeta_x)$$

$$\xi_{xz} = -(a_5\xi_x + b_5\xi_y + c_5\xi_z) \tag{4B.26}$$

$$\eta_{xz} = -(a_5\eta_x + b_5\eta_y + c_5\eta_z) \tag{4B.27}$$

$$\zeta_{xz} = -(a_5\zeta_x + b_5\zeta_y + c_5\zeta_z) \tag{4B.28}$$

where

$$a_5 = x_{\xi\xi}\xi_x\xi_z + x_{\eta\eta}\eta_x\eta_z + x_{\zeta\zeta}\zeta_x\zeta_z + x_{\xi\eta}(\xi_x\eta_z + \xi_z\eta_x) + x_{\xi\zeta}(\xi_x\zeta_z + \xi_z\zeta_x) + x_{\eta\zeta}(\eta_x\zeta_z + \eta_z\zeta_x)$$

$$b_5 = y_{\xi\xi}\xi_x\xi_z + y_{\eta\eta}\eta_x\eta_z + y_{\zeta\zeta}\zeta_x\zeta_z + y_{\xi\eta}(\xi_x\eta_z + \xi_z\eta_x) + y_{\xi\zeta}(\xi_x\zeta_z + \xi_z\zeta_x) + y_{\eta\zeta}(\eta_x\zeta_z + \eta_z\zeta_x)$$

$$c_5 = z_{\xi\xi}\xi_x\xi_z + z_{\eta\eta}\eta_x\eta_z + z_{\zeta\zeta}\zeta_x\zeta_z + z_{\xi\eta}(\xi_x\eta_z + \xi_z\eta_x) + z_{\xi\zeta}(\xi_x\zeta_z + \xi_z\zeta_x) + z_{\eta\zeta}(\eta_x\zeta_z + \eta_z\zeta_x)$$

$$\xi_{yz} = -(a_6\xi_x + b_6\xi_y + c_6\xi_z) \tag{4B.29}$$

$$\eta_{yz} = -(a_6\eta_x + b_6\eta_y + c_6\eta_z) \tag{4B.30}$$

$$\zeta_{yz} = -(a_6\zeta_x + b_6\zeta_y + c_6\zeta_z) \tag{4B.31}$$

where

$$a_6 = x_{\xi\xi}\xi_y\xi_z + x_{\eta\eta}\eta_y\eta_z + x_{\zeta\zeta}\zeta_y\zeta_z + x_{\xi\eta}(\xi_y\eta_z + \xi_z\eta_y) + x_{\xi\zeta}(\xi_y\zeta_z + \xi_z\zeta_y) + x_{\eta\zeta}(\eta_y\zeta_z + \eta_z\zeta_y)$$

$$b_6 = y_{\xi\xi}\xi_y\xi_z + y_{\eta\eta}\eta_y\eta_z + y_{\zeta\zeta}\zeta_y\zeta_z + y_{\xi\eta}(\xi_y\eta_z + \xi_z\eta_y) + y_{\xi\zeta}(\xi_y\zeta_z + \xi_z\zeta_y) + y_{\eta\zeta}(\eta_y\zeta_z + \eta_z\zeta_y)$$

$$c_6 = z_{\xi\xi}\xi_y\xi_z + z_{\eta\eta}\eta_y\eta_z + z_{\zeta\zeta}\zeta_y\zeta_z + z_{\xi\eta}(\xi_y\eta_z + \xi_z\eta_y) + z_{\xi\zeta}(\xi_y\zeta_z + \xi_z\zeta_y) + z_{\eta\zeta}(\eta_y\zeta_z + \eta_z\zeta_y)$$

Appendix 4C: Jacobian Matrices

The inviscid Jacobian matrices can be written as

$$\mathbf{A} = \frac{\partial \tilde{\mathbf{E}}}{\partial \hat{\mathbf{Q}}} \qquad \mathbf{B} = \frac{\partial \tilde{\mathbf{F}}}{\partial \hat{\mathbf{Q}}} \qquad \mathbf{C} = \frac{\partial \tilde{\mathbf{G}}}{\partial \hat{\mathbf{Q}}} \tag{4C.1}$$

$$\mathbf{A} = \frac{\partial}{\partial \hat{\mathbf{Q}}}\left[\frac{1}{J}(\xi_x \mathbf{E} + \xi_y \mathbf{F} + \xi_z \mathbf{G})\right]$$

$$= \xi_x \frac{\partial}{\partial \hat{\mathbf{Q}}}\left(\frac{\mathbf{E}}{J}\right) + \xi_y \frac{\partial}{\partial \hat{\mathbf{Q}}}\left(\frac{\mathbf{F}}{J}\right) + \xi_z \frac{\partial}{\partial \hat{\mathbf{Q}}}\left(\frac{\mathbf{G}}{J}\right) \tag{4C.2}$$

$$\mathbf{B} = \frac{\partial}{\partial \hat{\mathbf{Q}}}\left[\frac{1}{J}(\eta_x \mathbf{E} + \eta_y \mathbf{F} + \eta_z \mathbf{G})\right]$$

$$= \eta_x \frac{\partial}{\partial \hat{\mathbf{Q}}}\left(\frac{\mathbf{E}}{J}\right) + \eta_y \frac{\partial}{\partial \hat{\mathbf{Q}}}\left(\frac{\mathbf{F}}{J}\right) + \eta_z \frac{\partial}{\partial \hat{\mathbf{Q}}}\left(\frac{\mathbf{G}}{J}\right) \tag{4C.3}$$

$$\mathbf{C} = \frac{\partial}{\partial \hat{\mathbf{Q}}}\left[\frac{1}{J}(\zeta_x \mathbf{E} + \zeta_y \mathbf{F} + \zeta_z \mathbf{G})\right]$$

$$= \zeta_x \frac{\partial}{\partial \hat{\mathbf{Q}}}\left(\frac{\mathbf{E}}{J}\right) + \zeta_y \frac{\partial}{\partial \hat{\mathbf{Q}}}\left(\frac{\mathbf{F}}{J}\right) + \zeta_z \frac{\partial}{\partial \hat{\mathbf{Q}}}\left(\frac{\mathbf{G}}{J}\right) \tag{4C.4}$$

$$\mathbf{A} = \begin{bmatrix} \frac{C_p U}{Rh} & \rho\alpha_1 & \rho\alpha_2 & \rho\alpha_3 & -\frac{\rho U}{h} \\ \alpha_1 + \frac{C_p u U}{Rh} & \rho(U + \alpha_1 u) & \rho u\,\alpha_2 & \rho u\,\alpha_3 & -\frac{\rho u U}{h} \\ \alpha_2 + \frac{C_p v U}{Rh} & \rho v\,\alpha_1 & \rho(U + \alpha_2 v) & \rho v\,\alpha_3 & -\frac{\rho v U}{h} \\ \alpha_3 + \frac{C_p w U}{Rh} & \rho w\,\alpha_1 & \rho w\,\alpha_2 & \rho(U + \alpha_3 w) & -\frac{\rho w U}{h} \\ \frac{C_p H U}{Rh} & \rho(uU + \alpha_1 H) & \rho(vU + \alpha_2 H) & \rho(wU + \alpha_3 H) & \rho U\left(1 - \frac{H}{h}\right) \end{bmatrix} \tag{4C.5}$$

where

$$U = \alpha_1 u + \alpha_2 v + \alpha_3 w \tag{4C.6}$$

For the Jacobian matrix **A**, the $\alpha's$ are

$$\alpha_1 = \xi_x \qquad \alpha_2 = \xi_y \qquad \alpha_3 = \xi_z \tag{4C.7}$$

C_p denotes the specific heat at constant pressure. For the Jacobian **B**, $\alpha_1 = \eta_x$, $\alpha_2 = \eta_y$, and $\alpha_3 = \eta_z$. Similarly for the Jacobian **C**, the $\alpha's$ are $\alpha_1 = \zeta_x$, $\alpha_2 = \zeta_y$, and $\alpha_3 = \zeta_z$.

The viscous Jacobians, $\mathbf{R}_{\xi\xi} = \partial\tilde{\mathbf{E}}_v^{(1)}/\partial\hat{\mathbf{Q}}$, $\mathbf{R}_{\eta\eta} = \partial\tilde{\mathbf{F}}_v^{(1)}/\partial\hat{\mathbf{Q}}$, and $\mathbf{R}_{\zeta\zeta} = \partial\tilde{\mathbf{G}}_v^{(1)}/\partial\hat{\mathbf{Q}}$ can be written in the following forms:

$$\mathbf{R}_{\xi\xi} = \begin{bmatrix} 0 & 0 & 0 & 0 & 0 \\ 0 & \mu\left(\frac{1}{3}\alpha_1^2 + \Lambda\right) & \mu\alpha_{12} & \mu\alpha_{13} & 0 \\ 0 & \rho v\alpha_1 & \mu\left(\frac{1}{3}\alpha_2^2 + \Lambda\right) & \mu\alpha_{23} & 0 \\ 0 & \mu\alpha_{13} & \mu\alpha_{23} & \mu\left(\frac{1}{3}\alpha_3^2 + \Lambda\right) & 0 \\ 0 & \pi_u & \pi_v & \pi_w & \frac{\mu}{\mathrm{Pr}}\Lambda \end{bmatrix} \tag{4C.8}$$

where

$$\begin{aligned} &\alpha_1 = \xi_x \qquad \alpha_2 = \xi_y \qquad \alpha_3 = \xi_z \\ &\alpha_{12} = \frac{1}{3}\alpha_1\alpha_2 \qquad \alpha_{13} = \frac{1}{3}\alpha_1\alpha_3 \qquad \alpha_{23} = \frac{1}{3}\alpha_2\alpha_3 \end{aligned} \tag{4C.9}$$

$$\Lambda = \alpha_1^2 + \alpha_2^2 + \alpha_3^2$$

$$\begin{aligned} \pi_u &= \mu\left[u\left(\frac{1}{3}\alpha_1^2 + \Lambda\right) + v\alpha_{12} + w\alpha_{13}\right] \\ \pi_v &= \mu\left[v\left(\frac{1}{3}\alpha_2^2 + \Lambda\right) + u\alpha_{12} + w\alpha_{23}\right] \\ \pi_w &= \mu\left[w\left(\frac{1}{3}\alpha_3^2 + \Lambda\right) + u\alpha_{13} + v\alpha_{23}\right] \end{aligned} \tag{4C.10}$$

$\mathbf{R}_{\eta\eta}$ can obtained by setting $\alpha_1 = \eta_x$, $\alpha_2 = \eta_y$, and $\alpha_3 = \eta_z$ and $\mathbf{R}_{\zeta\zeta}$ by setting $\alpha_1 = \zeta_x$, $\alpha_2 = \zeta_y$, and $\alpha_3 = \zeta_z$.

References

Balakrishnan, R. and Agarwal, R.K., Entropy consistent formulations and numerical simulation of the BGK-Burnett equations for hypersonic flows in a continuum-transition regime, *Proceedings of the Sixteenth International Conference for the Numerical Method in Fluid Dynamics*, Springer, Monterey, CA, 1996.

Bhatnagar, P.L., Gross, E.P., and Krook, M., A model for collision processes in gases. I. Small amplitude processes in charged and neutral one-component systems, *Phys. Rev.*, Vol. 94, No. 3, p. 511, 1954.

Burnett, D., The distribution of velocities in a slightly non-uniform gas, *Proceedings of the London Mathematical Society*, Vol. 39, pp. 385–430, 1935.

Chapman, S. and Cowling, T.G., *The Mathematical Theory of Non-Uniform Gases*, 3rd ed., Cambridge University Press, Cambridge, 1970.

Fisco, K.A. and Chapman, D.R., Comparison of Burnett, Super-Burnett and Monte-Carlo solutions for hypersonic shock structure, *Proceedings of the* 16th *International Symposium on Rarefied Gas Dynamics*, Pasadena, CA, pp. 374–395, 1988.

Lockerby, D.A. and Reese, J.M., High-resolution Burnett simulations of micro Couette flow and heat transfer, *J. Computational Phys.*, Vol. 188, p. 333, 2003.

Shavaliev, M. S., The Burnett approximation of the distribution function and the super-Burnett contributions to the stress tensor and the heat flux, *J. Appl. Math. Mech.*, Vol. 42, pp. 656–702, 1978.

Shuen, J.S., Chen, K.H., and Choi, Y.H., A coupled implicit method for chemical non-equilibrium flow at all speeds, *J. Comput. Phys.*, Vol. 106, p. 306, 1993.

Vincenti, W.G. and Kruger, C.H., *Introduction to Physical Gas Dynamics*, Wiley, New York, 1965.

Woods, L.C., Transport process in dilute gases over the whole range of Knudsen numbers, Part 1: General theory, *J. Fluid Mech.*, Vol. 93, pp. 585–607, 1979.

Xue, H., Ji, H.M., and Shu, C., Analysis of micro-Couette flow using the Burnett equations, *Int. J. Heat Mass Transf.*, Vol. 44, p. 4139, 2001.

Yoon, S. and Jameson, A., Lower-Upper symmetric-Gauss-Seidel method for the Euler and Navier-Stokes equations, *AIAA J.*, Vol. 26., 1988.

Zhong, X., *Development and Computation of Continuum High Order Constitutive Relations for High-Altitude Hypersonic Flow*. Ph.D. Thesis, Stanford University, 1991.

Chapter

5

Statistical Method: Direct Simulation Monte Carlo Method and Information Preservation Method

5.1 Conventional DSMC

5.1.1 Overview

DSMC (direct simulation Monte Carlo) is a direct particle simulation method based on kinetic theory (Bird 1963; 1965*a*,*b*; 1976; 1978; 1994). The fundamental idea is to track thousands or millions of randomly selected, statistically representative particles, and to use their motions and interactions to modify their positions and states appropriately in time. Each simulated particle represents a number of real molecules. The primary approximation of DSMC is the uncoupling of the molecular motions and the intermolecular collisions over small time intervals that are less than the mean collision time. The DSMC computation is started from some initial conditions and followed in small time steps that can be related to physical time. Collision pairs of molecule in a small computational cell in physical space are randomly selected based on a probability distribution after each computation time step. Complex physics such as radiation, chemical reactions, and species concentrations can be included in the simulations without invoking nonequilibrium thermodynamic assumptions that commonly afflict nonequilibrium continuum flow calculations. The DSMC technique is explicit and time marching, and therefore can produce unsteady flow simulations. For steady flows, DSMC simulation proceeds until a stationary flow is established within a desired accuracy. The macroscopic flow quantities can then be obtained by time averaging the cell-based values. For unsteady flows, ensemble averaging of many independent Monte Carlo simulations can be carried out to obtain the final results within a prescribed statistical accuracy.

A significant advantage of DSMC is that the total computation required is proportional to the number of molecules simulated N, in contrast to N^2 for the molecular dynamics (MD) simulations. In essence, particle motions are modeled deterministically while collisions are treated statistically. The backbone of DSMC follows directly from classical kinetic theory, and hence the applications of this method are subject to the same limitations as the theory. These limitations are the fundamental assumption of molecular chaos and dilute gases. DSMC has been shown to be accurate at the Boltzmann equation level for monatomic gases undergoing binary collisions (Bird 1970, 1994).

The statistical error of a DSMC solution is inversely proportional to the square root of the sample size N. A large sample size is needed to reduce the statistical error. The primary drawback of DSMC is the significant computer resources required for simulating a practical flow.

Bird (1963) first applied the DSMC method to a homogeneous gas relaxation problem, and then to a study of shock structure (Bird 1965*b*). The detailed principles of DSMC can be found in *Molecular Gas Dynamics* (Bird 1976) and the recent edition *Molecular Gas Dynamics and the Direct Simulation of Gas Flows* (Bird 1994).

Since its introduction, significant research has been performed and reported on improving the accuracy of the DSMC method as well as expanding its application to various fluid flow problems. Improved and more complex physical molecular models have also been developed for DSMC. The original *hard sphere* (HS) molecular model has been extended to the more realistic *variable hard sphere* (VHS) model (Bird 1981), *variable soft sphere* (VSS) model (Koura and Matsumoto 1991, 1992), and *generalized hard sphere* (GHS) model (Hash and Hassan 1993; Hassan and Hash 1993). The Borgnakke-Larsen (BL) phenomenological model (Borgnakke and Larsen 1975) was introduced in 1975 to handle inelastic binary collisions of polyatomic gases. The BL collision model accounts for the vibrational and rotational nonequilibrium and it has been widely used. In 1987, Bird proposed a radiation model for use in radiative nonequilibrium flows. This model was used successfully in predicting the radiative heating on the Aero-Assist Flight Experiment (AFE) vehicle (Moss et al. 1988). A model for dissociation-recombination interactions was introduced by Nanbu (1991). Boyd et al. (1992) proposed a vibrational relaxation method for use with the VHS model. Methods for ternary reactions, or three-body collisions, have also been proposed, even through binary collisions prevail in most flow problems. Several collision-sampling techniques were also developed.

These include the *time counter* (TC) method (Bird 1976) and the *no time counter* (NTC) method (Bird 1989). Some of the molecular models were described in Chap. 2.

The accuracy of the gas-surface interaction model has also been improved. The boundary condition models initially used by Bird and earlier researchers are simple and provide only an approximation to real gas–surface interactions. More advanced boundary conditions for DSMC calculations, such as the *Maxwell* models, *Cercignani-Lampis* (C-L) and *Cercignani-Lampis-Lord* (CLL) models can be found in Maxwell (1879), Cercignani and Lampis (1971), Cercignani (1988), Lord (1991, 1992), and Collins and Knox (1994). DSMC has been combined with the *Monotonic-Lagrangian-Grid* (MLG) algorithm, allowing the grid to automatically adapt to the local number density of the flowfield (Oh et al. 1995). DSMC has also been improved with the information preservation (IP) method, effectively reducing the statistical scatter for low-speed flows (Fan and Shen 1998, 2001; Cai et al. 2000; Sun and Boyd 2002). The statistical scatter associated with DSMC has also been reduced by another technique of DSMC-filter (Kaplan and Oran 2002) where it involves a postprocessing operation by employing a filter to extract the solution from a noisy DSMC calculation. The filters called flux-corrected transport, uses a high-order, nonlinear monotone convection algorithm. Simulations show that filtering removes high-frequency statistical fluctuations and extract as solution from a noisy DSMC calculation for low-speed Couette flows. All the above efforts have improved the computational accuracy and versatility of DSMC. A summary of these works is listed in Table 5.1.1.

In the last three decades, DSMC has been primarily used in the simulations of hypersonic rarefied-gas flows. The results have been validated well with many experimental data and DSMC has been widely accepted

TABLE 5.1.1 Summary of Significant Research Efforts on DSMC Methodology

DSMC model	Author	Year
BL	Borgnakke et al.	1975
VHS	Bird	1981
Radiation	Bird; Moss et al.	1987; 1988
NTC	Bird	1989
Dissociation	Nanbu	1991
Vibrational relaxation	Boyd	1992
VSS	Koura and Matsumoto	1991; 1992
GHS	Hash and Hassan	1993
Boundary conditions	Collins and Knox	1994
DSMC-MLG	Oh et al.	1995
DSMC-IP	Fan and Shen; Sun and Boyd	1998; 2002
DSMC-filter	Kaplan and Oran	2002

as an important numerical method in rarefied-gas flows. DSMC has been used for different applications. Campbell (1991) used DSMC to investigate the plume/free stream interactions applicable to a rocket exhaust plume at high altitudes. Celenligil et al. (1989) simulated the flow for blunt AFE configuration. Boyd et al. (1992) studied the flow of small nitrogen nozzles and plumes using DSMC and compared the numerical results with Navier-Stokes equations solutions and experimental data. Carlson and Wilmoth (1992) used the DSMC to solve the type IV shock-interaction problem at a scramjet cowl lip. Drags on spheres at hypersonic speeds were calculated for flows in continuum-transition regime and compared to experimental data by Dogra et al. (1992). Celenligil and Moss (1992) used a three-dimensional version of DSMC to study a delta wing configuration and found good agreement with wind-tunnel data. A comprehensive review of the applications of DSMC can be found in Oran et al. (1998).

5.1.2 Methodology

The procedures involved in applying DSMC to steady or unsteady flow problems are presented in Fig. 5.1.1 (Bird 1976). Execution of the method requires the physical domain to be divided into computational cells. The cells provide geometric boundaries and volumes required to sample macroscopic properties. It is also used as a unit where only molecules located within the same cell, at a given time, are allowed for collision. The cell dimensions, Δx_d, must be such that the change in flow properties is small (Bird 1976). This requires that Δx_d must be much smaller than the characteristic length of the geometry, or less than $\lambda/3$ in general. It is also important that the selection of a time step, Δt_d, should be less than the molecular mean collision time. Simulation results have been found independent of the time step increment as long as the spatial and temporal requirements are satisfied. The core of the DSMC algorithm consists of four primary processes: move the particles, index and cross-reference the particles, simulate collisions, and sample the flow field. These procedures are uncoupled during each time step. A general implementation of the DSMC method can be summarized as follows.

1. *Initialization*: A DSMC simulation keeps track of the time-dependent movement of a huge amount of molecules. Like a continuum CFD calculation, the DSMC simulation proceeds from a set of prescribed initial condition. The molecules are initially distributed in the computational domain. These simulated molecules, each representing a large number of real gas molecules, are assigned random velocities, usually based on the equilibrium distribution. For an equilibrium stationary problem, the initial condition does not affect the final DSMC

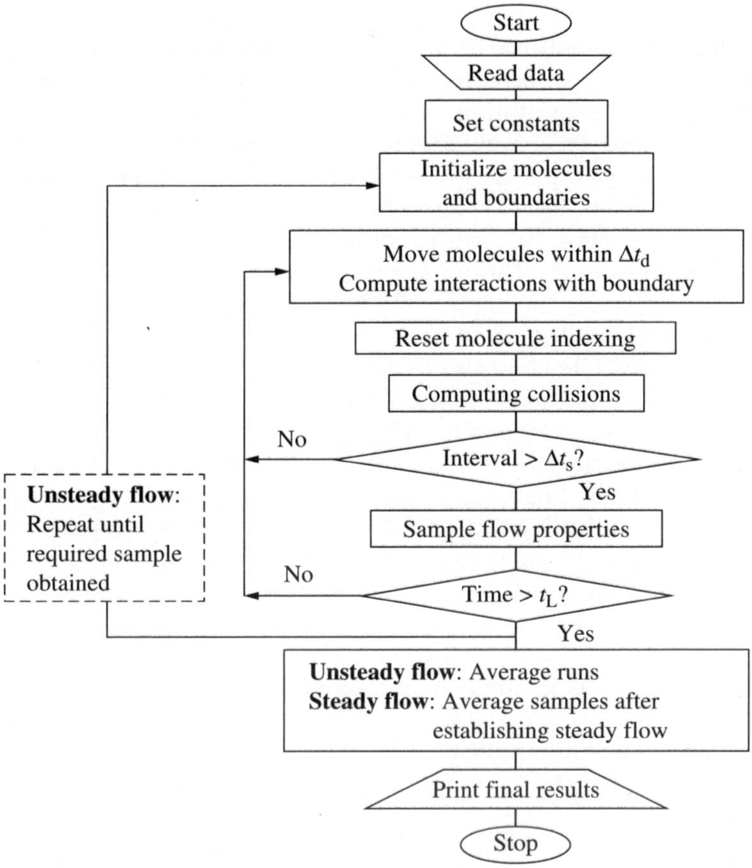

Figure 5.1.1 DSMC flow chart. (*Bird 1976*.)

solution, but will affect the computing time. Physically correct molecular sizes and weights are used in the calculation of gas properties. The position, velocity, collision cross section, and temperature of molecules, and the boundary conditions determine the subsequent evolution of the system. Therefore, for unsteady problems, the solution may be dependent on the particular choice of the initial, boundary, and input conditions (Oran et al. 1998).

2. *Movement*: The simulated representative molecules are moved for a convection time step of Δt_d. This molecular motion is modeled deterministically. This process enforces the boundary conditions and samples macroscopic flow properties along solid surfaces. Modeling molecule-surface interactions requires applying the conservation laws to individual molecules instead of using a velocity distribution function. Such treatment of the boundary conditions allows DSMC to be extended

to include physical effects such as chemical reactions, catalytic walls, radiation effects, three-body collisions, and ionized flows, without major modifications in the basic procedure. When a simulated molecule moves across boundary, computational boundary condition is employed. Specular reflection is used for symmetric boundary. Maxwellian diffusive model or Cercignani-Lampis-Lord model are usually used for the surface boundary, which are described in detail in Chap. 7. When a molecule moves out of an exit boundary, all its properties are taken out of the subsequent simulations. At the inlet/outlet boundary, new molecules are also allowed to enter the computational field. The molecule number and the location, velocity, and internal energy of the molecule are randomly determined based on time interval Δt_d and the macroscopic flow properties at the inlet/outlet boundary.

3. *Indexing and cross-referencing*: After all the simulated molecules have completed their movement of a time step Δt_d, they are in new locations in space and may be in a different field cell. New molecules may be introduced from an inlet boundary. Therefore, simulated molecules must be reindexed and tracked based on their spatial coordinates and associated field cells (computational grids). A scheme for molecular referencing is a prerequisite to the selection of collision partners and the sampling for the macroscopic properties of the flow field in the following steps. Efficient indexing and tracking schemes are keys to practical DSMC applications that involve large-scale information processing. To improve the accuracy and computing efficiency, subcells, such as virtual subcells, static subcells, and transient subcells, have been proposed for the DSMC method.

4. *Collision*: With the simulated particles appropriately indexed, the molecular collision is considered. The collision process is modeled statistically, which is different from that in deterministic simulation methods such as MD. Only the particles within a given computational cell are considered to be possible collision partners. Within each cell, a representative set of collisions occurs and collision pairs are selected randomly. The postcollision molecular velocities are determined and the particles move accordingly in the next time step. Uncoupling the molecular motions and the intermolecular collision requires that Δt_d be smaller than the mean collision time. There are several collision-sampling methods that have been applied successfully. The currently preferred model is the *no-time-counter* (NTC) technique (Bird 1994) used in conjunction with the subcell technique (Bird 1986). The subcell method calculates local collision rates based on the individual cell but restricts possible collision pairs to within subcells. The procedure improves accuracy by ensuring that collisions occur only between two near neighbors.

5. *Sampling*: The macroscopic flow properties are sampled after repeating step 2 to step 4 for a large time interval. The velocity of the molecules in a particular cell is used to calculate macroscopic quantities at the geometric center of the cell. Since these steps do not depend on the sampling process, the computational time can be reduced through sampling the flow properties every nth time step. The time-marching DSMC procedures are explicit and, therefore, always produce unsteady flow simulation. For a steady flow problem, a DSMC simulation proceeds until steady flow is established at a sufficiently large time t_L, and the desired steady result is the time average of all values sampled after the simulation has reached a steady state. For an unsteady flow, an ensemble of many independently performed simulations may be collected. An ensemble average property can then be obtained by averaging the instantaneous results in the sample over an area of a volume element.

5.1.3 Binary elastic collisions

The DSMC implementation of the binary elastic collision kinetics, discussed in Chap. 2, is described in this section. In such a collision, two scattering angles, χ and ε, are defined for DSMC simulations. χ denotes the deflection angle and, as has been defined previously, represents the angle between the precollision relative velocity and the postcollision relative velocity. ε is the azimuthal impact angle measured between the collision plane and some reference plane. For hard sphere collisions (HS and VHS), both scattering angles, χ and ε, are uniformly distributed. The molecule diameter d and the impact parameter b are related by $b/d = \cos(\chi/2)$. For VSS collisions, a parameter β is used to characterize the anisotropy of the deflection angle as

$$\left(\frac{b}{d}\right)^{1/\beta} = \cos\left(\frac{\chi}{2}\right) \tag{5.1.1}$$

The parameters must be used for computing the postcollision relative velocity $\mathbf{c}_r^*$ as a function of the precollision relative velocity $\mathbf{c}_r$ and the postcollision relative speed $\mathbf{c}_r^*$. For elastic collision, $\mathbf{c}_r^*$ is simply the magnitude of $\mathbf{c}_r$

$$c_r = |\mathbf{c}_1 - \mathbf{c}_2| \tag{5.1.2}$$

whereas, for inelastic collisions it may be different. The procedure for computing $\mathbf{c}_r^*$ is as follows.

First, the azimuthal impact angle ε is computed using a random fraction $R_{f,1}$ as

$$\varepsilon = 2\pi R_{f,1} \tag{5.1.3}$$

Second, the deflection angle χ is computed using another random fraction $R_{f,2}$

$$\cos\chi = 2R_{f,2}^{1/\beta} - 1 \tag{5.1.4}$$

$$\sin\chi = \sqrt{1-\cos^2\chi} \tag{5.1.5}$$

Next, a set of Cartesian coordinates (x', y', z') is introduced with x' in the direction of $\mathbf{c}_r$. The vector $\mathbf{g}_1$ representing $\mathbf{c}_r^*$ in this coordinate system (x', y', z') is

$$\mathbf{g}_1 = (c_r\cos\chi, c_r\sin\chi\cos\varepsilon, c_r\sin\chi\sin\varepsilon) \tag{5.1.6}$$

The precollision relative velocity $\mathbf{c}_r$ has components of (u_r, v_r, w_r) in the original Cartesian coordinates (x, y, z). The direction cosines of x' are $(u_r/c_r, v_r/c_r, w_r/c_r)$ in the original coordinates. Since the orientation of the reference plane is arbitrary, the y'-axis may be chosen to be normal to the x-axis. The direction cosines of y' are then $(0, w_r(v_r^2+w_r^2)^{-1/2}, -v_r(v_r^2+w_r^2)^{-1/2})$. Those of z' are $((v_r^2+w_r^2)^{1/2}/c_r, -u_r v_r(v_r^2+w_r^2)^{-1/2}/c_r, -u_r w_r(v_r^2+w_r^2)^{-1/2}/c_r)$. So, the coordinate transformation from (x', y', z') to (x, y, z) is a second-order tensor $\mathbf{g}_T$ of

$$\mathbf{g}_T = \begin{bmatrix} u_r/c_r & v_r/c_r & w_r/c_r \\ 0 & w_r\left(v_r^2+w_r^2\right)^{-1/2} & -v_r\left(v_r^2+w_r^2\right)^{-1/2} \\ \left(v_r^2+w_r^2\right)^{1/2}/c_r & -u_r v_r\left(v_r^2+w_r^2\right)^{-1/2}/c_r & -u_r v_r\left(v_r^2+w_r^2\right)^{-1/2}/c_r \end{bmatrix} \tag{5.1.7}$$

The components of $\mathbf{c}_r^*$ in the original coordinate system can be obtained through the product of vector $\mathbf{g}_1$ with the tensor $\mathbf{g}_T$, as

$$\mathbf{c}_r^* = \mathbf{g}_1 \cdot \mathbf{g}_T \tag{5.1.8}$$

and the required expressions of $\mathbf{c}_r^*$ are

$$u_r^* = \cos\chi u_r + \sin\chi\cos\varepsilon\left(v_r^2+w_r^2\right)^{1/2} \tag{5.1.9a}$$

$$v_r^* = \cos\chi v_r + \sin\chi(c_r w_r\cos\varepsilon - u_r v_r\sin\varepsilon)/\left(v_r^2+w_r^2\right)^{1/2} \tag{5.1.9b}$$

$$w_r^* = \cos\chi w_r - \sin\chi(c_r v_r\cos\varepsilon - u_r w_r\sin\varepsilon)/\left(v_r^2+w_r^2\right)^{1/2} \tag{5.1.9c}$$

From the conservations of momentum and energy, the postcollision velocities of the two collision molecules can be obtained as

$$\mathbf{c}_1^* = \mathbf{c}_m + \frac{m_2}{m_1 + m_2}\mathbf{c}_r^* \qquad (5.1.10\text{a})$$

$$\mathbf{c}_2^* = \mathbf{c}_m - \frac{m_1}{m_1 + m_2}\mathbf{c}_r^* \qquad (5.1.10\text{b})$$

where $\mathbf{c}_m$ is the velocity of the center of mass of the pair of collision molecules given by

$$\mathbf{c}_m = \frac{m_1\mathbf{c}_1 + m_2\mathbf{c}_2}{m_1 + m_2} \qquad (5.1.11)$$

5.1.4 Collision sampling techniques

Several collision-sampling techniques are discussed by Bird (1994). In this section, two of the most popular techniques, the time counter (TC) and the no time counter (NTC) sampling, are outlined. Both methods use *acceptance–rejection* statistics described by Bird (1976) to select the collision partners. The acceptance–rejection technique uses random numbers to determine whether a randomly selected pair of molecules will interact. The details of this technique will be described in App. 5A.

The probability of collision between two molecules in a homogeneous gas is proportional to the product of their relative speed c_r and total collision cross-section σ_T, as shown in Chap. 2. The total collision rate N_c per unit volume of gases is

$$N_c = \frac{1}{2}n^2\sigma_T\, c_r \qquad (5.1.12)$$

which could be used to establish the number of collisions in each cell at a time step Δt_d in a homogeneous gas is then $N_c\Delta t_d$. In the DSMC procedures, this number can be calculated. The mean value of the product of c_r and σ_T is calculated for each cell, and the maximum value could be recorded. The collision pairs could then be chosen by the acceptance-rejection method, with the probability P of a particular pair being given by the ratio of their product of c_r and σ_T to the maximum of the product in the cell.

$$P = \frac{\sigma_T\, c_r}{(\sigma_T\, c_r)_{\max}} \qquad (5.1.13)$$

The TC technique assigns an incremental time δt to each collision. For HS and VHS molecules, δt is given by

$$\delta t = \frac{2}{N_m \sigma_T c_r} \tag{5.1.14}$$

where N_m is the number of simulated molecules in a cell. Sufficient collisions are simulated in each cell for the sum of all δt to equal to the convection time step Δt_d. The TC technique is very efficient. Its computational cost is directly proportional to N_c. It has been found that TC technique predicts correct collision rates for moderate nonequilibrium flows. However, for flows with extreme nonequilibrium regions, such as strong shock fronts, it gives an inaccurate collision rate (Bird 1989). The TC technique allows the acceptance of unlikely collision pairs, resulting in an incorrect incremental time δt. This problem was corrected by the NTC technique. In NTC method, the procedures are similar to those in the TC technique, except that the summation of collision incremental time δt is replaced by the summation of the number of collision pairs until it reaches the number of allowed collisions, N_{cp}, which is given by

$$N_{\text{cp}} = \frac{N_m \bar{N}_m S_m (\sigma_T c_r)_{\max} \Delta t_d}{2V_c} \tag{5.1.15}$$

where S_m is the number of real molecules a simulated molecule represents, and V_c the cell volume. The computational cost of the NTC method remains directly proportional to N_m. It is the most widely used collision-sampling technique for DSMC and is also used in the IP1D program in App. 5C.

5.1.5 Cell schemes

The DSMC method uses field-cell systems for the sampling of the macroscopic flow properties and for the selection of collision partners. The macroscopic properties are assumed uniform in the cell and, as a result, the cell dimensions should be small in comparison with the length scale of the macroscopic flow gradients. The molecules in the cell are regarded as representative of those at the position of the cell, and the relative locations of molecules within the cell are ignored in the selection of collision partners.

Primitive implementations of the DSMC method choose the collision partners from any location in the same flow field cell and satisfactory results have been obtained as long as the fundamental constraints of DSMC are met. The sampled density is used to establish the collision rate and it is desirable to have the number of molecules per cell as large as possible, generally around 20. On the other hand, in the selection of collision partners, it is desirable to have this number as small as possible to reduce the mean separation of collision pairs. Meiburg (1986) questioned the validity of the method, claiming that, by allowing molecules of a cell to collide, DSMC could not support a rotational component within the cell. In response to the claim, Bird introduced the concept of *subcells*. In this scheme, subcells are created by further subdividing a flow-field cell and sorting molecules within that cell into the subcells. The standard collision methodology was then modified such that candidate collision pairs are formed at random from within the same subcell if possible. This reduces the separation distance between collision pairs, effectively increasing the spatial resolution of the simulation at minimal additional computational cost.

In widely used G2/A3 codes of Bird (1992), subcells were a fixed element of the flow field grid. The subcell resolution was the same for all flow field cells in a given region, and the subcell resolution was specified as a preprocessing input and held static during a simulation. This requires that the user specifies the ratio of the real-to-simulated molecules such that both the cells and the subcells are adequately populated. This is an important limitation on the number of subcells. The limitation is not about sufficient statistical representation, rather the ability to even form a collision pair from within the same subcell. When the number of molecules is small in a subcell, it becomes likely that there will be less than two molecules in the subcell. Searching in the neighboring subcells for a collision partner can add an additional layer of complexity to the collision algorithm. The problem can be more troublesome for three-dimensional problems. For example, for each direction being bisected, there will be a total of 2^3 subcells per cell. In a best-case scenario, 16 molecules are needed in a field cell to maintain an average of two molecules per subcell. Often, there will be subcells with fewer than two molecules. This would require the implementation of an even more complex search algorithm than that for two-dimensional problems.

As an alternative, Bird (2000) devised *transient-adaptive subcells* to address the under-populated problems associated with the use of subcell. The subcell resolutions are determined dynamically for each flow field cell based on the number of molecules in the cell. The subcell resolution is such that there are as many subcell as molecules. Dynamic subcells provide obvious advantages over static subcells, in terms of both usability and efficiency. However, the design of an appropriate search methodology to locate a second collision partner is still a relevant issue, since on an average there are fewer than two molecules in a subcell.

The static or transient-adaptive subcell scheme cannot guarantee that a subcell would have less than two molecules in it on occasion. LeBeau et al. (2003) proposed the use of *virtual subcells*. It simply involves picking the first collision partner at random, and then locating the nearest other molecule in the flow-field cell as the second partner. The method introduces a N^2-like search algorithm to the procedure and additional computational cost. Only in cases where the molecule number N is kept small in each flow-field cell the virtual subcell scheme is computationally competitive with the true subcell methods.

5.1.6 Sampling of macroscopic properties

The macroscopic properties of a flow field are sampled from field cells, and the value represents the flow property at the cell center. For a steady state problem of monatomic molecules, the flow field is usually sampled after flow has transitioned from the initial state to a stationary state. The transition usually takes a few thousands time steps. The flow density and velocity can be obtained through the following equations:

$$N = \sum_{i=1}^{N_t} N_i \tag{5.1.16}$$

$$\rho = \frac{1}{N_t} \frac{N F_{\text{num}}}{\Delta V} m \tag{5.1.17}$$

$$U = \frac{1}{N} \sum_{i=1}^{N_t} \sum_{j=1}^{N_i} c_{u,j} \tag{5.1.18}$$

$$V = \frac{1}{N} \sum_{i=1}^{N_t} \sum_{j=1}^{N_i} c_{v,j} \tag{5.1.19}$$

$$W = \frac{1}{N} \sum_{i=1}^{N_t} \sum_{j=1}^{N_i} c_{w,j} \tag{5.1.20}$$

Where N_i is the number of molecules in the sampled field cell at the i-th sampling time step, N the total number of molecules sampled during the total time steps of N_t in the field cell, ΔV the volume of the sampled field cell, F_{num} the number of real molecules each simulated particle represents (weight factor), m mass of the simulated molecule. $c_{u,j}$, $c_{v,j}$, and $c_{w,j}$ are the molecular velocity component in the x-, y-, and z-directions, respectively, of the jth molecule sampled at the ith time step. ρ is flow density. U, V, and W are the mean flow velocity in the x, y, and z directions, respectively. The thermal temperatures T_x, T_y, and T_z in the x-, y-, and z-directions can be obtained as

$$T_x = \frac{m}{k}\left\{\frac{1}{N}\sum_{i=1}^{N_t}\sum_{j=1}^{N_i} c_{u,j}^2 - \left(\frac{1}{N}\sum_{i=1}^{N_t}\sum_{j=1}^{N_i} c_{u,j}\right)^2\right\} \tag{5.1.21a}$$

$$T_y = \frac{m}{k}\left\{\frac{1}{N}\sum_{i=1}^{N_t}\sum_{j=1}^{N_i} c_{v,j}^2 - \left(\frac{1}{N}\sum_{i=1}^{N_t}\sum_{j=1}^{N_i} c_{v,j}\right)^2\right\} \tag{5.1.21b}$$

$$T_z = \frac{m}{k}\left\{\frac{1}{N}\sum_{i=1}^{N_t}\sum_{j=1}^{N_i} c_{w,j}^2 - \left(\frac{1}{N}\sum_{i=1}^{N_t}\sum_{j=1}^{N_i} c_{w,j}\right)^2\right\} \tag{5.1.21c}$$

The flow temperature and pressure can be obtained as

$$T = \frac{1}{3}(T_x + T_y + T_z) \tag{5.1.22}$$

$$p = \left(\frac{1}{N_t}\frac{NF_{\text{num}}}{\Delta V}\right)kT \tag{5.1.23}$$

From Eqs. (5.1.16–23), it can be seen that only the number of molecules N_i and the following six summations are necessary to be saved to obtain the macroscopic flow properties of flow property of density, pressure, temperature, and velocity during sampling procedure.

$$\begin{matrix} \sum_{i=1}^{N_t}\sum_{j=1}^{N_i} c_{u,j} & \sum_{i=1}^{N_t}\sum_{j=1}^{N_i} c_{v,j} & \sum_{i=1}^{N_t}\sum_{j=1}^{N_i} c_{w,j} \\ \sum_{i=1}^{N_t}\sum_{j=1}^{N_i} c_{u,j}^2 & \sum_{i=1}^{N_t}\sum_{j=1}^{N_i} c_{v,j}^2 & \sum_{i=1}^{N_t}\sum_{j=1}^{N_i} c_{w,j}^2 \end{matrix} \tag{5.1.24}$$

The aerothermodynamic properties on the surface are sampled from the momentum flux and energy flux on the surface. For example, the pressure on the surface p_w can be obtained by sampling the difference between the normal momentum flux on the surface before and after gas–surface interaction per area ΔA during sampling time t_s.

$$p_w = \frac{F_{\text{num}}}{t_s \cdot \Delta A}\sum_{j=1}^{N_s} m\left(c_{n,j}^r - c_{n,j}^i\right) \tag{5.1.25}$$

where c_n the normal component of molecular velocity, N_s the total number of molecules impacting the surface element during sampling time t_s.

The superscripts "i" and "r" denote the values of before and after impacting the wall element, respectively. The shear stress τ_w on the surface can be obtained as

$$\tau_w = \frac{F_{\text{num}}}{t_s \cdot \Delta A} \sum_{j=1}^{N_s} m\left(c_{t,j}^r - c_{t,j}^i\right) \tag{5.1.26}$$

where c_t is the molecular velocity of the tangential component. And heat transfer rate can be obtained by sampling the difference of energy fluxes. That is,

$$q_w = \frac{F_{\text{num}}}{t_s \cdot \Delta A} \left[\sum_{j=1}^{N_s} \left(\frac{1}{2} m c_j^2\right)^i - \sum_{j=1}^{N_s} \left(\frac{1}{2} m c_j^2\right)^r \right] \tag{5.1.27}$$

5.2 DSMC Accuracy and Approximation

5.2.1 Relationship between DSMC and Boltzmann equation

The general form of the Boltzmann equation for a simple dilute gas, Eq. (2.6.10) described in Chap. 2, defines the relationship between the velocity distribution function and its dependent variables. It is the governing equation for gases in the entire transition regime of interest in this study. The Boltzmann equation is derived from the fundamental principles of classical kinetic theory and is restricted to dilute gas flows in molecular chaos.

The DSMC method, on the other hand, is derived from the same first principles as the Boltzmann equation, but not from the equation itself. Due to its ties to classical kinetic theory, the DSMC method is subject to the same restrictions of dilute gas and molecular chaos. Unlike the Boltzmann equation, however, the DSMC method does not require the existence of inverse collisions that are dictated by symmetry considerations of binary dynamics. This allows the application of the method to some complex phenomena, such as ternary chemical reactions, that are inaccessible to the Boltzmann equation.

A derivation of the Boltzmann equation from DSMC procedures could be obtained for hard sphere molecules based on the NTC collision technique. The left-hand side of Eq. (2.6.10) states that the quantity nf remains constant in the phase space in the absence of collisions if one moves along with a group of molecules in a Lagrangian manner. Similarly, the DSMC procedures trace the paths of the simulated molecules in the phase space, and the processes are consistent with the Boltzmann formulation. Any discrepancy between DSMC and the Boltzmann equation would be from the collision term on the right-hand side.

The collision term conventionally comprises the gain and loss integrals. The loss term represents the rate of scattering by collision out of the phase space element $d\mathbf{c}\,d\mathbf{r}$ per unit volume of this element. This rate of scattering may be derived from the DSMC procedures by the following arguments.

The spatial cell in the simulated flow may be treated as the element $d\mathbf{v}$ of physical space, and the set of N_m molecules in the cell defines the velocity distribution function f. The number of molecules of class $\mathbf{c}$ within the cell at time t is $nf\,d\mathbf{c}\,d\mathbf{v}$, or $N_m\,f d\mathbf{c}$. Now consider a collision between a molecule of this class and one of class $\mathbf{c}_1$. The probability of finding a molecule in class $\mathbf{c}_1$ is $f_1 d\mathbf{c}_1$, with f_1 being of f at velocity $\mathbf{c}_1$, and the collision rate for such collisions is

$$\nu = \frac{(N_m f d\mathbf{c})(f_1 d\mathbf{c}_1)}{\Delta t} \tag{5.2.1}$$

For HS molecules, the increment of time Δt contributed by such a collision to a counter for class $\mathbf{c}$ molecules is given from Eq. (5.2.1) by

$$\Delta t = \frac{1}{n\sigma_T\, c_r} \tag{5.2.2}$$

and substituting Eq. (5.2.2) into Eq. (5.2.1) yields to

$$\nu = n^2 \sigma_T\, c_r\, f f_1 d\mathbf{c}_1\, d\mathbf{c}\, d\mathbf{v} \tag{5.2.3}$$

Finally, the rate of molecules per unit volume scattering out of class $\mathbf{c}$, i.e., the loss term, is obtained by substituting the definition of σ_T, integrating class $\mathbf{c}_1$ in Eq. (5.2.3) over the velocity space, and dividing by $d\mathbf{c}\,d\mathbf{v}$. The resulting expression

$$\int_{-\infty}^{\infty} \int_{0}^{4\pi} n^2 (f f_1) c_r \sigma\, d\,\Omega d\mathbf{c}_1 \tag{5.2.4}$$

is identical to the loss term of the Boltzmann equation. The gain term in the Boltzmann equation can be derived in a similar manner as that for the loss term through the existence of inverse collisions for the simple monatomic gas model. The DSMC method selects collision partners and replaces their precollision velocity components by appropriate postcollision values. Therefore, the correct gain term is automatically obtained without requiring the existence of inverse collision. This holds irrespective of the existence of inverse collisions, indicating that the DSMC method is less restrictive than the Boltzmann equation.

Since the procedures of the DSMC technique are consistent with the formulation of the Boltzmann equation, the results of DSMC provide accurate solutions to the Boltzmann equation as long as the numerical approximations are kept within allowable bounds.

5.2.2 Computational approximations and input data

Two distinct types of errors are associated with a DSMC simulation. The first type is caused by the computational approximations inherent to the method. These include errors due to the finite cell sizes in the physical space, the finite size of the time step, the ratio of actual to simulated molecules, and the various aspects in the implementation of the boundary conditions. The second type of error is the result of uncertainties, or inadequacies, in the physical model input parameters. These include uncertainties about the type of species modeled, their interaction cross-sections, and the other aspects of boundary conditions and interactions.

When the cell size in physical space is large, macroscopic gradients are typically underpredicted. For accurate DSMC simulations, cell dimensions must be smaller than λ in each spatial direction. Another computational approximation involves the time interval over which molecular motions and collisions are uncoupled. The effects of the time step are negligible if the time step Δt_d is smaller than the mean collision time of molecules. DSMC procedures are not subject to a stability criterion such as the *Courant-Friedrichs-Lewy* (CFL) condition of continuum CFD. Disturbances propagate at sound or shock speed throughout the DSMC computational domain, even though the ratio of spatial cell size to time step may be a smaller value.

Large values of the ratio S_m of the actual to simulated molecules are typical in most applications. This can lead to unacceptable levels of statistical scatter in a single independent computation. To reduce the scatter to acceptable levels, a large ensemble average is needed. Typical values of S_m used may range from 10^9 to 10^{18} for a three-dimensional computation. In problems involving chemical reactions or thermal radiation, large value of S_m is particularly problematic. Important physical effects, caused by only a few molecules, may not be simulated properly. In these cases, special remedies, such as those proposed by Bird (1994) and Boyd (1996) are required.

The errors associated with various models and input data are not easily summarized as those inherent to the method itself. Input data for the molecular models may consist of the type of molecular gas-surface interactions or the total collision cross-section. The errors associated with various molecular models and input data are complex and can not be analyzed readily.

Similar to continuum CFD, DSMC results are strongly influenced by the treatment of surface interactions. The DSMC method relies on gas-surface interaction models that are simple, fast, and accurate for a wide range of engineering problems. The primary types of molecular-surface interactions invoked in DSMC have been briefly described earlier.

The thermal accommodation coefficient a_T is an input parameter, which is usually determined experimentally. Another input data is the total collision cross-section σ_T, which is the equivalent of chemical reactions in continuum CFD. The models for σ_T may range from the simplest hard-sphere model to complex models for inelastic collisions that account for vibrational and rotational nonequilibrium (Borgnakke and Larsen 1975). The HS model is more or less adequate for idealized monotonic gas computations but not for real gases. The VHS model (Bird 1992) corrects the primary deficiency in the hard-sphere model, namely, an inaccurate representation of the total collision cross section, while retaining its simplicity in implementation. The VSS model (Koura and Matsumoto 1991; 1992) further refines the VHS model by improving postcollision scattering dynamics. The GHS model (Hash and Hassan 1993) extends the VHS model to allow relaxation of the internal modes.

5.3 Information Preservation Method

5.3.1 Overview

IP development and applications. The DSMC method (Bird 1994) is one of the most successful numerical approaches for simulating rarefied gas flows. With appropriate outflow boundary conditions, it can also be used to simulate microflows. However, the statistical scatter associated with DSMC prevents its further applications to microflow of extremely low speeds. The information preservation (IP) method has been developed to overcome this problem of DSMC.

Fan and Shen (1998) first proposed an IP scheme for low-speed rarefied gas flows. Their method uses the molecular velocities of the DSMC method as well as the preserved information velocities that record the collection of an enormous number of molecules that a simulated particle represents. The information velocity is based on an inelastic collision model, and sampled in the same manner as that for the macroscopic flow velocity. IP has been applied to low-speed Couette, Poiseuille, and Rayleigh flows in the slip, transition and free-molecular flow regimes with very good agreement with the corresponding analytical solutions (Fan and Shen 2001). They showed that the IP scheme could reduce computational time by several orders of magnitude compared with a regular DSMC simulation. This scheme has also been applied in the simulation of low-speed microchannel flows (Cai et al. 2000) and the flow around a NACA0012 airfoil (Fan et al. 2001) with isothermal wall conditions.

The model formulation described in the above IP applications is not general. Sun and Boyd (2002) developed a general two-dimensional IP method to simulate subsonic microflows. The preserved macroscopic information is first solved in a similar manner to that for the microscopic information in the DSMC method and is then modified to include pressure effects. With modeling for gas energy transport included, the IP method has been successfully applied to simulate high-speed Couette flow, Rayleigh flow, and the flow over a NACA0012 airfoil.

Statistical scatter in DSMC. IP addresses the statistical scatter inherent in the DSMC method. Statistical scatter also imposes serious limitation in the application of DSMC. To illustrate this fact, Fan and Shen (2001a) considered the DSMC simulation of a uniform flow with velocity $\mathbf{U}$. The macroscopic velocity $\mathbf{u}_c$, sampled in a field cell, can be decomposed as

$$\begin{aligned}\mathbf{u}_c &= \frac{1}{N}\sum_{i=1}^{N}\mathbf{c}_i \\ &= \mathbf{U} + \frac{1}{N}\sum_{i=1}^{N}\mathbf{c}_{t,i}\end{aligned} \tag{5.3.1}$$

where N is the sample size in the field cell, $\mathbf{c}_i$ is the molecular velocity of the ith particle. According to kinetic theory, $\mathbf{c}_i$ consists of the mean flow velocity $\mathbf{U}$ and the thermal velocity $\mathbf{c}_{t,i}$. The thermal part $\mathbf{c}_{t,i}$ is random and obeys the Maxwellian distribution in an equilibrium gas. The DSMC method collects $\mathbf{c}_i$ and uses it to compute both the molecular trajectory and field velocity $\mathbf{u}_c$. Equation (5.3.1) shows that it is the explicit term $\frac{1}{N}\sum_{i=1}^{N}\mathbf{c}_{t,i}$ that makes $\mathbf{u}_c$ different from the exact value of $\mathbf{U}$. This thermal velocity term is the major contributing source to statistical scatter in the DSMC method. As has been shown in Chap. 2, the most probable thermal speed is about 1.2 times that of the speed of sound for monatomic, ideal gas. Therefore, for high-speed hypersonic flows, the statistical scatter is small. For low-speed microflows, on the other hand, the DSMC scatter can become large. The IP techniques have been proposed to reduce the statistical scatter of DSMC.

IP of Fan and Shen. Based on the above kinetic analysis of molecular velocity $\mathbf{c}_i$, Fan and Shen (1998) first proposed an IP technique. They assigned each simulated molecule in the DSMC method two types of velocities. One is the molecular velocity $\mathbf{c}_i$ to compute the molecular motion following the same steps in the DSMC method. The other is the information velocity $\mathbf{v}_{\mathrm{ip},i}$. $\mathbf{v}_{\mathrm{ip},i}$ is the collective velocity of the enormous number of real molecules that a single simulated molecule represents, and therefore corresponds to $\mathbf{U}$. The information velocity is different from macroscopic velocity, which actually is the average of molecular velocities over all the real molecules represented by the simulated molecules in the DSMC method. The IP technique employs the information velocity to compute the macroscopic velocity,

$$\mathbf{u}_c = \frac{1}{N} \sum_{i=1}^{N} \mathbf{v}_{\text{ip},i} = \mathbf{U} \tag{5.3.2}$$

Therefore, the statistical scatter source $1/N \sum_{i=1}^{N} \mathbf{c}_{t,i}$ in the DSMC sampling Eq. (5.3.1) does not appear in the IP sampling Eq. (5.3.2). For instance, for a uniform flow of velocity $\mathbf{U}$, the macroscopic velocity calculated by using Eq. (5.3.2) is the exact value of $\mathbf{U}$ regardless of the sample size.

The IP technique of Fan and Shen (1998) used the DSMC simulation particles as a carrier and preserved the mean flow velocity $\mathbf{U}$ as the information velocity of the simulated molecules. This technique does not affect the DSMC procedures. Particles move along their trajectories and recalculate their translational velocities and internal energy with every collision or wall interaction. The IP method also uses the molecule distribution and molecular collision in the DSMC procedures to preserve the information velocity, although by different rules. In addition to molecular collision, the IP technique also considered the pressure gradient as an external force that caused changes to the particle velocities. Arguing that the field cell can be regarded as a control volume from a macroscopic point of view, Fan and Shen used the following equations for the conservation of mass and momentum of the preserved variables,

$$\frac{\partial}{\partial t} n_c + \nabla \cdot (n\bar{\mathbf{v}}_{\text{ip},i})_c = 0 \tag{5.3.3}$$

$$\frac{\partial}{\partial t}(\bar{\mathbf{v}}_{\text{ip},i}) + [\nabla p/(nm)]_c = 0 \tag{5.3.4}$$

where n_c is the number density of molecules in a cell control volume, ∇p the pressure gradient in the cell, and i the index to the molecules in the cell. In the IP molecular collisions, the preserved information velocities are divided equally. This collision model is *primitive*. Sun and Boyd (2002) has suggested an improved model and it is described in Sec. 5.3.4.

The original IP technique of Fan and Shen (1998) was based on an isothermal assumption that variations of the translational temperature are small and can be neglected. This assumption holds well for low-speed microflows. Sun and Boyd (2002) modified the approach to consider the translational temperature change in a general form and developed an improved collision model. The following discussions will focus on the IP method of Sun and Boyd (2002).

5.3.2 IP governing equations

Gas flows can generally be described by the density, velocity, and temperature in the flow field. In the IP scheme of Sun and Boyd (2002), each simulated particle has the following information: location, microscopic velocity, internal energy, macroscopic density, macroscopic velocity, and macroscopic temperature. Because the particle's location, microscopic velocity, and internal energy are handled by the DSMC method, only the preserved macroscopic information requires modeling.

To clarify the connection between the macroscopic information and the microscopic properties of particles in gas flows, two sets of velocities are defined for a particle i : the molecular velocity $\mathbf{c}_i$ and the preserved macroscopic velocity $\mathbf{v}_{\text{ip},i}$. With the mean velocity of the flow field written as $\mathbf{c}_0$, they hold the following relationship with scatter $\mathbf{c}'_i$, $\mathbf{c}''_i$, and $\mathbf{c}'''_i$ as

$$\mathbf{c}_i = \mathbf{c}_0 + \mathbf{c}'_i \tag{5.3.5}$$

$$\mathbf{v}_{\text{ip},i} = \mathbf{c}_0 + \mathbf{c}''_i \tag{5.3.6}$$

$$\mathbf{c}'''_i = \mathbf{c}'_i - \mathbf{c}''_i \tag{5.3.7}$$

Therefore,

$$\mathbf{c}_i = \mathbf{v}_{\text{ip},i} + \mathbf{c}'''_i \tag{5.3.8}$$

The information preserved equations (IPE) can be obtained by substituting Eq. (5.3.8) into the moment Eqs. (4.2.7), (4.2.10), and (4.2.14) without an external force.

$$\frac{\partial}{\partial t}\rho + \nabla\cdot(\rho\bar{\mathbf{v}}_{\text{ip},i}) = 0 \tag{5.3.9}$$

$$\frac{\partial}{\partial t}(\rho\bar{\mathbf{v}}_{\text{ip},i}) + \nabla\cdot(\rho\overline{\mathbf{c}_i}\bar{\mathbf{v}}_{\text{ip},i}) = -\nabla p + \nabla\cdot\boldsymbol{\tau}_{\text{ip}} \tag{5.3.10}$$

$$\frac{\partial}{\partial t}\left(\frac{1}{2}\rho\overline{\left(v^2_{\text{ip},i} + \xi RT_{\text{ip},i}\right)}\right) + \nabla\cdot\left(\frac{1}{2}\rho\overline{\left(v^2_{\text{ip},i} + \xi RT_{\text{ip},i}\right)\mathbf{c}_i}\right)$$
$$= \nabla\cdot\mathbf{h}_{\text{ip}} - \nabla\cdot(p\bar{\mathbf{v}}_{\text{ip},i}) + \nabla\cdot(\bar{\mathbf{v}}_{\text{ip},i}\cdot\boldsymbol{\tau}_{\text{ip}}) \tag{5.3.11}$$

where

$$\boldsymbol{\tau}_{\text{ip}} = -(\rho\overline{\mathbf{c}'_i\mathbf{c}'''_i} - p\mathbf{I}) \tag{5.3.12}$$

$$p = nk\bar{T}_i \tag{5.3.13}$$

$$\mathbf{h}_{\text{ip}} = -\frac{1}{2}\rho\overline{\mathbf{c}'_i c'''^2_i} \tag{5.3.14}$$

Equation (5.3.12) gives the viscous stress tensor, Eq. (5.3.13) the pressure, and Eq. (5.3.14) the heat transfer vector. Equations (5.3.9–11) correspond to the transfer equation for molecular quantity Q in the conservative form for the preserved mass ($Q = m$), the preserved momentum ($Q = m\mathbf{v}_{\mathrm{ip},i}$), and the preserved energy $\left[Q = \frac{1}{2}m(\mathbf{v}_{\mathrm{ip},i}^2 + \xi R T_{\mathrm{ip},i})\right]$. Note that $T_{\mathrm{ip},i}$ in Eq. (5.3.11) is defined by

$$T_{\mathrm{ip},i} = c_i'''^2/(\xi R) \tag{5.3.15}$$

which is different from the thermal temperature $T = \overline{c_i'^2}/(\xi R)$. The thermal temperature T can be obtained from the preserved information $T_{\mathrm{ip},i}$,

$$T = \bar{T}_{\mathrm{ip},i} + \frac{\overline{v_{\mathrm{ip},i}^2} - \bar{v}_{\mathrm{ip},i}^2}{\xi R} \tag{5.3.16}$$

by writing the average of the total molecular energy as $\frac{1}{2}\overline{c_i^2} = \frac{1}{2}(c_0^2 + \xi R T)$ from Eq. (5.3.5) and $\frac{1}{2}\overline{c_i^2} = \frac{1}{2}(\overline{v_{\mathrm{ip},i}^2} + \xi R \bar{T}_{\mathrm{ip},i})$ from Eq. (5.3.8), with the assumption that $\bar{v}_{\mathrm{ip},i} = c_0$.

5.3.3 DSMC modeling for the information preservation equations

Equations (5.3.9) to (5.3.11) are the governing equations for the IP variables. They are, however, not a closed set of equations and closure models are needed. Direct mathematical modeling for some of the terms in the IPE Eqs. (5.3.9)–(5.3.11), such as the viscous stresses and the heat fluxes, are very difficult. Attempts to model the information temperature have been made (Fan and Shen et al. 2001; Sun 2001). Sun and Boyd (2002) have reported some success in high-speed Couette flows and NACA0012 flows.

In the DSMC procedures, the transport of particle mass, momentum, and energy are caused by particle movement, particle collision, and external force (if it exists). As was shown earlier, these processes are consistent with the Boltzmann equation. According to IPE, these mechanisms also cause the transport of the preserved mass, momentum, and energy for each simulated particle. The convective terms in IPE Eqs. (5.3.9) to (5.3.11) are in the same form as the convective terms in Eqs. (4.2.7), (4.2.10), and (4.2.13), which describe the particle movement in the DSMC procedure. That means, the particle movement in IP can be described by the particle movement in the DSMC procedures. As a result, the IP technique uses the DSMC movement as its carrier for the preserved information in the simulation of gas flows. This, however, is not to say that the amount of the transfer of the preserved information as a result of the particle movement is the same as that of the molecular momentum and energy transfer.

The molecular collision gives rise to the viscous stress and the heat flux effects in DSMC as well as IP. As shown in kinetic theory, the viscous stress

$$\tau_{\mathrm{DSMC}} = -(\rho\overline{\mathbf{c}_i'\mathbf{c}_i'} - p\mathbf{I}) \tag{5.3.17}$$

and heat transfer flux

$$\mathbf{h}_{\mathrm{DSMC}} = -\frac{1}{2}\rho\overline{\mathbf{c}_i' c_i'^2} \tag{5.3.18}$$

are accurately calculated in DSMC simulations. To account for their difference, the viscous stress and heat transfer flux in the IP technique can be written as

$$\tau_{\mathrm{ip}} = \tau_{\mathrm{DSMC}} + \Delta\tau_{\mathrm{ip}} \tag{5.3.19}$$

$$\mathbf{h}_{\mathrm{ip}} = \mathbf{h}_{\mathrm{DSMC}} + \Delta\mathbf{h}_{\mathrm{ip}} \tag{5.3.20}$$

Fan and Shen (1998, 2001) showed that, for low-speed rarefied gas flows where only factors affecting the momentum transport were considered, the preserved information velocities obtained in the IP simulation agreed well with the DSMC solutions. Therefore, the difference in the viscous stress between DSMC and IP approaches could be neglected in momentum transport. That is, $\Delta\tau_{\mathrm{ip}} = 0$. The difference in the energy transport $\Delta\mathbf{h}_{\mathrm{ip}}$ has been considered by Sun and Boyd (2002) and their model is described in section 5.3.4.

With the use of the particle movement simulated in the DSMC procedure as the information carrier, the convection term of preserved momentum can be completely accounted for in the DSMC procedure if $\Delta\tau_{\mathrm{ip}} = 0$. The convection term of preserved energy contains third-order correlations, much more complicated than that for the preserved momentum. Usually, it is different from the convection term of particle energy in Eq. (4.2.13). If we take the difference between the convection terms of Eqs. (5.3.11) and (4.2.13) and combined it into $\Delta\mathbf{h}_{\mathrm{ip}}$ as energy flux, the second term in the left-hand side of Eq. (5.3.11) can be taken off in the DSMC procedures. Therefore, IPE for the correction of the preserved mass, momentum, and preserved energy transport can be reduced to

$$\frac{\partial}{\partial t}\rho = -\nabla\cdot(\rho\bar{\mathbf{v}}_{\mathrm{ip},i}) \tag{5.3.9a}$$

$$\frac{\partial}{\partial t}(\rho_c\mathbf{v}_{\mathrm{ip},i}) = -\nabla p \tag{5.3.21}$$

$$\frac{\partial}{\partial t}\left(\frac{1}{2}\rho_c\left(v_{\mathrm{ip},i}^2 + \xi RT_{\mathrm{ip},i}\right)\right) = \nabla\cdot(\Delta\mathbf{h}_{\mathrm{ip,c}}) - \nabla\cdot(p\bar{\mathbf{v}}_{\mathrm{ip,c}}) \tag{5.3.22}$$

where the viscous dissipation term has been assumed negligible in energy transport. Equations (5.3.9a), (5.3.21), and (5.3.22) are referred to as the control equations in IP. Equation (5.3.9a) forces the mass conservation of preserved mass based on cell average. Equations (5.3.21) and (5.3.22) account for the microscopic mechanisms for simulated particles in a sense of preserved information that are not included in the DSMC procedure, yet still contribute to the rate of change of the preserved momentum and energy. It can be seen that the transport of the preserved momentum and energy in IP results from additional factor of the pressure gradient, which is true because the pressure gradient effect is always considered from macroscopic point of view. In these control equations, the quantities on the right-hand side and the density ρ_c on the left-hand side are field properties of the cell where the simulated particles stay at the calculation time t. All these variables can be calculated, except for the first term on the right-hand side of Eq. (5.3.22), $\nabla \cdot (\Delta \mathbf{h}_{\text{ip,c}})$. The energy flux $\Delta \mathbf{h}_{\text{ip,c}}$ needs physical modeling. In principle, modeling $\Delta \mathbf{h}_{\text{ip,c}}$ should include the difference in the convection terms of energy Eqs. (5.3.11) and (4.2.13) and the difference in disturbance energy diffusion based on thermal fluctuation. That is,

$$\Delta \mathbf{h}_{\text{ip,c}} = -\frac{1}{2}\rho\overline{\left(v_{\text{ip},i}^2 - c_0^2\right) + \xi R(T_{\text{ip},i} - T)} \cdot \overline{c_i} - \frac{1}{2}\rho\left(\overline{c_i' c_i''^2} - \overline{c_i' c_i'^2}\right) \quad (5.3.23)$$

It can be seen from Eqs. (5.3.22) and (5.3.23) that the particle movement has an additional effect on the preserved energy transport in IP.

5.3.4 Energy flux model for the preserved temperature

Kinetic theory for stationary monatomic gases shows that the average internal energy of the molecules crossing a surface element exceeds that in a spatial element by a factor 4/3 (see App. 5B for details). That is, these molecules have an average translational energy of $2kT$ for stationary gases with temperature T. However, the average translational energy is $\frac{3}{2}kT$ for all the spatial molecules. This is because the probability for fast molecules to cross a surface element per unit time is greater than the corresponding probability for slower molecules. Physically, it can be understood that molecules crossing a surface element carry an additional energy of $\frac{1}{2}kT$ on an average. That is, their preserved average translational energy can be modeled as

$$\frac{1}{2}\overline{c_i'''^2} = \frac{3}{2}R(T + T_a) \quad (5.3.24)$$

where

$$T_a = \frac{1}{3}T \tag{5.3.25}$$

for the preserved velocity $\bar{\mathbf{v}}_{\mathrm{ip},i} = 0$. In the DSMC procedure, the field is sampled based on field cells. Therefore, the term $\nabla \cdot (\Delta \mathbf{h}_{\mathrm{ip,c}})$ in Eq. (5.3.22) corresponds to the additional energy transport carried by the molecule crossing its cell interface from the viewpoint of preserved information.

Sun and Boyd (2002) proposed an approximate model to describe the additional energy transport for the IP technique, where a flow of two monatomic gases separated by a plate is considered as shown in Fig. 5.3.1. The temperature on the left side of the plate is T_1 and the right side is T_2. The particles preserve the macroscopic temperature information with T_1 or T_2, and the averaged translational energy with $\frac{3}{2}kT_1$ and $\frac{3}{2}kT_2$ for the two groups of particles, respectively. The two gases mix together after the plate is suddenly removed. Some molecules described by T_1 move to the other side, transfer energy of $\frac{3}{2}kT_1$ plus an additional $\frac{1}{2}kT_1$ to the other side on an average, and vice versa. Assuming that number of particles crossing in each direction is equal, the net energy flux based on the moving particles is $\frac{3}{2}k(T_1 - T_2)$ plus the additional $\frac{1}{2}k(T_1 - T_2)$. This process is modeled by assuming that the T_1 particles transfer $\frac{3}{2}kT_1$ plus an additional $\frac{1}{2}k(T_1 - T_{\mathrm{ref}})$ to the other side, and the T_2 particle transfer $\frac{3}{2}kT_2$ plus an additional $\frac{1}{2}k(T_2 - T_{\mathrm{ref}})$ to the other side, where T_{ref} is a reference temperature. Therefore, the IP approach captures the net energy flux across an interface from a viewpoint of macroscopic average.

In the energy model, a particle carries an additional energy when it crosses a field cell interface. The additional energy is "borrowed" from all the other particles in the same cell, which means other particles

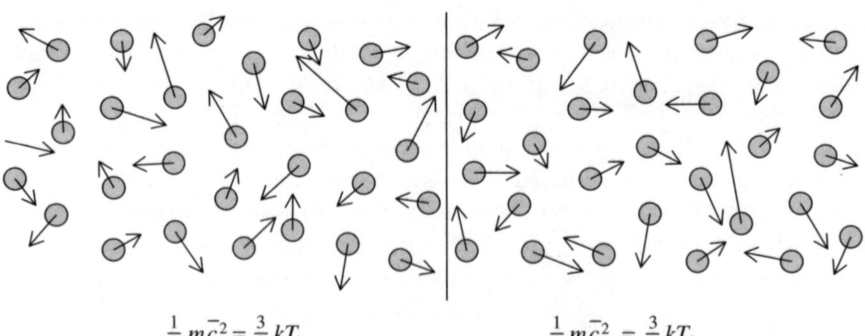

Figure 5.3.1 Stationary monatomic gases separated by a plate. (*Sun and Boyd 2002.*)

need to share this energy to satisfy the conversion of energy. The additional energy is taken as $\frac{1}{2}k(T - T_{\text{ref}})$, where T_{ref} is very close to the preserved temperature T of the particle. The part $\frac{1}{2}kT_{\text{ref}}$, in the additional energy of one particle, is balanced by another particle such that the net flux across an interface can be appropriately modeled. For the stationary gas flow, the number of particles crossing each direction is the same, and the reference temperature can be taken as the temperature on the interface that a particle crosses, which can be interpolated from the preserved temperatures of two neighboring cells. Then the effect of the additional energy $\frac{1}{2}kT_{\text{ref}}$ of a particle can be balanced by another particle that crosses the same interface from the other direction. For the steady flow with small bulk velocity, however, the number of particles crossing in each direction is not the same, but very close. The additional energy $\frac{1}{2}kT_{\text{ref}}$ of most particles crossing an interface can be balanced by other particles crossing the same interface from the other side, and the additional energy $\frac{1}{2}kT_{\text{ref}}$ of the particles that are not balanced needs to be balanced by another means. Statistically, there is another molecule leaving a computational cell from one interface when a molecule enters the cell through another interface for steady flows. Therefore the additional energy $\frac{1}{2}kT_{\text{ref}}$ of the particles that enter a cell from one interface and are not balanced by the above means can be approximately balanced by particles leaving the cell from other interfaces because the difference of the reference temperatures (the flow temperatures on different interfaces) is relatively small. Hence, the additional energy transfer model can model the net energy flux approximately for steady flows. For unsteady flows, the model can also be a good approximation if the frequency is low or the temperature variation is small. In the implementation, the IP method dedicates an additional variable T_a for particles to describe the additional energy $\frac{1}{2}k(T - T_{\text{ref}})$ as $\frac{1}{2}\xi kT_a$. As stated earlier, the additional energy is borrowed from other particles, so a new variable $T_{\text{a,c}}$ is preserved for cells to record the borrowed energy as $\frac{1}{2}\xi kT_{\text{a,c}}$. At the end of each time step, the borrowed energy is evenly provided by all the particles in the cell to maintain the conservation of energy.

5.3.5 IP implementation procedures

In general, the IP method preserves macroscopic information in the individual particles simulated in the DSMC procedures, and updates the preserved information every several time steps of particle movement and particle collision. IP obtains flow field information by sampling the preserved information. This procedure is illustrated in Fig. 5.3.2.

In the IP technique, each particle has the following information: particle's location, particle's microscopic velocity, internal energy, preserved

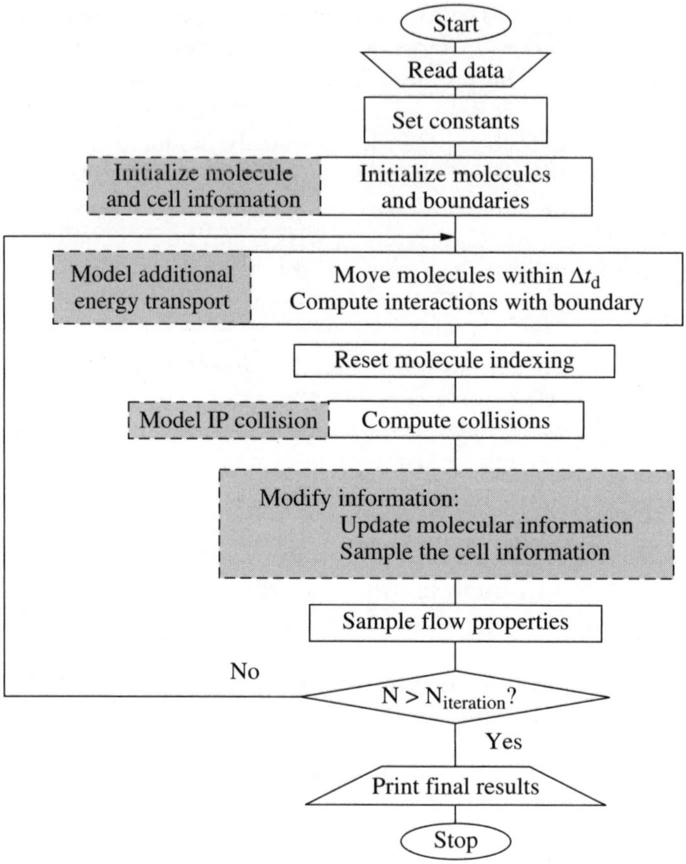

Figure 5.3.2 DSMC-IP flowchart.

velocity $\mathbf{v}_{ip,i}$, preserved temperature $T_{ip,i}$, and preserved additional temperature $T_{a,i}$, as well as the following information for each computational field cell: the preserved density ρ_c, preserved velocity $\mathbf{v}_c$, preserved temperature T_c, and preserved additional cell temperature $T_{a,c}$. They are initialized by the ambient condition with the preserved additional temperature set as zero. Next, all particles update preserved information according to Eqs. (5.3.91), (5.3.21), and (5.3.22), along with the kinetic processes modeled by the DSMC procedure. Basically, the IP implementation is based on the DSMC procedures, as shown in Fig. 5.3.2. In the DSMC-IP flowchart, the key steps for IP are shown in dash-lined boxes. In each time step, one movement, collision, and modification step are executed to update the information preserved in the simulated particles and field cells. The collision and movement step are decoupled as a particle collision step and a particle movement step according to the regular DSMC procedures. A general implementation of the IP technique can be summarized as follows:

1. *Initialization*: The information velocity $\mathbf{v}_{\text{ip},i}$ and temperature $T_{\text{ip},i}$ of simulated particles and the information density ρ_c, temperature T_c, and velocity $\mathbf{v}_c$ of computational field cells are set as the initial flow condition, while the particle additional temperature $T_{\text{a},i}$ and cell additional temperature $T_{\text{a,c}}$ are assigned to zero. The molecular velocity, position, and internal energy are initialized as in the usual DSMC procedures.

2. *Particle movement*: Particles move with their molecular velocity in the usual DSMC procedures. However, the preserved information of particles may change when they interact with interfaces. Possible types of particle–interface interaction are:

2*a*. *Migration between cells.* When particle i moves from cell k to another, momentum and energy transfer occur, and additional energy transfer is required as described in the above energy flux model. The preserved additional energy for the particle i and the cell k is adjusted as

$$T'_{\text{a},i} = (T_i - T_{\text{ref}})/\xi \tag{5.3.26}$$

$$T'_{\text{a,c},k} = T_{\text{a,c},k} + T_{\text{a},i} - T'_{\text{a},i} \tag{5.3.27}$$

where T_{ref} is the interface temperature interpolated from the preserved cell temperatures of neighboring cells, and $T_{\text{a,c},k}$ the additional temperature in cell k.

2*b*. *Inflow and outflow.* If a particle leaves the computational domain, its preserved information are discarded. New particles that enter the computational domain are assigned information according to the boundary condition with $T_{a,i} = 0$.

2*c*. *Reflection from a wall.* The preserved information of particles colliding with a wall is assigned with the collective behavior of a large number of real molecules. That is, the normal velocity component is reversed in a specular reflection, and the preserved velocity and temperature of the reflected particles are set as the wall velocity and temperature in a diffuse reflection. Also, the preserved additional temperature is changed.

Specular reflection

$$T'_{\text{a},i} = -T_{\text{a},i} \tag{5.3.28}$$

$$T'_{\text{a,c},k} = T_{\text{a,c},k} + T_{\text{a},i} - T'_{\text{a},i} \tag{5.3.29}$$

Diffuse reflection

$$T''_{\mathrm{a},i} = (T_i - T_{\mathrm{ref}})/\xi \tag{5.3.30}$$

$$T'_{\mathrm{a,c},k} = T_{\mathrm{a,c},k} + T_{\mathrm{a},i} - T''_{\mathrm{a},i} \tag{5.3.31}$$

$$T'_{\mathrm{a},i} = (T_w - T_{\mathrm{ref}})/\xi \tag{5.3.32}$$

Here, $T''_{\mathrm{a},i}$ represents the additional temperature of the particle before impacting the wall, and $T_{\mathrm{ref}} = \sqrt{T_i \cdot T_w}$ the constant gas temperature of collisionless flow between two plates with one at T_i and the other at T_w (Sun and Boyd 2002).

2d. *Reflection from a symmetric boundary.* The normal velocity component is reversed, and the parallel velocity component remains unchanged, when a particle reflects from a symmetric boundary.

3. *Particle collision*: When particles collide with their collision partner, their molecular momentum and energy transfer is modeled in the usual DSMC collision procedure. In the algorithm, the indices of the simulated particles are reset before the collision step. In the collisions, changes in the preserved velocity and temperature should be appropriately considered. A collision results in an equilibrium state for the particles; hence the preserved information of these particles tends to be the same after collisions. A simple collision model is proposed as

$$\mathbf{v}''_{\mathrm{ip},1} = \mathbf{v}''_{\mathrm{ip},2} = \left(\mathbf{v}'_{\mathrm{ip},1} + \mathbf{v}'_{\mathrm{ip},2}\right)/2 \tag{5.3.33}$$

$$T''_{\mathrm{ip},1} = T''_{\mathrm{ip},2} = \left(T'_{\mathrm{ip},1} + T'_{\mathrm{ip},2}\right)/2 + \left(v'_{\mathrm{ip},1} - v'_{\mathrm{ip},2}\right)^2/(4\xi R) \tag{5.3.34}$$

$$T''_{\mathrm{a},1} = T''_{\mathrm{a},2} = \left(T'_{\mathrm{a},1} + T'_{\mathrm{a},2}\right)/2 \tag{5.3.35}$$

Equations (5.3.33) and (5.3.34) conserve the information momentum and energy, respectively, for collision molecules. To improve the accuracy in calculating the flow viscosity and thermal conductivity, a phenomenological collision model for IP has been proposed by Sun and Boyd (2002).

After the procedures of particle movement and collision are considered, the additional energy preserved by the cell k is shared by all the particles (N_p) in the cell, and the additional energy preserved will be reset. That is,

$$T'_{\mathrm{ip},i} = T_{\mathrm{ip},i} + T_{\mathrm{a,c},k}/N_p \tag{5.3.36}$$

$$T'_{\mathrm{a,c},k} = 0 \tag{5.3.37}$$

4. *Particle information modification*: The preserved information of particles is modified with the pressure field during the modification step. Equations (5.3.21) and (5.3.22) are solved by a finite volume method.

$$\mathbf{v}_{\text{ip},i}^{t+\Delta t} - \mathbf{v}_{\text{ip},i}^{t} = -\frac{\Delta t}{\rho_c \Delta V} \oint\limits_{\text{cell}} p_c \, d\mathbf{s} \tag{5.3.38}$$

$$\left(\frac{v_{\text{ip},i}^2}{2} + \frac{\xi R T_{\text{ip},i}}{2}\right)^{t+\Delta t} - \left(\frac{v_{\text{ip},i}^2}{2} + \frac{\xi R T_{\text{ip},i}}{2}\right)^{t} = -\frac{\Delta t}{\rho_c \Delta V} \oint\limits_{\text{cell}} p_c \bar{\mathbf{v}}_{\text{ip},c} \cdot d\mathbf{s} \tag{5.3.39}$$

Here, ΔV is the volume of a field cell, and $d\vec{\mathbf{s}}$ the surface area vector of the cell. The information on the cell surface is linearly interpolated using the information of the neighboring cells. To avoid statistical effects due to number density fluctuation of particles in a cell in Eqs. (5.3.38) and (5.3.39), the density ρ_c is replaced with the ratio of real mass of total represented molecules in the cell to the volume of the cell.

Note that Eqs. (5.3.38) and (5.3.39) are the Lagrangian inviscid fluid dynamics equations. Therefore, this step of IP updates the macroscopic information of simulated particles according to the inviscid fluid dynamics with Lagrangian method.

5. *Cell information update*: After modifying particles with their preserved information, the preserved information for cells is updated by averaging the information of all the N_p particles in the cell.

$$\mathbf{v}_c' = \sum_{i=1}^{N_p} \left(\frac{\mathbf{v}_{\text{ip},i}}{N_p}\right) \tag{5.3.40}$$

$$T_c' = \sum_{i=1}^{N_p} \left(\frac{T_{\text{ip},i} + T_{a,i}}{N_p}\right) \tag{5.3.41}$$

$$\rho_c' = \rho_c - \frac{\Delta t}{\Delta V} \oint\limits_{\text{cell}} \rho_c \bar{\mathbf{v}}_{\text{ip},c} \cdot d\mathbf{s} \tag{5.3.42}$$

6. *Flow property sampling*: The flow properties are obtained by using time average of the preserved information. The flow velocity, temperature, and density are obtained as

$$\mathbf{V}_f = \sum_{t=1}^{N_{\text{step}}} \left(\frac{\mathbf{v}_{c,t}}{N_{\text{step}}} \right) \tag{5.3.42}$$

$$T_f = \frac{1}{N_{\text{step}}} \sum_{t=1}^{N_{\text{step}}} \left[T_{c,t} + \frac{1}{\xi R} \left(\sum_{i=1}^{N_{p,t}} \frac{\mathbf{v}_{\text{ip},i}^2}{N_{p,t}} - \left(\sum_{i=1}^{N_{p,t}} \frac{\mathbf{v}_{\text{ip},i}}{N_{p,t}} \right)^2 \right) \right] \tag{5.3.43}$$

$$\rho_f = \sum_{t=1}^{N_{\text{step}}} \left(\frac{\rho_{c,t}}{N_{\text{step}}} \right) \tag{5.3.44}$$

And the flow fluxes properties on the wall, such as the pressure, shear stress, and heat transfer, can be given as

$$p_w = p_c + \frac{1}{t_s \cdot \Delta A} \sum_{j=1}^{N_s} m \left(v_{\text{ip,n},j}^r - v_{\text{ip,n},j}^i \right) \tag{5.3.45}$$

$$\tau_w = \frac{1}{t_s \cdot \Delta A} \sum_{j=1}^{N_s} m \left(v_{\text{ip},\tau,j}^r - v_{\text{ip},\tau,j}^i \right) \tag{5.3.46}$$

$$q_w = \frac{1}{t_s \cdot \Delta A} \left[\sum_{j=1}^{N_s} \left(\frac{1}{2} m v_{\text{ip},j}^2 + \frac{\xi}{2} k (T_{\text{ip},j} + T_{\text{a},j}) \right)^i - \sum_{j=1}^{N_s} \left(\frac{1}{2} m v_{\text{ip},j}^2 + \frac{\xi}{2} k (T_{\text{ip},j} + T_{\text{a},j}) \right)^r \right] \tag{5.3.47}$$

where N_s is the total number of molecules compacting the wall element during the sampling time t_s, ΔA the area of the wall element, p_c the cell pressure close to the wall element. The subscript "n" represents the normal component of the molecule information velocity, and "τ" the tangential component. The superscript "r" and "i" denote the values before and after the collision with the wall element, respectively.

Steps 2 to 5 are repeated until the flow reaches a stationary state, following the similar procedures of the usual DSMC method. Step 6 is used to obtain the macroscopic properties of the simulated flow.

5.4 DSMC-IP Computer Program and Applications

5.4.1 IP1D program

In App. 5C, a Fortran 90 code, called IP1D.F90, has been written to apply Fan's IP, Sun's IP, and a conventional DSMC method to Couette flows in a monatomic gas. The geometry is one-dimensional with a unit depth in the z-direction and two plane, diffusively reflecting walls that are normal to the x-axis. When compiling, there are two options to choose the different IP techniques, macro "IPFAN" for Fan's IP and macro "IPSUN" for Sun's IP. The conventional DSMC method is used if neither option is selected. Again, the IP technique does not affect the DSMC solution. IP1D.F90 becomes a DSMC solver when the lines inside the IP macros are removed.

In the implementation, the HS molecule model is used for the DSMC elastic binary collision process. In the gas-surface interaction, the Maxwellian diffuse reflection model is considered. In the calculation, this program employs normalized variables with length normalized by the mean free path in the initial gas $\lambda_0 = 1/(\sqrt{2}\pi d^2 n)$. A Knudsen number is defined as the ratio of λ_0 to the distance h between the lower and upper boundaries so that the distance is effectively the inverse Knudsen number of the flow. Temperature and density are normalized by their values in the initial gas. Velocity is normalized by $\sqrt{2R}$. And the time step Δt is a value that factor, dtr, times the mean collision time of molecules in the initial gas, $t_{c,0}$. That is,

$$\Delta t = \text{dtr} \cdot t_{c,0} \tag{5.4.1}$$

In the demonstration code, dtr is set 0.2. The mean collision time is

$$t_{c,0} = \left(\frac{1}{2}\sqrt{\pi}/\sqrt{T_0}\right)(\lambda_0/\sqrt{2R}) \tag{5.4.2}$$

for HS molecule model. To improve the DSMC efficiency and accuracy, the transient subcells (Bird 2000) is used.

Parameters **mnm**, **mcell**, **mmc**, and **ndiv** are defined in the module **dimHeader** that sets the maximum number of simulated molecules, the maximum number of field cells, the maximum number of molecules in a cell, and the number of subcells in each field cell, respectively. Global variables associated with the calculation are declared in module **calculation**, molecular species properties in module **species**, other molecular properties in module **molecules**, flow properties of each field cell in module **flowField**, and variables for field and surface sampling in module **sampling**. The variables in subroutine **inputData** set the data for a particular run of the program. The main program **Couette** contains an inner loop over the number **nis** of time step between flow samples and a second loop over the number **nso** of sample between the output files for the flow field, wall properties, and the updating of the restart file **couette.res**. Files **FieldS.dat** and **FieldU.dat** are the DSMC long-time averaged and instantaneous solution, respectively. **FieldIPS.dat** and **FieldIPU.dat** are the IP long-time averaged and instantaneous solutions, respectively. The program stops after 10***NPT** cycles have been completed. Apart from these loops, the main program calls subroutines that are modularized for the standard DSMC and IP procedures.

Subroutine **setInitialState** is called with the parameter **nql** equals to 1 for a new run, which is set in the subroutine **inputData**. Or, subroutine **readRestart** is called with **nql** equals to 0. With the program restarting from a previous calculation, subroutine **clearSteadySample** is called to clear and initialize the field and wall samplings for long-time average quantity when variable **nqls** equals to 1. The subroutine **setInitialState** sets the physical constants and normalizing parameters for length, time, and velocity. It also initializes the flow field and individual molecule. The flow field is divided into **ncy** cells of uniform width **cy**, and each of these is divided into **ndiv** equal-width subcells. The locations and velocity components of the simulated molecules are set randomly according to the initial state of the gas. Finally, the molecules are sorted in the order of field cells and subcells by calling **indexMols**. Because the flow is one-dimensional, only the y-coordinate is stored for each molecule. All three-velocity components are saved, since the collisions are computed as three-dimensional events.

Subroutine **moveMols** moves the molecules through distances appropriate to the time interval **dtm**. The information velocity **pvip** and information temperature **ptip** of the simulated molecules evolve with the momentum and energy transport, respectively, while molecular velocities do not change. When a molecule collides with the surface at $y = 0$ and $y =$ ybn, the Maxwellian reflection model is implemented in subroutine **reflect**. Assuming all the impacting molecules are diffusively reflected with Maxwellian distribution, the velocity components of the reflected molecule are

$$v = [-\ln(R_f)]^{1/2}\sqrt{2kT/m} \tag{5.4.3}$$

$$u = a \sin b \tag{5.4.4}$$

$$w = a \cos b \tag{5.4.5}$$

where

$$a = [-\ln(R_f)]^{1/2}\sqrt{2kT/m} \tag{5.4.6}$$

$$b = 2\pi R_f \tag{5.4.7}$$

where R_f denotes random fraction. The gas properties on the surface are sampled before and after the reflection. The information velocities of reflected molecules are set to zero. The information temperature and the additional temperature for the reflected molecule and field cell are set according to the general IP step 2c. After the molecules complete their movement for the time interval **dtm**, the molecule numbers are re-sorted in a cross-referencing array **ir** in the order of the field cells. Molecules from the same cells are potential collision pairs. Based on the new cross-reference, subroutine **aEnergy1** calculates the additional information temperature for the molecules and cells according to Eqs. (5.3.26) and (5.3.27) in the general IP step 2a.

The NTC method and transient subcells technique are employed in subroutine **collisions** to determine the appropriate set of collisions. In the selection of collision pairs, the acceptance-rejection method is applied. The acceptance rate of the collision pairs is sampled as,

$$\frac{c_r}{(c_r)_{\max}} \tag{5.4.8}$$

based on the HS molecular collision model. The maximum value of the relative speed is initially set to the most probable speed of the molecules at the initial temperature. Binary collision Eqs. (5.1.3) to (5.1.5) and (5.1.9) to (5.1.11) for HS model are used to compute the postcollision components of the molecular velocity. There is a parameter **ipcol** that determines the collision model for IP technique. With **ipcol** equal to 1, the simple collision model shown in Eqs. (5.3.33) to (5.3.35) is applied to calculate the information velocity, information temperature, and additional information temperature of the collision molecules. The improved collision model proposed by Sun and Boyd (2002) is also implemented with **ipcol** equal to 2.

Subroutine **aEnergy2** distributes the additional information energy of a cell to all molecules in the cell according to Eqs. (5.3.36) and (5.3.37) in the general IP step 3, after the procedures of particle movement and collision. Subroutine **IPupdate** updates the right-hand-side terms of Eqs. (5.3.40) and (5.3.42), which will be used to calculate mass, momentum, and energy transport effects on the number density, information velocity, and information temperature of molecules in the next time step.

The flow properties in each cell are sampled in subroutine **sampleFlow**, which includes **IPsample**. The output files for the flow field and wall properties are generated in subroutine **output** according to Eqs. (5.3.43) and (5.3.45), and the equations are identified in the comment. Double precision is used for variables such as the total number of collisions and long-time average samplings that are expected to vary beyond the limit of single precision arithmetic.

As mentioned earlier, the data that affect the dimensions of the arrays is set in the module **dimHeader**. The test case sets the maximum number of molecules to 1M (**mnm** = 1,000,000), maximum number of cells 60 (**mcell** = 60) with 10 subcells per cell, and maximum number of molecules per cell 0.2M (**mmc** = 200,000). The input parameters are read from a file **couette.in**. There, the total number of the simulated molecules is 1M. The initial flow has the Knudsen number of 0.01 at temperature 300K and pressure 1 atm. The gas properties of argon are set in subroutine **inputData**. The flow is sampled every other time step and the output files are updated at intervals of 100 samples.

5.4.2 IP1D applications to microCouette flows

The first application of the program is to validate the DSMC solutions by using a continuum slip flow model (Liou et al. 2003). The model solves the Navier-Stokes (NS) equations with Maxwell-Smoluchowski slip boundary conditions to determine the slip velocity components and the temperature jump at the upper and lower surface. A microCouette flow between two plane surfaces is simulated. The upper and lower plate move at a speed of $0.5U_w$ in the opposite directions. U_w is set to the most probable molecular speed at the surface temperature $T_w = 1000$K. The gap between the two plates is 60 λ, where λ denotes the mean free path of the initial argon gas. The Knudsen number based on the initial gas state is about 0.017. Figure 5.4.1 shows a comparison of the vertical distributions of velocity u normalized by U_w. The velocity profiles agree well and both show a slight departure from a linear distribution.

Both the DSMC and NS solutions exhibit the expected compressibility effects. Because of viscous heating, the initial pressure increases to a steady value around 1.05 atm, an increase of about 5 percent, as shown in Fig. 5.4.2. The pressure is uniform across the flow, while the temperature increases from the surface to the center of the channel. The rise in the central portion of the flow is over 7 percent of wall temperature. The comparison of the DSMC and NS temperature profiles are shown in Fig. 5.4.3. The temperature jump on both surfaces is about 0.8 percent over the wall temperature for both the DSMC and NS simulations. Generally, the pressure and temperature distributions of DSMC agree well with those of NS solution. The compressibility effects calculated by the DSMC method are somewhat higher than those by the continuum slip flow model.

The program has also been applied to compare the microCouette solutions of the DSMC and Sun's IP method. Fan's IP does not calculate the temperature, with the assumption of isothermal flow fields. As a result, the velocity solutions obtained by Fan's IP approach do

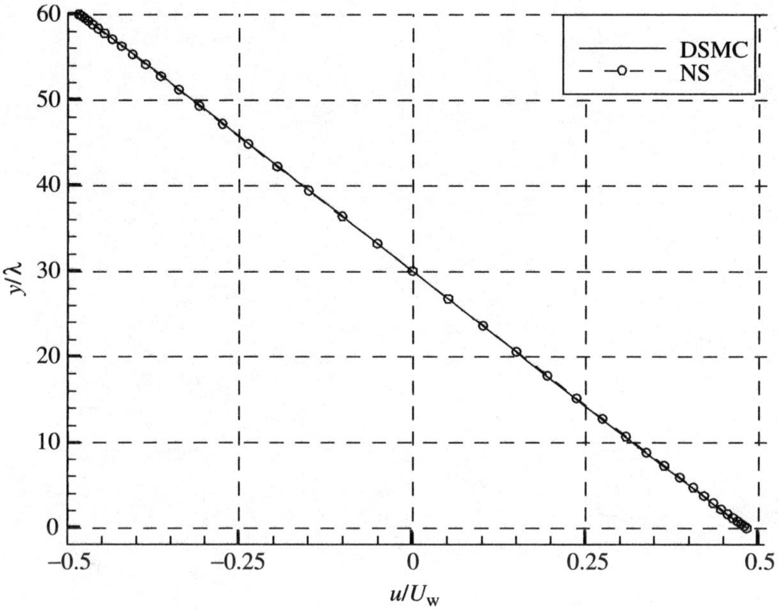

Figure 5.4.1 Comparisons of the velocity profiles in the microCouette flow simulated by DSMC and NS, at Kn = 0.017.

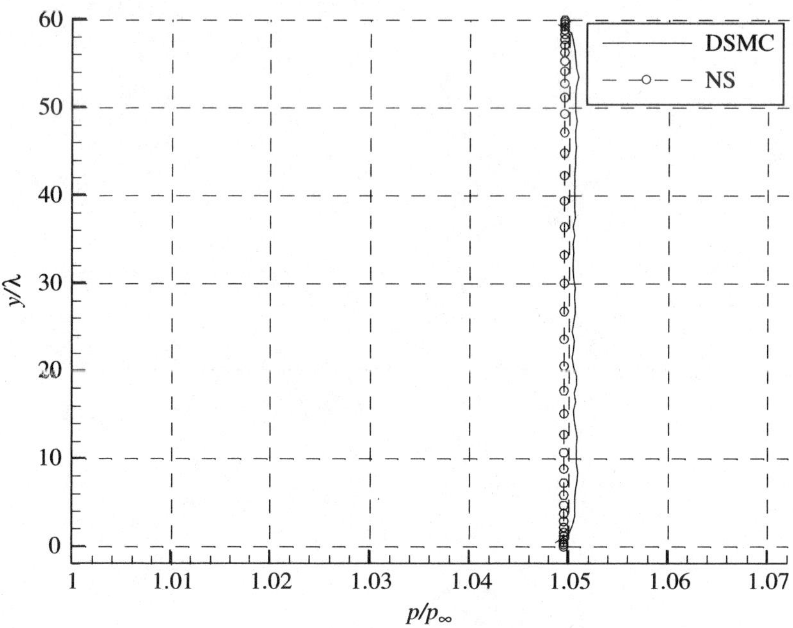

Figure 5.4.2 Comparisons of the pressure profiles in the microCouette flow simulated by DSMC and NS, at Kn = 0.017.

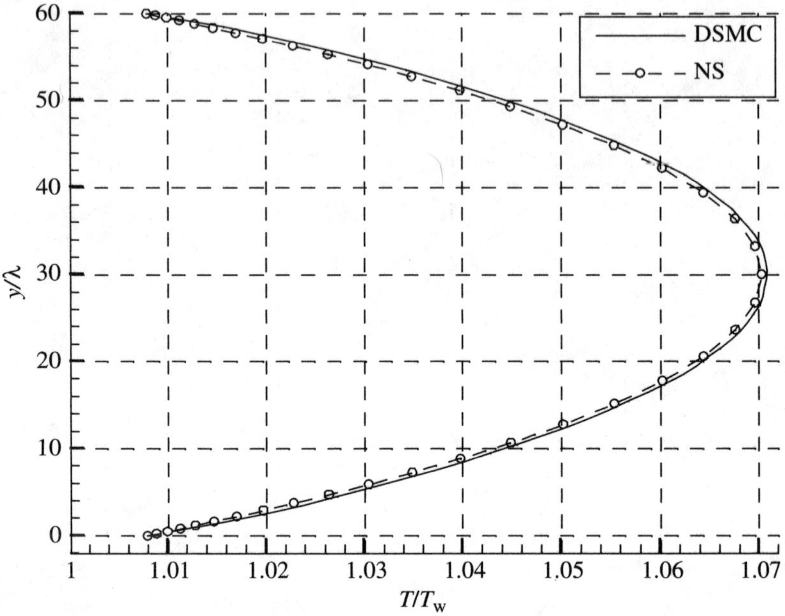

Figure 5.4.3 Comparisons of the temperature profiles in the microCouette flow simulated by DSMC and NS, at Kn = 0.017.

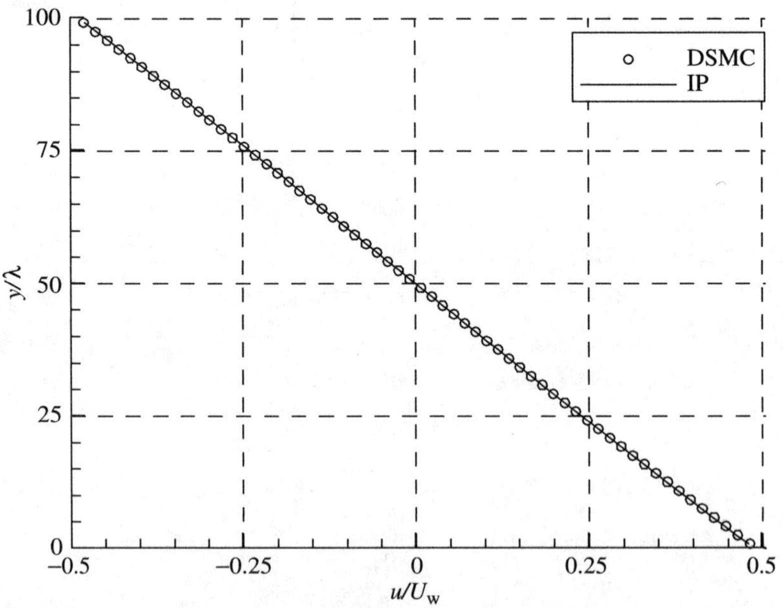

Figure 5.4.4 Comparisons of the velocity profiles in the microCouette flow simulated by DSMC and IP, at Kn = 0.01.

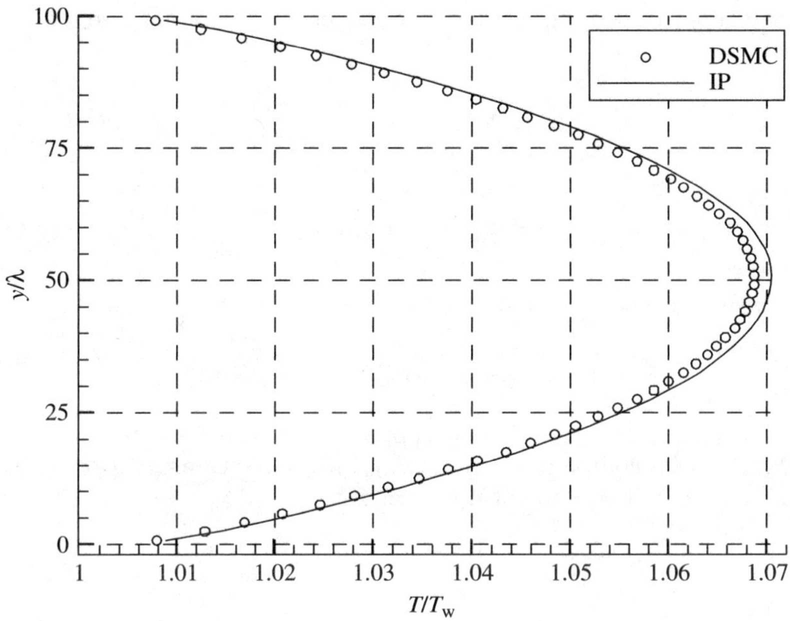

Figure 5.4.5 Comparisons of the temperature profiles in the microCouette flow simulated by DSMC and IP, at Kn = 0.01.

not differ significantly from those by Sun's IP. Therefore, only Sun's IP solutions are shown in the following numerical experiments. The flow parameters are largely the same as the simulated cases described earlier, except that the wall temperature and the initial gas temperature are both set to 300K and the plate gap is 100 $\lambda(\mathrm{Kn} = 0.01)$. Figure 5.4.4 shows the comparison of velocity profiles. It can be seen that the velocity profiles obtained by the IP method agree very well with those by the DSMC method. As shown in Fig. 5.4.5, the temperature distribution simulated by the IP method agrees well with the DSMC method near the surfaces. Near the center, the temperature obtained by the IP technique is about 0.2 percent higher than that of the DSMC method. Apparently, the collision model used in the current IP technique may need further improvement.

5.5 Analysis of the Scatter of DSMC and IP

To compare the velocity scatter of DSMC and IP, monatomic gas is considered for convenience and the velocity is normalized by a factor $\sqrt{R}$. The mean x-component velocity $\tilde{U}$ and the standard deviation $\sigma_{\tilde{u}}$

of the molecular velocity $\tilde{u}_i$ of the total number N of particles in a field cell are

$$\tilde{U} = \sum_{i=1}^{N} \tilde{u}_i / N \tag{5.5.1}$$

$$\sigma_{\tilde{u}}^2 = \sum_{i=1}^{N} (\tilde{u}_i - \tilde{U})^2 / N \tag{5.5.2}$$

In the DSMC method, the temperature is sampled by

$$T = \frac{1}{3}\left(\sigma_{\tilde{u},\mathrm{DSMC}}^2 + \sigma_{\tilde{v},\mathrm{DSMC}}^2 + \sigma_{\tilde{w},\mathrm{DSMC}}^2\right) \tag{5.5.3}$$

where $\sigma_{\tilde{u},\mathrm{DSMC}}$, $\sigma_{\tilde{v},\mathrm{DSMC}}$, and $\sigma_{\tilde{w},\mathrm{DSMC}}$ represent the standard deviation of the DSMC velocity components in the x, y, and z directions, respectively. From Eq. (5.3.16), it can be obtained that

$$T = \bar{T}_{\mathrm{ip},i} + \frac{1}{3}\left(\sigma_{\tilde{u},\mathrm{ip}}^2 + \sigma_{\tilde{v},\mathrm{ip}}^2 + \sigma_{\tilde{w},\mathrm{ip}}^2\right) \tag{5.5.4}$$

where $\sigma_{\tilde{u},\mathrm{ip}}$, $\sigma_{\tilde{v},\mathrm{ip}}$, $\sigma_{\tilde{w},\mathrm{ip}}$ are the standard deviation of the IP information velocity components in the x, y, and z directions, respectively. In comparison, the standard deviation of the DSMC velocity is basically from the thermal fluctuations and its value is in the order of gas temperature. The standard deviation of the IP velocity can be interpreted as the statistical error deducted by the thermal fluctuation. It can be very small value if the information temperature $\bar{T}_{\mathrm{ip},i}$ is well modeled.

In the following, the statistical scatters are observed in numerical experiments with the micro-Couette flows of Kn $= 0.01$ introduced above. All the data is obtained after 34,300 time steps, when the flow approaches the stationary state. Table 5.5.1 gives the mean velocities U and V, and the standard deviations σ_u and σ_v of the molecular velocity of the particles in the different field cells from the DSMC method. These

TABLE 5.5.1 Mean Velocity and Standard Deviation in Different Cells from the DSMC Method, $T = 300$K, Kn $= 0.01$

y/λ	N	U (m/s)	V (m/s)	σ_u (m/s)	σ_v (m/s)
0.8333	0.174E + 05	171.0540	2.4154	355.3231	354.0095
7.5000	0.169E + 05	143.3956	0.0586	354.9523	358.5170
14.1667	0.166E + 05	123.8581	3.1054	359.4852	360.6309
20.8333	0.166E + 05	102.4277	0.9865	357.8456	356.9699
27.5000	0.165E + 05	80.0715	1.1626	360.3816	363.7546
34.1667	0.166E + 05	54.0752	3.3344	365.5974	360.1658
40.8333	0.165E + 05	28.4963	−2.3811	366.5619	364.0483
49.1667	0.164E + 05	7.5551	0.1943	363.1191	363.5411

data are sampled in one time step. From the table it can be seen that the standard deviations of the molecular velocity are in the same order of magnitude as the flow temperature in the cells. In the middle cells, the deviations are at least two orders of magnitude higher than the signals. Many more particles are needed in these cells in one time step to suppress the noise. Table 5.5.2 shows the counterpart of Table 5.5.1 from the IP method. It is seen that the noise is decreased by two orders of magnitude in the x direction, and even higher orders of magnitude in the y direction.

Figure 5.5.1 compares the particle velocity phases obtained by the DSMC and IP methods at the cell near the lower surface ($y/\lambda = 0.8333$). Notice that the velocity phase from the IP method has been magnified by 40 times to compare with that from the DSMC method. The velocity phase for the IP method is compact, more than 40 times smaller and narrower than the DSMC results. In the DSMC method, it spreads almost homogeneously from its center as shown in Fig. 5.5.1*b*.

Figure 5.5.2 gives the statistical fluctuations of the mean velocity U in the cell close to the lower plate ($y/\lambda = 0.8333$). The mean velocities are obtained by 100 independent runs. Twelve hundred particles are sampled for each run in Fig. 5.5.2*a*, while the sample size in Fig. 5.5.2*b* is 10,000 per cell. In both Figs. 5.5.2*a* and 5.5.2*b*, the IP velocities are very smooth, while there are large fluctuations in the DSMC velocities. However, the DSMC fluctuations decrease significantly with the larger sample size. Eight more sets of velocity fluctuations obtained with eight more different sample sizes from 100 to 10,000 particles per cell. By using statistical analysis of these velocity fluctuations, the velocity statistical scatter is obtained as 25.04 and 0.54 m/s with sample size of 100 particles per cell, in the DSMC and IP solution, respectively. Sampling with 10,000 particles per cell, the statistical scatters become 2.64 and 0.05 m/s, for the DSMC and IP solutions, respectively. The properties of velocity statistical scatter associated with the DSMC and IP

TABLE 5.5.2 Mean Velocity and Standard Deviation in Different Cells from the IP Method, $T = 300$K, Kn = 0.01

y/λ	N	U (m/s)	V (m/s)	σ_u (m/s)	σ_v (m/s)
0.8333	0.174E + 05	169.6835	0.2722	7.0119	0.7678
7.5000	0.169E + 05	146.0909	0.0591	7.2623	1.0833
14.1667	0.166E + 05	123.0724	1.7150	7.3453	1.8157
20.8333	0.166E + 05	100.1639	0.0286	7.3023	0.8141
27.5000	0.165E + 05	77.2525	−2.6594	7.4039	1.0565
34.1667	0.166E + 05	54.5685	0.7335	7.4090	1.7231
40.8333	0.165E + 05	31.6614	−0.9568	7.3574	1.1672
49.1667	0.164E + 05	3.2754	−2.4789	7.3608	1.9621

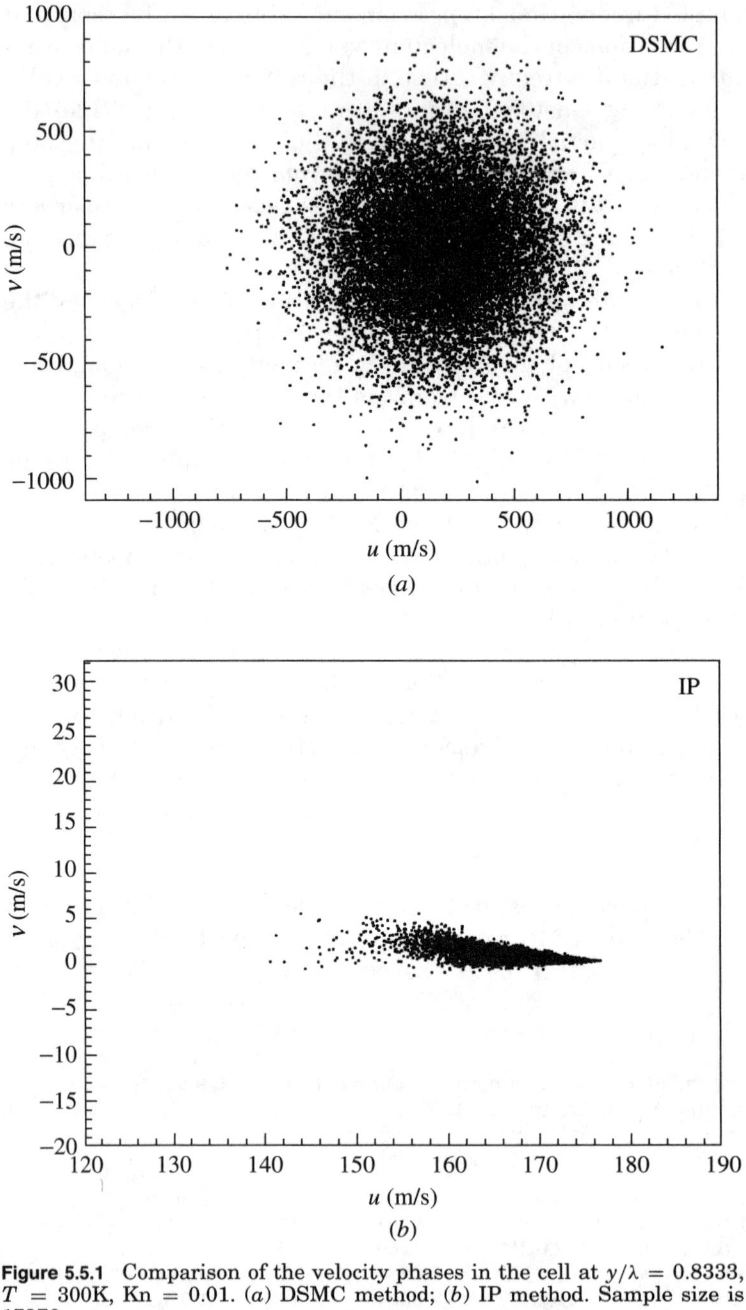

Figure 5.5.1 Comparison of the velocity phases in the cell at $y/\lambda = 0.8333$, $T = 300\text{K}$, Kn $= 0.01$. (*a*) DSMC method; (*b*) IP method. Sample size is 17370.

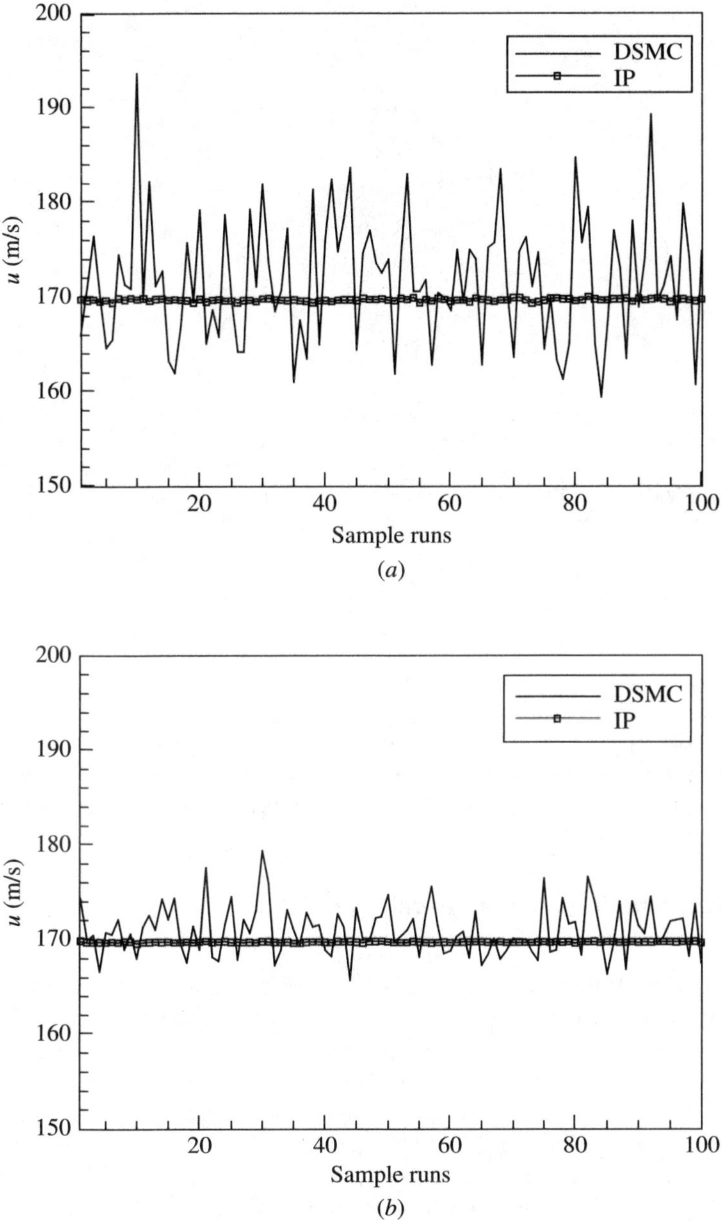

Figure 5.5.2 Comparison of the statistical fluctuation in DSMC and IP velocities in the cell at $y/\lambda = 0.8333$, $T = 300\text{K}$, $\text{Kn} = 0.01$. (*a*) Sample size is 1200 particles per cell; (*b*) Sample size is 10,000 particles per cell.

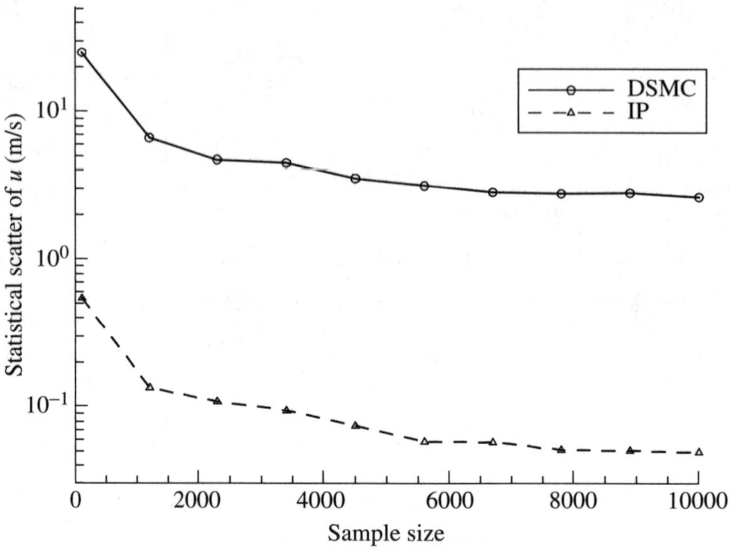

Figure 5.5.3 Comparison of the statistical scatter associated with the DSMC and IP method obtained from microCouette flows. $T = 300\text{K}$, $\text{Kn} = 0.01$.

method are plotted in Fig. 5.5.3. It can be seen that velocity scatter in the IP solutions is always about two orders of magnitude lower than that associated with the DSMC solutions, when sampled with the equal number of particles per cell.

Appendix 5A: Sampling from a Probability Distribution Function

As general probabilistic modeling of physical processes, the DSMC simulation requires the generation of representative values of variables that are distributed in a prescribed manner. This is done through random numbers, which is assumed to be a set of successive random fractions R_f and uniformly distributed between 0 and 1. If the *probability distribution function* (PDF) has a *cumulative distribution function* (CDF), which can be inversed into an explicit form, *the inverse-cumulative* method can be applied to sample its representative values; otherwise, the *acceptance-rejection* method may be used.

Inverse-cumulative method

It is assumed that the distribution of variable x may be described by a normalized probability distribution function $f(x)$, and the variable ranges from a to b. The total probability is

$$\int_a^b f(x)dx = 1 \tag{5A.1}$$

The cumulative distribution function is defined as

$$F(x) = \int_a^x f(x')dx' \tag{5A.2}$$

The representative value of x can be sampled by generating a random fraction R_f and setting it equal to the cumulative density function $F(x)$ in the form of

$$F_x = R_f \tag{5A.3}$$

Usually, the value x is obtained through an explicit form of the inverse function $F^{-1}(x)$, if it is possible to invert Eq. (5A.3) to get an explicit function for x.

In the DSMC simulations, there are usually three examples using the inverse-cumulative method to sample the value x.

Example 5A.1 The variable x is uniformly distributed between a and b. This distribution is encountered when setting a random initial or inflow condition for the space and velocity distribution of particles in equilibrium flow. For this case $f(x)$ is a constant and the normalization condition of Eq. (5A.2) gives

$$f(x) = \frac{1}{b-a} \tag{5A.4}$$

Therefore, from Eq. (5A.2),

$$F(x) = \frac{x-a}{b-a} \tag{5A.5}$$

and Eq. (5A.3) gives

$$\frac{x-a}{b-a} = R_f \tag{5A.6}$$

Rearrange Eq. (5A.6) to give

$$x = a + R_f(b-a) \tag{5A.7}$$

Example 5A.2 The variable x is distributed between a and b such that the probability of x is proportional to x. This distribution is encountered when setting a random radius in an axially symmetric flow. For this case $f(x)$ is

$$f(x) \propto x \tag{5A.8}$$

and, again the normalization condition of Eq. (5A.2) gives

$$f(x) = \frac{2x}{b^2 - a^2} \tag{5A.9}$$

Therefore, from Eq. (5A.2),

$$F(x) = \frac{x^2 - a^2}{b^2 - a^2} \tag{5A.10}$$

and the representative value of x is taken as

$$x = [a^2 + R_f(b^2 - a^2)]^{1/2} \tag{5A.11}$$

The similar distribution when x is proportional to x^2 in a spherical flow is

$$x = [a^3 + R_f(b^3 - a^3)]^{1/3} \tag{5A.12}$$

Example 5A.3 The variable x is distributed between 0 and ∞, with its probability distribution function

$$f(\beta^2 x^2) = \exp(-\beta^2 x^2) \tag{5A.13}$$

where β is a coefficient that varies with temperature. This distribution is used in the gas-interface diffusive reflection model where to sample the reflected velocities. The cumulative density function is

$$F(\beta^2 x^2) = 1 - \exp(-\beta^2 x^2) \tag{5A.14}$$

And, noting that R_f is equivalent to $1 - R_f$, Eq. (5A.13) gives

$$x = [-\ln(R_f)]^{1/2}/\beta \tag{5A.15}$$

Acceptance–rejection method

Unfortunately, it is impossible to invert Eq. (5A.3) to obtain an explicit form of function like Eq. (5A.7) for x in many realistic distributions of function $f(x)$. Alternatively, a representative value of x can be generated by using the *acceptance–rejection* algorithm through the following five steps:

Step 1: Normalize the distribution function $f(x)$ by its maximum value f_{max} to give

$$f_x' = f(x)/f_{max} \tag{5A.16}$$

Step 2: Choose a value of x at random on the basis of x being uniformly distributed between its limits; i.e., from Eq. (5A.7).

Step 3: Calculate the normalized distribution function f_x' with the chosen x.

Step 4: Generate a second random fraction R_f.

Step 5: If $f_x' > R_f$, then accept the chosen x and stop; otherwise, go back to step 2 and repeat the procedures until x is accepted.

In a DSMC simulation, the thermal velocity components and internal energy of an equilibrium gas are generated by the acceptance–rejection method from its specified velocity distribution function and internal energy distribution function, respectively. The acceptance–rejection method is also used to determine the choosing of collision pairs in the procedure of molecular collision.

Appendix 5B: Additional Energy Carried by Fast Molecules Crossing a Surface

The flux of molecular quantities across a surface element is considered in an equilibrium gas. The stream velocity $\mathbf{c}_0$ is inclined at the angle θ to the unit normal vector $\mathbf{e}$ to the surface element, as shown in Fig. 5B.1. Without loss of generality, Cartesian coordinates are chosen such that the stream velocity lies in the yz-plane, with the x-axis in the negative $\mathbf{e}$ direction.

Each molecule has velocity components

$$u = c_0 \cos\theta + u' \qquad v = c_0 \sin\theta + v' \qquad w = w' \tag{5B.1}$$

The flux of quantity Q across the surface element along the positive x-direction is

$$n\overline{Qu} = n \int_{-\infty}^{\infty} \int_{-\infty}^{\infty} \int_{0}^{\infty} Quf \, du \, dv \, dw \tag{5B.2}$$

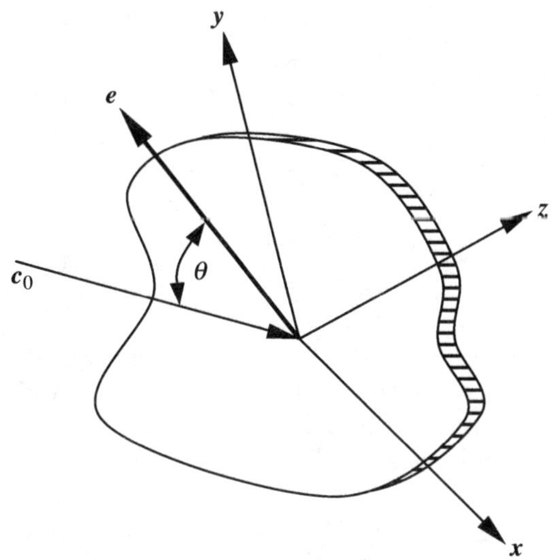

Figure 5B.1 Coordinate system for the analysis of molecular flux across a surface element.

For an equilibrium gas, the Maxwellian distribution function f_0 is substituted into Eq. (5B.2), and the Q-flux across the element per unit area per unit time can be obtained as

$$n\overline{Qu} = \frac{n\beta^3}{\pi^{3/2}} \int_{-\infty}^{\infty} \int_{-\infty - c_0}^{\infty} \int_{\cos\theta}^{\infty} Q(c_0 \cos\theta + u') \times \exp[-\beta^3(u'^2 + v'^2 + w'^2)] du' dv' dw' \tag{5B.3}$$

The number flux $\dot{N}_x$ of molecules across the surface element is obtained with $Q = 1$ in Eq. (5B.3).

$$\dot{N}_x = \frac{n\sqrt{kT/m}}{\sqrt{2\pi}} [\exp(-s^2) + \sqrt{\pi} s(1 + \mathrm{erf}(s))] \tag{5B.4}$$

where

$$s = c_0 \cos\theta \sqrt{m/(2kT)} \tag{5B.5}$$

The translational energy flux $\dot{q}_{x,\mathrm{tr}}$ to the element is obtained by setting $Q = \frac{1}{2}mc^2$ in Eq. (5B.3) to give

$$\dot{q}_{x,\mathrm{tr}} = \frac{nm(\sqrt{2kT/m})^3}{4\sqrt{\pi}} \left\{ \left(2 + \frac{s^2}{\cos^2\theta}\right) [\exp(-s^2) + \sqrt{\pi} s(1 + \mathrm{erf}(s))] + \frac{1}{2}\sqrt{\pi} s(1 + \mathrm{erf}(s)) \right\} \tag{5B.6}$$

And the averaged translational energy per molecule crossing a surface element is obtained through dividing Eq. (5B.6) by Eq. (5B.4) to give

$$\overline{\left(\frac{1}{2}mc_i^2\right)}_{c_i \cdot n > 0} = \frac{\dot{q}_{x,\mathrm{tr}}}{\dot{N}_x} = \frac{3}{2}kT + \frac{1}{2}mc_0^2 + (1 + \xi_a)\frac{1}{2}kT \tag{5B.7}$$

where

$$\xi_a = \left(1 + \frac{\exp(-s^2)}{\sqrt{\pi} s[1 + \mathrm{erf}(s)]}\right)^{-1} \tag{5B.8}$$

Comparing with kinetic theory equation $\overline{\left(\frac{1}{2}mc_i^2\right)} = \frac{3}{2}kT + \frac{1}{2}mc_0^2$ for all the molecules in an equilibrium state with mean flow velocity c_0 and temperature T, the additional internal energy E_a carried by a molecule crossing the surface on average can be obtained as

$$E_a = (1 + \xi_a)\frac{1}{2}kT \tag{5B.9}$$

For stationary equilibrium monatomic gases, ξ_a becomes 0 with $c_0 \to 0$. The additional internal energy is

$$E_a = \frac{1}{2}kT \tag{5B.10}$$

and

$$\overline{\left(\frac{1}{2}mc_i^2\right)}_{c_i \cdot n > 0} = 2kT \qquad \overline{\left(\frac{1}{2}mc_i^2\right)} = \frac{3}{2}kT$$

Therefore the ratio of $\overline{\left(\frac{1}{2}mc_i^2\right)}_{c_i \cdot n > 0}$ to $\overline{\left(\frac{1}{2}mc_i^2\right)}$ becomes 4/3. That is, the averaged internal energy of the molecules crossing a surface element exceeds that in a spatial element by a factor 4/3 for stationary monatomic gases.

Appendix 5C: One-Dimensional DSMC-IP Computer Program

IP1D input data: couette.in

```
1,        0                  !nql,nqls
1000000                      !nmi
0.01,      300,    1.0 !kn,ftmp,pref
1.0,      1.0          !svf,tr
0.2                    !dtr
60                     !ncy
2                      !nis
50                     !nso
10                     !nsi
300                    !npt
```

IP1D makefile

```
#=========================================================
# -DIPFAN  Fan's IP only
# -DIPSUN  Sun's IP, compiled together with -DIPFAN
#=========================================================
CC=pgf90
OPTFLAGS := -O -r8
EE=IP1D
#----------------------------------------------------------
DFLAG1 = -DIPFAN
DFLAG2 = -DIPSUN
#----------------------------------------------------------
$(EE): $(EE).o
       ${CC} ${OPTFLAGS} ${DFLAG1} ${DFLAG2} -o $(EE).x $(EE).o
       mv ${EE}.x ../
$(EE).o: $(EE).F90
       ${CC} ${OPTFLAGS} ${DFLAG1} ${DFLAG2} -c $(EE).F90
clean:
      rm -r *.o *.mod

run:
      ./${EE}.x > output &
```

Source code of IP1D.F90

```
!_____________________________________________________________________
!|                                                                   |
!|  program CouetteIP                                                |
!|                                                                   |
!|          : written for micro-Couette flows by DSMC-IP method      |
!|          : 1D version 1.0                                         |
!|          : June, 2004                                             |
!|                                                                   |
!|          : by Dr. Yichuan Fang and Dr. William W. Liou            |
!|          : Department of Mechanical and Aeronautical Engineering  |
!|          : Western Michigan University                            |
!|          : Kalamazoo, MI 49008                                    |
!|___________________________________________________________________|
!=====================================================================

module dimHeader

!--DECLARE THE DIMENSION SIZE

integer,parameter :: mnm = 1000000, mcell = 60, mmc = 200000, ndiv = 10

!--MNM the maximum number of molecules
!--MCELL the maximum number of cells
!--MMC the maximum number of moleucles in a cell
!--NDIV
!  the optimum NDIV is set when there are about 2 molecules per subcell
end module dimHeader

!-----------------------

module calculation

!--DEFINES THE VARIABLES ASSCOCIATED WITH THE CALCULATION

real,parameter :: pi = 3.141592654, spi = 1.77245385, dpi = 6.283185308
double precision,parameter :: boltz = 1.3805d-23

integer :: npr
real :: dtm,time,dtr,fmct,ranf
double precision :: totcol,clsep

!--NPR the number of output intervals
!--DTM the time step
!--TIME the flow time
!--TOTCOL the cumulative number of collisions
!--CLSEP the cumulative collision separation
!--FMCT the mean collision time in reference gas
!--RANF random fraction
!--BOLTZ Boltzmann constant K (J/K)
end module calculation

!-----------------------

module species

!--DECLARE THE SPECIES PROPERTIES OF SIMULATED MOLECULES

real,dimension(5) :: sp
real,dimension(6) :: spm
integer :: ispr,ispf
```

```
real :: mfp,gasR
!--SP(I) I=1 the molecule diamter
!        I=2 the reference temperature (K)
!        I=3 the viscosity-temperature index
!        I=4 the reciprocal of the VSS scattering parameter
!        I=5 the molecule mass
!--SPM   silimar with SP
!        I=6 the GAMA function part
!--ISPR the number of degrees of rotational freedom
!--ISPF the number of degrees of freedom
end module species

!------------------------

module molecules

!--DECLARES THE VARIABLES REPRESENT THE SIMULATED MOLECULES
!  AND SETS THE NUMBER OF SIMULATED MOLECULES

use dimHeader
real,dimension(mnm)   :: pp
real,dimension(3,mnm) :: pv
integer,dimension(2,mnm) :: ir,iro
integer,dimension(3,mmc) :: irs
integer :: nm,nmi,msc
real :: cxs
#ifdef IPFAN
  real,dimension(3,mnm) :: pvip
#ifdef IPSUN
  real,dimension(2,mnm) :: ptip
#endif
#endif

!--PP(N) y position coorddinates
!--PV(1,N),PV(2,N),PV(3,N) u,v,w velocity components
!--IR(1,N) the cell in which the molecule lies
!--IR(2,L) the molecule numbers in order of cells(cross-reference array)
!--NM the number of molecules
!--NMI,MNM initial and maximum number of molecules
!--MSC the number of sub-cells in a cell
!--CXS the collision cross-section
!--PVIP(J,N)
!       J=1,2,3 u,v,w IP velocity components
!--PTIP(J,N)
!       J=1,2 for T and Ta of IP-energy preserver
end module molecules

!----------------------------

module flowField

!--DEFINES THE VARIABLES ASSOCIATED WITH THE FLOW

use dimHeader
!
real,dimension(3,mcell)    :: cell
integer,dimension(2,mcell) :: icell
integer,dimension(2,ndiv)  :: isc
integer,dimension(2) :: nscn
integer :: ncy,ncell,nis,nso,nsi,npt
```

```
real :: cvol,area,ybn,cy,rcy,dcrit,tr,tw,svf,fvel,vmpf,vmpw, &
        ftmp,fden,pref,kn
#ifdef IPFAN
  real,dimension(3,mcell) :: caip,cvip
  real,dimension(mcell)   :: cpip,cnip
#ifdef IPSUN
  real,dimension(2,mcell) :: ctip
  real,dimension(mcell)   :: ceip
#endif
#endif

!--CELL(N,M) relates to cell N
!    N=1 the lower y coordinates of the cell
!    N=2 the maximum value of relative velocity in collisions
!    M=3 the remainder when collisions are computed
!--ICELL(N,M)
!    N=1 the start address -1 in IR( 2 of molecules in cell N
!    N=2 the number of molecules in cell M
!--ISC(N,J)
!    N=1 the start address -1 of transient subcell J molecules in IRS(1
!    N=2 the number of molecules in transient subcell J
!--NSCN(M)
!    the relative row of the subcells in the first layer of adjacent &
     sub cells
!--NCY the number of cells in the y direction (cell rows)
!--NCELL the number of cells
!--CY the cell dimensions in the y directions
!--RCY the reciprocal cell dimension in the y directions
!--CVOL the volume of a cell
!--YBN data items to set magnitude of flowfield
!--FDEN,FTMP initial number density and temperature
!--VMPF,MFP reference molecular speed and free path
!--TR,TW temperature ratio, surface temperature
!--FVEL velocity of upper wall
!--VMPW most probable mulecular speed at the surface
!--DCRIT critical distance to cope with round-off error
!--CVIP(J,M) IP velocity in cell M
!--CAIP(J,M) IP velocity acceleration in cell M
!--CEIP(M) IP energy increase by pressure work
!--CPIP(M) IP pressure in cell M
!--CNIP(M) number density in cell M
!--CTIP(J,M) IP temperature in cell M
!      J=1,2 for T and Ta
end module flowField

!------------------------

module sampling

!--DECLARES THE VARIABLES USED FOR SAMPLING THE FLOW PROPERTIES

use dimHeader

integer :: nsamp,nsamps,nstep
real :: timi,timis
double precision,dimension(7,mcell) :: cs,scs
real,dimension(2,7) :: surf
double precision,dimension(2,7) :: ssurf
#ifdef IPFAN
  double precision,dimension(8,mcell) :: csip,scsip
  real,dimension(2,2:7) :: surfip
```

```
  double precision,dimension(2,2:7) :: ssurfip
#endif

!--NSAMP the number of samples in the unsteady sampling
!--NSAMPS the number of samples in the steady sampling
!--CS(I,N) the samples in cell N
!    I=1 the sampled sum of molecules
!    I=2,3,4 the sampled sums of u,v,w
!    I=5,6,7 the sampled sums of the square of u,v,w
!--SCS similar to CS for the steady sampling
!--SURF(L,I) sampled properties on the wall
!    L=1,2 the lower, upper wall
!    I=1 the number sum to the wall
!--SSURF similar to SURF for steady sampling
!--TIMI,TIMIS start time for unsteady, steady sampling
!--NSTEP the number of time steps
end module sampling

!------------------------

module wcontrol
integer :: iwrite,ires
character(len=7) :: fp = './RUN1/'
character(len=40) :: fname
end module wcontrol

!=========================================================================
!                                MAIN PROGRAM
!=========================================================================

program COUETTE

use molecules
use flowField
use calculation
use sampling

implicit none

integer :: n,m,nql,nqls,ierr
!--N,M working integers
!--NQL 0,1 for continuing, new calculation

call inputData(nql,nqls)
if(nql == 1) then
  call setInitialState
  call clearSteadySample
else
  call readRestart
  if(nqls == 1) call clearSteadySample
endif

do while (npr < 10*npt)
  npr = npr + 1
  write(*,*) 'OUTPUT INTERVAL:',npr
  do n = 1, nso
    do m = 1, nis
      nstep = nstep + 1
      write(*,*) ' TIME',time/fmct,' COLS',totcol,' MOLS',nm
      write(*,*) ' SEP',clsep/totcol
      time = time + dtm
```

```
      call moveMols
      call indexMols
#ifdef IPSUN
      call aEnergy1
#endif
      call collisions
#ifdef IPFAN
#ifdef IPSUN
      call aEnergy2
#endif
      call IPupdate
#endif
    end do
    if(n == nsi) call clearSample
    call sampleFlow
  end do
  call output
  call writeRestart
end do

end program COUETTE

!------------------------------------------------------------------------

subroutine setInitialState

use species
use molecules
use flowField
use calculation
use sampling

implicit none

real :: a,b,eta,yheight,rmu,re
integer :: j,n,nrow

!eta = 999999.9   !HS MOLECULE
!eta = 7.452      !VHS MOLECULE
!eta = 5.0        !MAXWELL MOLECULE
spm(1) = pi*sp(1)**2
spm(2) = sp(2)
spm(3) = sp(3)
!spm(3) = 0.5*(eta+3)/(eta-1)
spm(4) = sp(4)
spm(5) = sp(5)/2.0
!spm(6) = ggam(2.5-spm(3))  ! USED IN VHS OR VSS

nm    = nmi
ncell= ncy
tw    = ftmp*tr
pref = pref*101325.0
fden = pref/(boltz*ftmp)
vmpf = sqrt(ftmp)
vmpw = sqrt(tw)
fvel = svf*vmpw

!--MFP mean free path of molecules at the initial state
mfp      = 1.0/(sqrt(2.0)*fden*spm(1))
yheight = mfp/kn
```

```
area     = nm/(fden*yheight)
gasR     = boltz/sp(5)
rmu      = 5.0/16.0*spi*mfp*vmpf*sqrt(2.0*gasR)*fden*sp(5)
re       = sp(5)*fden*fvel*sqrt(2.0*gasR)*yheight/rmu

!--                    -- NORMALIZED PARAMETERS --
!- cross-section
cxs    = spm(1)/(mfp**2)
fmct   = 0.5*spi/vmpf
dtm    = dtr*fmct
ybn    = 1.0/kn
cy     = ybn/float(ncy)
rcy    = 1.0/cy
dcrit  = cy/3.0e+6
area   = area/mfp**2
fden   = fden*mfp**3
cvol   = area*cy

open(2,file='couette.out',status='unknown')
write(2,*) 'Length scale: ',yheight
write(2,*) 'Pressure: ',pref/101325.0
write(2,*) 'Kn :',kn
write(2,*) 'Re :',re
write(2,*) 'Viscosity: ',rmu
write(2,*) 'R:          ',gasR
write(2,*) 'Temperature: ',ftmp
write(2,*) 'Mean free path: ',mfp
write(2,*) 'Molecules per cube mean free path: ',fden
write(2,*) 'Molecule mass: ',sp(5)
close(2)

time    = 0.
npr     = 0.
nsamp   = 0
nsamps  = 0
nstep   = 0
totcol  = 1.
clsep   = 0.
surf    = 0.
ssurf   = 0.

#ifdef IPFAN
  cnip = 1.
  cpip = 0.5*cnip*ftmp
  caip = 0.
  cvip = 0.
  pvip = 0.
#ifdef IPSUN
  ctip(1,:) = ftmp
  ctip(2,:) = 0.
  ceip      = 0.
  ptip(1,:) = ftmp
  ptip(2,:) = 0.
#endif
#endif

do n = 1, ncy
  cell(2,n) = 4.0*vmpf
  cell(3,n) = 0.0
end do
```

```
!--SET THE INITIAL STATE OF THE MOLECULES
do n = 1, nm
  call random_number(ranf)
  pp(n) = ranf*ybn
  call random_number(ranf)
  a = sqrt(-log(ranf))
  call random_number(ranf)
  b - dpi*ranf
  pv(2,n) = a*sin(b)*vmpf
  pv(3,n) = a*cos(b)*vmpf
  call random_number(ranf)
  a = sqrt(-log(ranf))
  call random_number(ranf)
  b = dpi*ranf
  pv(1,n) = a*sin(b)*vmpf
#ifdef IPFAN
  do j = 1, 3
    pvip(j,n) = 0.
  end do
#endif
end do

call indexMols

msc     = ndiv
nscn(1) = -1
nscn(2) = 1

do n = 1, ncy
  cell(1,n) = (n-1)*cy
end do

write(*,*) 'INITIAL STATE SET'

end subroutine setInitialState

!------------------------------------------------------------------------

subroutine inputData(nql,nqls)

use species
use molecules
use flowField
use calculation

implicit none

integer :: nql,nqls,rsize
integer,allocatable :: rseed(:)

open(2,file="couette.in",status="old")
read(2,*) nql,nqls
if(nql==0) then
  close(2)
  return
endif
read(2,*) nmi
read(2,*) kn,ftmp,pref
read(2,*) svf,tr
read(2,*) dtr
read(2,*) ncy
```

```
read(2,*) nis
read(2,*) nso
read(2,*) nsi
read(2,*) npt
close(2)

!--PROPERTIES FOR AR
sp(1) = 3.659e-10     ! DSMC: DIAMETER FOR HS  MODEL
!sp(1) = 4.170e-10    ! DSMC: DIAMETER FOR VHS MODEL
!sp(1) = 3.963e-10    ! IP: DIAMETER FOR HS  MODEL
!sp(1) = 4.283e-10    ! IP: DIAMETER FOR VHS MODEL
sp(2) = 273.0         ! REFERENCE TEMPERATURE
sp(3) = 0.81          ! VISCOSITY-TEMPERATURE INDEX
sp(4) = 1.0           ! RECIPROCAL OF THE VSS SCATTERING PARAMETER
sp(5) = 6.63e-26      ! MASS

ispr = 0
ispf = 3

!--INITIALIZE RANDOM SEED
call random_seed(SIZE = rsize)
allocate(rseed(rsize))
rseed = 100
call random_seed(PUT = rseed(1:rsize))

end subroutine inputData

!------------------------------------------------------------------------

subroutine moveMols

use molecules
use flowField
use calculation
use sampling
use species

implicit none

real :: a,dti,dts,y,yi,dy
integer :: n,k,m,is

n = 0
do while (n < nm)
  n   = n+1
  dts = dtm
  yi  = pp(n)
  dy  = pv(2,n)*dts
  y   = yi+dy
  if(y > ybn) then
    is  = 2
    dti = dts*(ybn-yi)/dy
  else if(y <= 0.0) then
    is=1
    dti = dts*(-yi)/dy
  else
    is  = 0
    dti = dts
  endif

#ifdef IPFAN
```

```
  m = ir(1,n)
  a = pvip(1,n)**2 + pvip(2,n)**2 + pvip(3,n)**2
  do k = 1, 3
    pvip(k,n) = pvip(k,n) + caip(k,m)*dti
  end do
#ifdef IPSUN
  a = pvip(1,n)**2 + pvip(2,n)**2 + pvip(3,n)**2 - a
  ptip(1,n) = ptip(1,n) + (ceip(m)*dti-a)*(2.0/ispf)
#endif
#endif

  if(is == 0) then
    pp(n) = y
    dts   = 0.0
  else
    call reflect(is,n)
    dts = dts - dti
  endif
end do
end subroutine moveMols

!------------------------------------------------------------------------

subroutine reflect(nd,n)

use molecules
use flowField
use sampling
use calculation
use species

implicit none

integer,intent(in) :: nd,n
real :: a,b,yi,sgn,Tref,Tabs
integer :: k,m

!-- ND 1 for lower wall, 2 for upper wall
!
if(nd == 1) then
  yi  = 0.
  sgn = 1.
else if(nd == 2) then
  yi  = ybn
  sgn = -1.0
end if

!-                           --SAMPLE WALL INCIDENT--
if(npr >= 10) then
  pv(1,n)      = pv(1,n) - sgn*fvel/2.0
  surf(nd,1)   = surf(nd,1)  + 1.0
  ssurf(nd,1)  = ssurf(nd,1) + 1.0
  surf(nd,2)   = surf(nd,2)  + pv(1,n)
  surf(nd,3)   = surf(nd,3)  - pv(2,n)
  ssurf(nd,2)  = ssurf(nd,2) + pv(1,n)
  ssurf(nd,3)  = ssurf(nd,3) - pv(2,n)
  a = 0.5*( pv(1,n)**2 + pv(2,n)**2 + pv(3,n)**2 )
   surf(nd,4) =  surf(nd,4) + a
  ssurf(nd,4) = ssurf(nd,4) + a
#ifdef IPFAN
  pvip(1,n)      = pvip(1,n)      - sgn*fvel/2.0
```

```
  ssurfip(nd,2) = ssurfip(nd,2) + pvip(1,n)
  ssurfip(nd,3) = ssurfip(nd,3) - pvip(2,n)
  a = 0.5*( pvip(1,n)**2 + pvip(2,n)**2 + pvip(3,n)**2 )
#ifdef IPSUN
  a = 0.5*( pvip(1,n)**2 + pvip(2,n)**2 + pvip(3,n)**2 ) &
     + 0.25*ispf*( ptip(1,n) + ptip(2,n) )
#endif
  ssurfip(nd,4) = ssurfip(nd,4) + a
#endif
end if
!-                          --GAS-SURFACE MODEL--
call random_number(ranf)
a = sqrt(-log(ranf))*vmpw
call random_number(ranf)
b = dpi*ranf
pv(1,n) = a * sin(b)
pv(3,n) = a * cos(b)
call random_number(ranf)
pv(2,n) = sgn*sqrt(-log(ranf))*vmpw

#ifdef IPFAN
  pvip(1,n) = 0.0
  pvip(2,n) = 0.0
  pvip(3,n) = 0.0

#ifdef IPSUN
  m    = ir(1,n)
  Tref = sqrt(ptip(1,n)*Tw)
  Tabs = (ptip(1,n)-Tref)/ispf
  ctip(2,m) = ctip(2,m)+ptip(2,n)-Tabs
 !ctip(2,m) = ctip(2,m)-Tabs
  ptip(2,n) = (Tw-Tref)/ispf
  ptip(1,n) = Tw
#endif
#endif

call boundaryIR(nd,n)

!-                           --SAMPLE WALL REFLECTION--
if(npr >= 10) then
   surf(nd,5) =  surf(nd,5) + pv(1,n)
   surf(nd,6) =  surf(nd,6) + pv(2,n)
  ssurf(nd,5) = ssurf(nd,5) + pv(1,n)
  ssurf(nd,6) = ssurf(nd,6) + pv(2,n)
  a = 0.5*( pv(1,n)**2 + pv(2,n)**2 + pv(3,n)**2 )
   surf(nd,7) =  surf(nd,7) + a
  ssurf(nd,7) = ssurf(nd,7) + a
#ifdef IPFAN
  ssurfip(nd,5) = ssurfip(nd,5) + pvip(1,n)
  ssurfip(nd,6) = ssurfip(nd,6) + pvip(2,n)
  a = 0.5*( pvip(1,n)**2 + pvip(2,n)**2 + pvip(3,n)**2 )
#ifdef IPSUN
  a = 0.5*( pvip(1,n)**2 + pvip(2,n)**2 + pvip(3,n)**2 ) &
    + 0.25*ispf*( ptip(1,n) + ptip(2,n) )
#endif
  ssurfip(nd,7) = ssurfip(nd,7) + a
#endif
end if

pv(1,n)    =   pv(1,n) + sgn*fvel/2.0
#ifdef IPFAN
```

```
pvip(1,n) = pvip(1,n) + sgn*fvel/2.0
#endif

end subroutine reflect

!-----------------------------------------------------------------------

subroutine boundaryIR(nd,n)

!-             -- Avoid changing the additional information --
!              -- for particles interacting with boundaries --

use molecules
use flowField

implicit none

integer,intent(in) :: nd,n

!-                         -Surface B.C.-
if(nd==1) ir(1,n) = 1
if(nd==2) ir(1,n) = ncy

!-                         -I/O B.C.-
end subroutine boundaryIR

!-----------------------------------------------------------------------

subroutine indexMols

use molecules
use flowField
use calculation

implicit none

integer :: l,m,n,nrow

iro = ir
do m = 1, ncell
  icell(2,m) = 0
end do
do n = 1, nm
  nrow = (pp(n)+dcrit)*rcy + 1
  if(nrow < 1) then
   write(*,*) 'nrow =',nrow
   write(*,*) n,pp(n),dcrit,rcy
   pause
  end if
  if(nrow > ncy) nrow = ncy
  m = nrow
  ir(1,n) = m
  icell(2,m) = icell(2,m) + 1
end do

n = 0
do m = 1, ncell
  if(icell(2,m) > mmc) then
    write(*,*) ' THE',icell(2,m),'MOLECULE IN CELL',m, &
      'EXCEEDS SPECIFIED MMC'
    do n = 1, icell(2,1)
```

```
      l = ir(2,n)
      if(n < 1000 ) write(72,*) n, pp(l),pv(2,l)
    end do
    stop
  endif
  icell(1,m) = n
  n = n + icell(2,m)
  icell(2,m) = 0
end do
do n = 1, nm
  m = ir(1,n)
  icell(2,m) = icell(2,m) + 1
  l = icell(1,m) + icell(2,m)
  ir(2,l) = n
end do

end subroutine indexMols

!-------------------------------------------------------------------------
#ifdef IPSUN

subroutine aEnergy1
!--       -- Calculate the additional energy for particles and --
!         -- cells owing to the particle movement(convection)   --
use molecules
use flowField
use species
implicit none
integer :: l,m,n
real :: a,Tref

do n = 1, nm
  m = iro(1,n)
  l = ir(1,n)
  if(m.ne.l) then
    call refEnergy(m,l,Tref)
    a = ptip(2,n)
    ptip(2,n) = (ptip(1,n)-Tref) / ispf
    ctip(2,m) = ctip(2,m) + a - ptip(2,n)
  end if
end do
end subroutine aEnergy1

!---------------------------

subroutine aEnergy2
!--            -- Distribute the additional energy of a cell --
use molecules
use flowField
implicit none
integer :: n,nc,k,k1,k2
real :: a

do nc = 1, ncell
  a = ctip(2,nc) / icell(2,nc)
  k1 = icell(1,nc) + 1
  k2 = icell(1,nc) + icell(2,nc)
  do k = k1, k2
    n = ir(2,k)
    ptip(1,n) = ptip(1,n) + a
  end do
```

```
  ctip(2,nc) = 0.0
end do
end subroutine aEnergy2

!----------------------------

subroutine refEnergy(c1,c2,Tref)
!--                -- Reference energy at the inter\kern.5ptface where --
!                  --  a particle leaves its cell through     --
use flowField
implicit none
integer, intent(in) :: c1,c2
real, intent(out) :: Tref

Tref = ( ctip(1,c1) + ctip(1,c2) ) / 2.0

end subroutine refEnergy

#endif

!------------------------------------------------------------------------

subroutine collisions

use molecules
use flowField
use calculation
use sampling
use species
implicit none
integer :: i,j,jj,jjj,k,kk,l,m,mm,n,nc,nsel,nrow,nmol,nsc,kc,nns,ipcol
real :: a,b,c,avn,asel,vr,vrr,sep,cu,ck
real, dimension(3) :: vrc,vccm,vrcp
real, dimension(2) :: trc,tccm,trcp
!--I,J,JJ,JJJ,K,L,M,MM,N,NNS,KC working integers
!--NC cell number
!--NSC sub-cell number
!--NMOL the number of molecules in a cell
!--IPCOL IP collision type control parameter
!--AVN average number of molecules in a cell
!--ASEL,NSEL the real<integer number of selections
!--VRC components of the pre-collision relative velocity
!--VRCP components of the post-collision relative velocity
!--VCCM components of the center of mass velocity
!--VR,VRR relative speed and square of relative speed

do nc = 1, ncell
  nmol = icell(2,nc)
  if(nsamp > 0) then
    avn = cs(1,nc) / nsamp
  else
    avn = nmol
  end if
  asel = 0.5*nmol*avn*cxs*cell(2,nc)*dtm/cvol + cell(3,nc)
  nsel = asel
  cell(3,nc) = asel - nsel
!--                      SET THE TRANSIENT SUB-CELLS IN THIS CELL
  do n = 1, nmol
    m  = icell(1,nc) + n
    mm = ir(2,m)
    irs(3,n) = mm
```

```
      nrow = ndiv*(pp(mm)-cell(1,nc)+dcrit)*rcy + 1
      if(nrow > ndiv) nrow = ndiv
      if(nrow < 1) nrow = 1
      irs(2,n) = nrow
    end do
!--                          SUB-CELL HAS BEEN DETERMINED
!                            SET THE INDEXING TO THE SUB-CELLS
    do n = 1, msc
      Isc(2,n) = 0
    end do
    do n = 1, nmol
      mm = irs(2,n)
      isc(2,mm) = isc(2,mm) + 1
    end do
    m = 0
    do n = 1, msc
      isc(1,n) = m
      m = m + isc(2,n)
      isc(2,n) = 0
    end do
    do n = 1, nmol
      mm = irs(2,n)
      isc(2,mm) = isc(2,mm) + 1
      k = isc(1,mm) + isc(2,mm)
      irs(1,k) = n
    end do

    do i = 1,nsel
      call random_number(ranf)
      k = int(ranf*(icell(2,nc)-0.0001)) + icell(1,nc) + 1
      l = ir(2,k)
!-                         L HAS BEEN CHOSEN RANDOMLY FROM THE CELL
      kc = k - icell(1,nc)
      nsc = irs(2,kc)
!-                         NSC IS THE SUB-CELL
      m = 1
      if(isc(2,nsc) > 1) then
!-                         CHOOSE COLLISION PARTNER FROM THE SAME SUB-CELL
        do while(m == 1)
          call random_number(ranf)
          k  = int(ranf*(isc(2,nsc)-0.0001)) + isc(1,nsc) + 1
          kk = irs(1,k)
          m  = irs(3,kk)
!--                        M HAS BEEN CHOSEN
        end do
      else
!-                         TRY FROM ADJACENT SUB-CELLS
        call random number(ranf)
        jj = int(ranf*1.999999)
!-         JJ IS THE RAMDOM ENTRY POINT TO THE ARRAY OF ADJACENT SUB-CELLS
        j = 1
        do while(j < 3)
!-                      CHOOSE COLLISION PARTNER FROM AN ADJACENT SUB-CELL
          jjj = j + jj
          if(jjj > 2) jjj = jjj - 2
          nrow = nsc + nscn(jjj)
          if((nrow > 0).and.(nrow <= ndiv)) then
            nns = nrow
            if(isc(2,nns) > 0) then
              if(isc(2,nns) == 1) then
                k = isc(1,nns) + 1
```

```
              else
                call random_number(ranf)
                k = int(ranf*(isc(2,nns)-0.0001)) + isc(1,nns) + 1
              end if
              kk = irs(1,k)
              m  = irs(3,kk)
            end if
          end if
          j = j + 1
        end do
        if(j == 3) then
!-                    CHOOSE COLLISION PARTNER FROM ANYWHERE IN THE CELL
          do while(m == 1)
          call random_number(ranf)
          k = int(ranf*(icell(2,nc)-0.0001)) + icell(1,nc) + 1
          m = ir(2,k)
          end do
        end if
      end if

!-                       POSSIBLE COLLISION PARTNERS HAVE BEEN SELECTED

      do k = 1, 3
        vrc(k) = pv(k,1) - pv(k,m)
      end do
      vrr = vrc(1)**2 + vrc(2)**2 + vrc(3)**2
      vr  = sqrt(vrr)
      if(vr > cell(2,nc)) cell(2,nc) = vr
      call random_number(ranf)
      if(ranf < vr/cell(2,nc)) then
!-                                      COLLISION OCCURS
        totcol = totcol + 1.0d0
        clsep  = clsep + abs(pp(1)-pp(m))
        do k=1,3
          vccm(k) = 0.5 * ( pv(k,1) + pv(k,m) )
        end do
        call random_number(ranf)
        b = 2.0 * ranf - 1.0
        a = sqrt(1.0 - b*b)
        call random_number(ranf)
        c = dpi * ranf
        vrcp(1) = b * vr
        vrcp(2) = a * cos(c) * vr
        vrcp(3) = a * sin(c) * vr
        do k=1,3
          pv(k,1) = vccm(k) + 0.5*vrcp(k)
          pv(k,m) = vccm(k) - 0.5*vrcp(k)
        end do

#ifdef IPFAN
!-      IP velocity
!-      changes with Collision
        ipcol = 2

        if(ipcol == 1) then
          vrr = 0.0
          do k = 1, 3
            vrr = vrr + (pvip(k,1)-pvip(k,m))**2
          end do
          c = vrr / (2.0*ispf)
          do k = 1, 3
```

```
          a = 0.5*( pvip(k,l) + pvip(k,m) )
          pvip(k,l) = a
          pvip(k,m) = a
        end do
#ifdef IPSUN
        do k = 1, 2
          a = 0.5 * ( ptip(k,l) + ptip(k,m) )
          ptip(k,l) = a
          ptip(k,m) = a
        enddo
        ptip(1,l) = ptip(1,l) + c
        ptip(1,m) = ptip(1,m) + c

      else if(ipcol == 2) then

        cu = -0.25; ck = 0.87
        vrr = 0.0
        do k = 1, 3
          vrc(k) = pvip(k,l) - pvip(k,m)
          vrr = vrr + vrc(k)**2
        end do
        c = vrr / (2.0*ispf)
        do k = 1, 3
          vccm(k) = 0.5*( pvip(k,l) + pvip(k,m) )
        end do
        do k = 1, 3
          vrcp(k) = cu * b * vrc(k)
        end do
        do k = 1, 3
          pvip(k,l) = vccm(k) + 0.5*vrcp(k)
          pvip(k,m) = vccm(k) - 0.5*vrcp(k)
        end do
        do k = 1, 2
          trc(k) = ptip(k,l) - ptip(k,m)
        end do
        do k = 1, 2
          tccm(k) = 0.5 * ( ptip(k,l) + ptip(k,m) )
        end do
        do k = 1, 2
          trcp(k) = ck * b * trc(k)
        end do
        do k = 1, 2
          ptip(k,l) = tccm(k) + 0.5*trcp(k)
          ptip(k,m) = tccm(k) - 0.5*trcp(k)
        end do
        ptip(1,l) = ptip(1,l) + c
        ptip(1,m) = ptip(1,m) + c*(1.0-(cu*b)**2)
#endif
      endif
#endif

    end if
  end do
end do

end subroutine collisions

!-----------------------------------------------------------------------

subroutine clearSample
```

```
use molecules
use flowField
use calculation
use sampling
implicit none

nsamp = 0
timi = time
cs(:,:) = 0.0
surf(:,:) = 0.0
#ifdef IPFAN
csip(:,:) = 0.0
surfip(:,:) = 0.0
#endif

end subroutine clearSample

!------------------------------------------------------------------------

subroutine clearSteadySample

use molecules
use flowField
use calculation
use sampling
implicit none

nsamps = 0
timis = time
scs(:,:) = 0.0
ssurf(:,:) = 0.0
#ifdef IPFAN
scsip(:,:) = 0.0
ssurfip(:,:) = 0.0
#endif

end subroutine clearSteadySample

!------------------------------------------------------------------------

subroutine sampleFlow

use molecules
use flowField
use calculation
use sampling

implicit none
integer :: n,m,nc
real :: a

nsamp = nsamp + 1
nsamps = nsamps + 1.0d0
do n = 1, nm
  nc = ir(1,n)
  if((nc < 1).or.(nc > ncell)) then
    write(*,*) 'MOLECULE',n,' IN ILLEGAL CELL',nc,pp(n)
    stop
  end if
   cs(1,nc) =  cs(1,nc) + 1.0
  scs(1,nc) = scs(1,nc) + 1.0d0
```

```
  do m = 1, 3
     cs(m+1,nc) =  cs(m+1,nc) + pv(m,n)
    scs(m+1,nc) = scs(m+1,nc) + pv(m,n)
    a = pv(m,n) * pv(m,n)
     cs(m+4,nc) =  cs(m+4,nc) + a
    scs(m+4,nc) = scs(m+4,nc) + a
  end do
end do
close(70)

#ifdef IPFAN
  call IPsample
#endif

end subroutine sampleFlow

!---------------------------------------------

subroutine IPsample

#ifdef IPFAN
use molecules
use flowField
use calculation
use sampling
use species

implicit none

integer :: n,m,nc,k
real,dimension(3,mcell) :: ccip

ccip = 0.0

!- SAMPLE V*V
do n = 1, nm
  nc = ir(1,n)
  do m = 1, 3
    ccip(m,nc) = ccip(m,nc) + pvip(m,n)*pvip(m,n)
  end do
end do

!--                    -- UPDATE IP FOR CELLS --

!-                  Sample ip flowfield density,velocities and temperature
do nc = 1, ncell
   csip(1,nc) =  csip(1,nc) + cnip(nc)
  scsip(1,nc) = scsip(1,nc) + cnip(nc)
  do k = 1, 3
   csip(k+1,nc) =  csip(k+1,nc) + cvip(k,nc)
  scsip(k+1,nc) = scsip(k+1,nc) + cvip(k,nc)
   csip(k+4,nc) =  csip(k+4,nc) + ccip(k,nc)/icell(2,nc) - cvip(k,nc)**2
  scsip(k+4,nc) = scsip(k+4,nc) + ccip(k,nc)/icell(2,nc) - cvip(k,nc)**2
  enddo
#ifdef IPSUN
   csip(8,nc) =  csip(8,nc) + ctip(1,nc)
  scsip(8,nc) = scsip(8,nc) + ctip(1,nc)
#endif
end do

#endif
```

```
end subroutine IPsample

!-------------------------------------------

subroutine IPupdate

#ifdef IPFAN
!--                   -- UPDATE CAIP,CEIP AND CNIP FOR NEXT TIMESTEP --
use molecules
use flowField
use calculation
use sampling

implicit none

real,allocatable :: v(:,:),p(:),n(:)
integer :: ecell,nc,ipc,nrow,np,nn,k,m
real :: dts,w,dn,de,nref,ps(3)
real,dimension(4,mcell) :: ccip

ecell = ncell + 2
allocate( v(3,ecell), p(ecell), n(ecell) )

ccip = 0.0

!-                   Sample v and t
do m = 1, nm
  nc = ir(1,m)
  do k = 1, 3
  ccip(k,nc) = ccip(k,nc) + pvip(k,m)
  end do
#ifdef IPSUN
  ccip(4,nc) = ccip(4,nc) + ptip(1,m) + ptip(2,m)
#endif
end do

!-                   Update ip for cells

!nref = fden * cvol
nref = nm / ncell
do nc = 1, ncell
  do k = 1, 3
  cvip(k,nc) = ccip(k,nc) / icell(2,nc)
  end do
  cnip(nc) = icell(2,nc) / nref
#ifdef IPSUN
  ctip(1,nc) = ccip(4,nc) / icell(2,nc)
  cpip(nc) = 0.5 * cnip(nc) * ctip(1,nc)
#else
  cpip(nc) = 0.5 * cnip(nc) * ftmp
#endif
end do
do nrow = 1, ncy
  nc = nrow
  ipc = nrow + 1
  n(ipc) = cnip(nc)
  p(ipc) = cpip(nc)
  v(1,ipc) = cvip(1,nc)
  v(2,ipc) = cvip(2,nc)
  v(3,ipc) = cvip(3,nc)
end do
```

```
!--                    Wall B.C.
do ipc = 1, ncy+2, ncy+1
  if(ipc == 1) nc = 2
  if(ipc == ncy+2) nc = ncy - 1
  n(ipc) = cnip(nc)
  p(ipc) = cpip(nc)
  v(1,ipc) = cvip(1,nc)
  v(2,ipc) = 0.0
  v(3,ipc) = 0.0
end do

dts = dtm
w = 0.5
ps = 0.0
do nrow = 1, ncy
  nc = nrow + 1
  np = nc - 1
  nn = nc + 1
  dn = -0.25 * rcy * ( n(nc)*(v(2,nn)-v(2,np)) + v(2,nc)*(n(nn)-n(np)) &
                      +n(nn)*v(2,nn) - n(np)*v(2,np) )
#ifdef IPSUN
  de = -0.25 * rcy * ( p(nc)*(v(2,nn)-v(2,np)) + v(2,nc)*(p(nn)-p(np)) &
                      +p(nn)*v(2,nn) - p(np)*v(2,np) )
#endif
  ps(2) = -0.5 * rcy * ( p(nn) - p(np) )

  do k = 1, 3
  caip(k,nrow) = caip(k,nrow)*(1.-w) + w*ps(k)/n(nc)
  end do
#ifdef IPSUN
  ceip(nrow) = ceip(nrow)*(1.-w) + w*2.0*de/n(nc)
#endif
  cnip(nrow) = cnip(nrow) + w*dn*dts
end do

#endif
end subroutine IPupdate

!------------------------------------------------------------------------

subroutine output

use species
use molecules
use flowField
use calculation
use sampling
use wcontrol

implicit none
double precision,dimension(8) :: op
double precision :: denn,denns,dennd,tt,tts,ttd
double precision,dimension(3) :: vels,vells
double precision,dimension(2,2:7) :: wall
integer :: j,n,nc,nrow,jj,kk
real :: y,a,ma,vref,pw,tao,ht

fname = fp//'wcontrol.in'
open(3,file = fname,status = 'old')
read(3,*) nso
read(3,*) nis
```

```
read(3,*) nsi
read(3,*) iwrite
read(3,*) ires
close(3)
if(iwrite == 0) return

kk = 1
#ifdef IPFAN
  kk = 2
#endif
ma = sqrt(2.0/1.667)*svf

vref = vmpf
a = 2.0/((time-timis)*area*fden*vref**2)
do jj = 1, kk
  if(jj == 1) then
    fname = fp//'Wall.dat'
    do j = 1, 2
    do n = 2, 7
    wall(j,n) = ssurf(j,n)
    end do
    end do
  end if
#ifdef IPFAN
  if(jj == 2) then
    fname=fp//'WallIP.dat'
    do j = 1, 2
    do n = 2, 7
    wall(j,n) = ssurfip(j,n)
    end do
    end do
  end if
#endif

  open(2,file = fname,status = 'unknown')
  write(2,10) kn,ma
  10 format(1x,'Kn =',f6.3,' Ma =',f6.3/)

  do j = 1, 2
    op(1) = ssurf(j,1)
    op(2) = ssurf(j,1) * a * vref
    op(3) = wall(j,3) * a
    op(4) = wall(j,6) * a
    op(5) = spi * wall(j,2) * a
    op(6) = spi * wall(j,5) * a
    op(7) = wall(j,4) * a / vref
    op(8) = wall(j,7) * a / vref
    if(svf > 0.001) then
      op(5) = op(5) * vref**2 / fvel
      op(6) = op(6) * vref**2 / fvel
    endif
    pw  = abs( op(3) + op(4) )
    tao = abs( op(5) - op(6) )
    ht  = op(7) - op(8)

    if(j == 1) write(2, 11)
    if(j == 2) write(2, 12)
 11 format(1x,'Properties on the LOWER surface'/)
 12 format(1x,'Properties on the UPPER surface'/)
    write(2, 13)
 13 format(1x,'  sample size number flux pressure(in,re)',1x, &
```

```
             '  shear stress(in,re)     Heat flux(in,re)')
    write(2, 14) op
    write(2, 15) pw
    write(2, 16) tao
    write(2, 17) ht
 14 format(1x,8e13.5)
 15 format(1x,' Wall pressure coef. =',e10.4)
 16 format(1x,' Wall Shear Stress coef. =',e10.4)
 17 format(1x,' Wall Heat Transfer coef. =',e10.4)
  end do
  close(2)
end do

!--                            -- FLOW FIELD --
!--                            -- Steady Flow--
!-                               DSMC RESULTS
fname = fp//'FieldS.dat'
open(2,file = fname,status = 'unknown')
write(2,*) 'cell, Y, Samples, N, u, v, Tx, Ty, Tz, T'
do nc = 1, ncy
  y = (nc-0.5) * cy

  do n = 1, 3
    vels(n)  = scs(n+1,nc) / scs(1,nc)
    vells(n )= scs(n+4,nc) / scs(1,nc)
  end do
  op(1) = scs(1,nc)
  op(2) = scs(1,nc) / (nsamps*cvol*fden)
  op(3) = vels(1) / vref
  op(4) = vels(2) / vref
  op(5) = 2.0 * (vells(1)-vels(1)**2)
  op(6) = 2.0 * (vells(2)-vels(2)**2)
  op(7) = 2.0 * vells(3)
  op(8) = ( op(5) + op(6) + op(7) ) / 3.0
  op(8) = op(8) / ftmp
  write(2,20) nc,y,op
end do
20 format(1x,I4,F9.4,E12.3,2F9.4,F10.4,4F9.4)
close(2)

#ifdef IPFAN
!-                               IP RESULTS
fname = fp//'FieldIPS.dat'
open(2,file = fname,status='unknown')
write(2,*) 'cell, Y, Samples, N, u, v, Tx, Ty, Tz, T'
do nc = 1, ncy
  y = (nc-0.5) * cy

  do n = 1, 3
  vels(n) = scsip(n+1,nc) / nsamps
  end do
  op(1) = scs(1,nc)
  op(2) = scsip(1,nc) / nsamps
  op(3) = vels(1) / vref
  op(4) = vels(2) / vref
  op(5) = 2.0 * scsip(5,nc) / nsamps
  op(6) = 2.0 * scsip(6,nc) / nsamps
  op(7) = 2.0 * scsip(7,nc) / nsamps
  op(8) = ( op(5) + op(6) + op(7) ) / 3.0
#ifdef IPSUN
  op(8) = op(8) + scsip(8,nc)/nsamps
```

```
#endif
  op(8) = op(8) / ftmp
  write(2,20) nc,y,op
end do
close(2)
#endif

!--                               -- Instantaneous Flow--
!-                                   DSMC RESULTS
fname = fp//'FieldU.dat'
open(2,file = fname,status = 'unknown')
write(2,*) 'cell, Y, Samples, N, u, v, Tx, Ty, Tz, T'
do nc = 1, ncy
  y = (nc-0.5) * cy

  do n = 1, 3
    vels(n)  = cs(n+1,nc) / cs(1,nc)
    vells(n) = cs(n+4,nc) / cs(1,nc)
  end do
  op(1) = cs(1,nc)
  op(2) = cs(1,nc) / (nsamp*cvol*fden)
  op(3) = vels(1) / vref
  op(4) = vels(2) / vref
  op(5) = 2.0 * (vells(1)-vels(1)**2)
  op(6) = 2.0 * (vells(2)-vels(2)**2)
  op(7) = 2.0 * vells(3)
  op(8) = ( op(5) + op(6) + op(7) ) / 3.0
  op(8) = op(8) / ftmp
  op(3) = op(3) * sqrt(2.0*gasR)
  op(4) = op(4) * sqrt(2.0*gasR)
  op(5) = sqrt(op(5)*2.0*gasR)
  op(6) = sqrt(op(6)*2.0*gasR)
  op(7) = sqrt(op(7)*2.0*gasR)
  write(2,20) nc,y,op
end do
close(2)

#ifdef IPFAN
!-                                IP RESULTS
fname = fp//'FieldIPU.dat'
open(2,file = fname,status='unknown')
write(2,*) 'cell, Y, Samples, N, u, v, Tx, Ty, Tz, T'
do nc = 1, ncy
  y = (nc-0.5) * cy

  do n = 1, 3
    vels(n) = csip(n+1,nc) / nsamp
  end do
  op(1) = cs(1,nc)
  op(2) = csip(1,nc) / nsamp
  op(3) = vels(1) / vref
  op(4) = vels(2) / vref
  op(5) = 2.0 * csip(5,nc) / nsamp
  op(6) = 2.0 * csip(6,nc) / nsamp
  op(7) = 2.0 * csip(7,nc) / nsamp
  op(8) = ( op(5) + op(6) + op(7) ) / 3.0
#ifdef IPSUN
  op(8) = op(8) + csip(8,nc)/nsamp
#endif
  op(8) = op(8) / ftmp
  op(3) = op(3) * sqrt(2.0*gasR)
```

```
    op(4) = op(4) * sqrt(2.0*gasR)
    op(5) = sqrt(op(5)*2.0*gasR)
    op(6) = sqrt(op(6)*2.0*gasR)
    op(7) = sqrt(op(7)*2.0*gasR)
    write(2,20) nc,y,op
  end do
  close(2)
  #endif

  end subroutine output

  !-----------------------------------------------------------------------

  subroutine writeRestart

  use species
  use molecules
  use flowField
  use calculation
  use sampling
  use wcontrol
  implicit none
  integer :: rsize
  integer,allocatable :: rseed(:)

  if(ires == 0) return

  call random_seed(SIZE = rsize)
  allocate(rseed(rsize))
  call random_seed(GET = rseed(1:rsize))

  fname = fp//'couette.res'
  open(7,file = fname,form = 'unformatted')
  write(7)npr,dtm,time,dtr,fmct,totcol,clsep,sp,spm,ispr,ispf,mfp,gasR,  &
         pp,pv,ir,irs,nm,nmi,msc,cxs,cell,icell,isc,nscn,ncy,ncell,nis,  &
         nso,nsi,npt,cvol,area,ybn,cy,rcy,dcrit,tr,tw,svf,fvel,vmpf,vmpw,&
         ftmp,fden,pref,kn,nsamp,nsamps,nstep,cs,scs,surf,ssurf,rseed
  #ifdef IPFAN
          write(7)pvip,caip,cvip,cpip,cnip,csip,scsip,surfip,ssurfip
  #ifdef IPSUN
          write(7)ptip,ctip,ceip
  #endif
  #endif
  close(7)
  end subroutine writeRestart

  !-----------------------------------------------------------------------

  subroutine readRestart

  use species
  use molecules
  use flowField
  use calculation
  use sampling
  use wcontrol
  implicit none
  integer :: rsize
  integer,allocatable :: rseed(:)

  call random_seed(SIZE = rsize)
```

```
allocate(rseed(rsize))

fname = fp//'couette.res'
open(7,file = fname,form='unformatted')
read(7) npr,dtm,time,dtr,fmct,totcol,clsep,sp,spm,ispr,ispf,mfp,gasR,  &
       pp,pv,ir,irs,nm,nmi,msc,cxs,cell,icell,isc,nscn,ncy,ncell,nis,  &
       nso,nsi,npt,cvol,area,ybn,cy,rcy,dcrit,tr,tw,svf,fvel,vmpf,vmpw,&
       ftmp,fden,pref,kn,nsamp,nsamps,nstep,cs,scs,surf,ssurf,rseed
#ifdef IPFAN
        read(7)pvip,caip,cvip,cpip,cnip,csip,scsip,surfip,ssurfip
#ifdef IPSUN
        read(7)ptip,ctip,ceip
#endif
#endif
close(7)
call random_seed(PUT = rseed(1:rsize))

fname = fp//'wcontrol.in'
open(3,file = fname,status = 'old')
read(3,*) nso
read(3,*) nis
read(3,*) nsi
read(3,*) iwrite
read(3,*) ires
close(3)

end subroutine readRestart
```

References

Bird, G.A., Approach to translational equilibrium in a rigid sphere gas, *Phys. Fluids*, Vol. 6, pp. 1518–1519, 1963.

Bird, G.A., The velocity distribution function within a shock wave. *J. Fluid Mech.*, Vol. 30, pp. 479–487, 1965*a*.

Bird, G.A., Shock wave structure in a rigid sphere gas, In *Rarefied Gas Dynamics*, J.H. de Leeuw (ed.), Vol. 1, pp. 216–222, Academic Press, NY, 1965*b*.

Bird, G.A., *Molecular Gas Dynamics*, Clarendon Press, Oxford, United Kingdom, 1976.

Bird, G.A., Direct simulation of the Boltzmann equations, *Phys. Fluids*, Vol. 13, pp. 2676–2681, 1970.

Bird, G.A., Monte Carlo simulation of gas flows, *Ann. Rev. Fluid Mech.*, Vol. 10, pp. 11–31, 1978.

Bird, G.A., Monte Carlo simulation in an engineering context, *Prog. Astronaut. Aeronaut.*, Vol. 74, pp. 239–255, 1981.

Bird, G.A., Nonequilibrium radiation during reentry at 10 km/s, AIAA Paper 87-1543, 1987.

Bird, G.A., Monte Carlo simulation in an engineering context, *Progress in Comut. Mech.*, Austin, TX, 1986.

Bird, G.A., *The G2/A3 Program System User's Manual*, G.A.B. Consulting, Killara, Australia, 1992.

Bird, G.A., *Molecular Gas Dynamics and the Direct Simulation of Gas Flows*, Clarendon Press, Oxford, United Kingdom, p. 209, 1994.

Bird, G.A., Perception of numerical methods in rarefied gas dynamics, *Prog. Astronaut. Aeronaut.*, Vol. 118, pp. 211–226, 1989.

Bird, G.A., "Forty Years of DSMC, and Now?" in *Rarefied Gas Dynamics: 22nd International Symposium*, Sydney, Australia, 9–14 July 2000, American Institute of Physics, 2001.

Borgnakke, C., and Larsen, P.S., Statistical collision model for Monte Carlo simulation of polyatomic gas mixtures, *J. Comp. Phys.*, Vol. 18, pp. 405–420, 1975.

Boyd, I.D., Penko, P.F., Meissner, D.L., and Witt, K.J.D., Experimental and numerical investigations of low-density nozzle and plume flows of nitrogen, *AIAA J.*, Vol. 30, pp. 2453–2461, 1992.

Boyd, I.D., "Conservative species weighting scheme for the direct simulation Monte Carlo method," *J. Thermophys. Heat Transfer*, Vol. 10, pp. 579–585, 1996.

Cai, C., Boyd, I.D., Fan, J., and Candler, G.V., Direct simulation methods for low-speed microchannel flows, *J. Thermophys. Heat Trans.*, Vol. 14, No. 3, pp. 368–378, 2000.

Campbell, D.H., DSMC analysis of plume-freestream interactions and comparison of plume flowfield predictions with experimental measurement, AIAA paper 91-1362, 1991.

Carlson, A.B., and Wilmoth, R.G., Shock interference prediction using direct simulation Monte Carlo, AIAA paper 92-0492, 1992.

Celenligil, M.C., and Moss, J.N., Hypersonic rarefied flow about a delta wing—direct simulation and comparison with experiment, *AIAA J.*, Vol. 30, p. 2017, 1992.

Celenligil, M.C., Moss, J.N., and Blanchard, R.C., Three-dimensional flow simulation about the AFE vehicle in the transitional regime, AIAA paper 89-0245, 1989.

Cercignani, C., *The Boltzmann Equation and Its Applications*, Springer, New York/Berlin, 1988.

Cercignani, C., and Lampis, M., Kinetic models for gas-surface interactions. *Transp. Theory Stat. Phys.*, Vol. 1, No. 2, 1971.

Collins, F.G., and Knox, E.C., Method for determining wall boundary conditions for DSMC calculations at high speed ratios, AIAA paper 94-0036, 1994.

Dogra, V.K., Moss, J.N., Wilmoth, R.G., and Price, J.M., Hypersonic rarefied flow past sphere including wake structure, AIAA paper 92-0495, 1992.

Fan, J., Boyd, I.D., and Cai, C., Statistical simulations of low-speed rarefied gas flows, *J. Computation. Phys.*, Vol. 167, pp. 393–412, 2001.

Fan, J., and Shen, C., Statistical simulation of low-speed unidirectional flows in transition regime, In *Proceeding of the 21st International Symposium on Rarefied Gas Dynamics*, Marseille, France, pp. 245–252, 1998.

Fan, J. and Shen, C., Statistical simulation of low-speed rarefied gas flows, *J. Computation. Phys.*, Vol. 167, pp. 393–412, 2001.

Hash, D.B. and Hassan, H.A., Direct simulation of diatomic gases using the generalized hard sphere model, AIAA paper 93-0730, 1993.

Hassan, H.A., and Hash, D.B., A generalized hard-sphere model for Monte Carlo simulation, *Phys. Fluids* A, Vol. 5, pp. 738–744, 1993.

Kaplan, C.R., and Oran, E.S., Nonlinear filtering for low-velocity gaseous microflows, *AIAA J.*, Vol. 40, pp. 82–90, 2002.

Koura, K. and Matsumoto, H., Variable soft sphere molecular model for inverse-power-law or Lennard-Jones potential, *Phys. Fluids* A, Vol. 3, pp. 2459–2465, 1991.

Koura, K., and Matsumoto, H., Variable soft sphere molecular model for air species, *Phys. Fluids* A, Vol. 4, pp. 1083–1085, 1992.

LeBeau, G.J., Boyles, K.A., and Lumpkin III, F.E., Virtual sub-cells for the direct simulation Monte Carlo method, AIAA Paper 2003-1031, 2003.

Liou, W.W., Liu, F., Fang, Y., and Bird, G.A., Navier-Stokes and DSMC simulations of forced chaotic microflows, *33rd AIAA Fluid Dynamic Conference and Exhibit*, AIAA paper 2003-3583, 2003.

Lord, R.G., Application of the Cercignani-Lampis scattering kernel to direct simulation Monte Carlo calculation, In *Rarefied Gas Dynamics*, A.E. Beylich (ed.), VCH, Aachen, Germany, pp. 1427–1433, 1991.

Lord, R.G., Direct simulation Monte Carlo calculations of rarefied flows with incomplete surface accommodation, *J. Fluid Mech.*, Vol. 239, p. 449, 1992.

Meiburg, E., Comparison of the molecular dynamics method and the direct simulation technique for flows around simple geometries, *Phys. Fluids*, Vol. 29, pp. 3107–3113, 1986.

Moss, J.N., Bird, G.A., and Dogra, V.K., Nonequilibrium thermal radiation for an aeroassist flight experiment vehicle, AIAA paper 88-0081, 1988.

Nanbu, K., Numerical simulation of Boltzmann flows of real gases—accuracy of models used in the Monte Carlo method, *Hypersonic Flows for Reentry Problems*, Springer, New York, pp. 120–138, 1991; *Inst. Fluid Sci.*, Vol. 4, pp. 93–113, 1992.
Oh, C.K., Oran, E.S., and Cybyk, B.Z., Microchannel flow computed with DSMC-MLG, AIAA paper 95-2090, 1995.
Oran, E.S., Oh, C.K., and Cybyk, B.Z., Direct simulation Monte Carlo: Recent advances and applications, *Annu. Rev. Fluid Mech.*, Vol. 30, pp. 403–441, 1998.
Sun, Q., and Boyd, I.D., A direct simulation method for subsonic microscale gas flows, *J. Computation. Phys.*, Vol. 179, pp. 400–425, 2002.
Sun, Q., Boyd, I.D., and Fan, J., Development of an information preservation method for subsonic, micro-scale gas flows, in T.J. Bartel and M.A. Gallis (eds.), 22nd *International Symposium on Rarefied Gas Dynamics*, Sydney, pp. 547–553, 2001.

Chapter

6

Parallel Computing and Parallel Direct Simulation Monte Carlo Method

6.1 Introduction

The DSMC method has been one of the most widely used tools for analyzing hypersonic rarefied gas flows (Moss et al. 1994). The method has been applied to a range of problems, for example, contaminant pollution over space platforms (Rault and Woronowicz 1995). Recent advances have resulted in procedures more efficient for dealing with complex three-dimensional geometry as well as significant reductions in the computational effort. Such advances have made practical the applications of DSMC to the calculations of the rarefied gas flow over the full Space Shuttle geometry (Bird 1990).

For microfluid flows in MEMS, the operating pressure can be near or even higher than the atmospheric pressure. Using DSMC to calculate such flows of high density can be quite demanding in both the computer memory and the computational time. Methods to deal with large variations of density have been proposed and applied successfully to many problems. These schemes generally require a significant amount of computer time for near-continuum conditions due to, partly, the increase of the collision frequency at high gas density. As such, further reductions in computational time are still needed to make the calculations of high-density, large-scale microfluid flow problems more affordable.

Parallel implementations of DSMC for high-speed rarefied gas applications have been reported in the literature (Wilmoth and Carlson 1992; Dietrich and Boyd 1995; LeBeau 1999). The computer platforms used include, for example, IBM SP-2 and CRAY-T3E. In recent years, hardware for personal computer (PC) has dramatically improved. This, when coupled with efficient algorithms for parallel implementation, offers a tremendous potential for a significant reduction of computation time for the DSMC of microfluid flows in a cost-effective manner. In fact, massive parallelization can achieve and even surpass the performance of the current supercomputers for certain types of problems.

This chapter presents the development of high-performance computing cluster (HPCC) and the basic implementation of a parallel algorithm for DSMC. The parallel efficiency has been benchmarked with applications to microfluid flows. The implementation of the parallel algorithm does not require any major changes to the procedures of DSMC and can, in principle, be applied to the general codes. Such an implementation will allow a wider range of flow conditions to be studied and make possible a direct comparison of the DSMC results with the continuum Navier-Stokes or Burnett equations solutions for microfluid flow problems.

6.2 Cost-Effective Parallel Computing

Microfluidic flow simulations using either the Burnett equations or DSMC involve intensive numerical computing. A large-scale unsteady microflow simulation by DSMC in particular will need to use state-of-the-art supercomputers to fully resolve the physics. On the other hand, an efficient parallization of the DSMC procedure can also bring down the computer run time on parallel computer platforms. Parallel implementation of DSMC for high-speed rarefied gas applications has been reported in the literature (Boyd 1991; Wilmoth 1992; Wilmoth et al. 1992; Dietrich and Boyd 1994; Dietrich and Boyd 1995; LeBeau 1999). The computer platforms used include, for example, IBM SP-1, IBM SP-2, Cray C-90, and Cray T3E. These supercomputers are expensive, beyond the reach of all but a few of the most prestigious universities and laboratories. Fortunately, high-performance computing on HPCC such as the Beowulf system (Sterling et al. 1999) is accelerating the development of CFD and other computational sciences, and it is briefly described in the following paragraphs.

6.2.1 Parallel architecture

The computing architecture (Flynn 1966) can be classified into four types in terms of instruction and data streams: single instruction and single data (SISD), multiple instruction and single data (MISD), single instruction and multiple data (SIMD), and multiple instruction and multiple data (MIMD). For example, IBM PC and IBM RS/6000 workstations belong to SISD; MP-2 belongs to SIMD; CRAY T3E and Beowulf system belong to MIMD; CM-5 is a hybrid of the SIMD and the MIMD. MISD is seldom used. MIMD is the dominant architecture in current high-performance computing community because of its flexibility.

The MIMD-type machines can be further classified into shared memory parallel computer and distributed memory parallel computer in terms of the type of memory access mode (Bell 1994). The shared memory is accessible for multiple processors. Programming for shared memory computer resembles that for vector computers, which is easier than that for distributed memory computers. However, its scalability is limited because of the limitation of system bus and memory access. Typical shared memory parallel computers are the CRAY C90, SGI Power Challenger, and PCs or workstations based on symmetric multiprocessors (SMP). In distributed memory parallel computers memory is physically distributed among processors; each local memory is only accessible by its specific processor. Message passing among processors is required for accessing remote memory, which saves the data at the request of a local processor. Because of its unlimited scalability, it is widely implemented in massively parallel processing system (MPP), which is the dominant trend in current high-performance computing. On the other hand, this kind of parallel computer has two disadvantages. First, time latency for message passing is usually long because of a software layer required to access remote memory. Second, directing message passing is the responsibility placed on a programmer. The programmer must explicitly implement the schemes of data distribution and processor mapping, all intraprocess communication, and synchronization.

The parallel computer system can be classified into highly specialized vector parallel machines and MPP system, based on the arrangement of computing processors. The vector parallel computers use ECL (emitter-coupled logic) processors, which were specially built and offered at a very high price. The MPP system, where high-performance microprocessors are connected through high-speed network, is relatively less expensive.

6.2.2 Development of HPCC

The history of the various supercomputers in the last six decades shows that the performances of the fastest computers have been increased exponentially by two orders of magnitudes every decade on an average, as shown by the dash line in Fig. 6.2.1. Most of the data presented in Fig. 6.2.1 before 1994 are taken from the book *Parallel Computing Works* (Fox et al. 1998). This tremendous rate of growth was accurately predicted by Moore's law and is expected to continue in the future. But it does not hold for the growth rate of the current parallel computer systems.

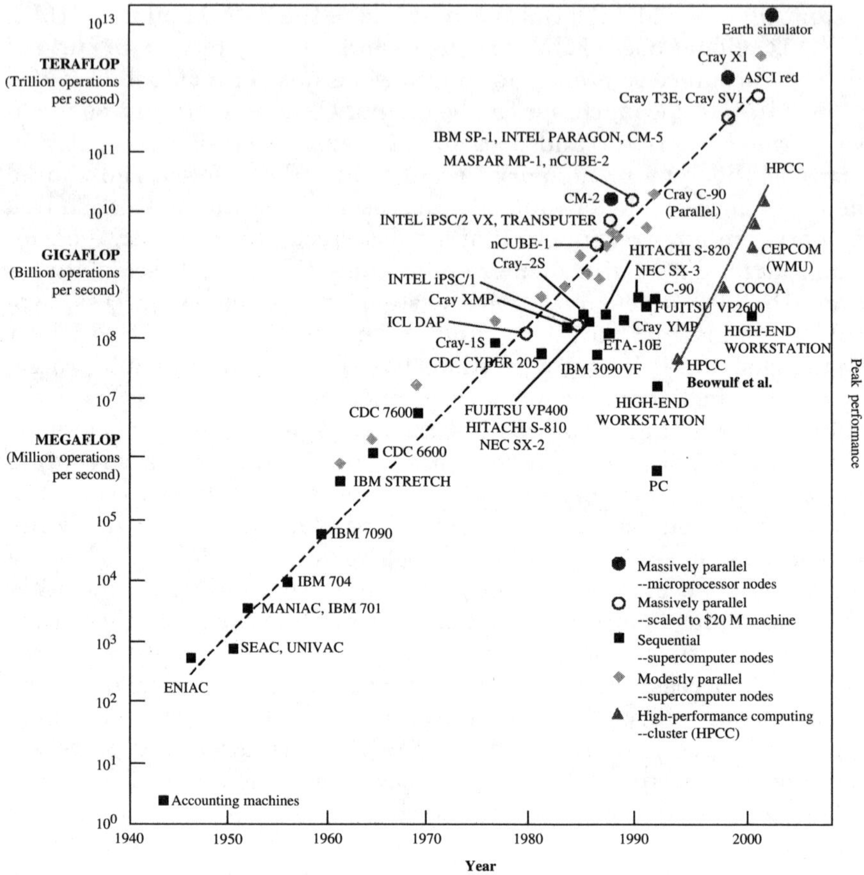

Figure 6.2.1 Historical trends of peak performance for sequential, parallel computers as well as high-performance computing cluster (HPCC).

Over the past 20 years, the performance of microcomputers has grown much faster relative to that of highly specialized vector parallel machines. Ten years ago, the best vector processor outperformed the best microprocessor by a factor of 20,000. Today the difference is a factor of less than 10. The vector parallel computers were very expensive, and their market was too small to drive rapid technological growth. In contrast, workstations and high-end PCs use microprocessors, which provide lower performance than ECL but at a much lower price. In the early 1990s, the microprocessor-based highly parallel computers were introduced into the high-performance market by Thinking Machines, Kendall Square, and Intel. These machines compensated for somewhat slower performance per processor by using more of the less expensive processors. Mainstream vendors like Cray, IBM, and Convex introduced their own such systems a few years later. Right now many of the high-performance parallel systems in the TOP500 list (2003), which ranks the performance of the most powerful supercomputer systems, use microprocessors. With the data taken from the Cray Inc. presentation (Cray Inc. 2003) and Intel website (Intel 2003), the comparison of peak performance for the fastest processors available in 2003 is shown in Fig. 6.2.2. The peak performance is counted in the 32-bit-floating-point operations per second (flop/s). It can be seen that the fastest microprocessor at the time, Intel Itanium2, can be only about three times slower than the fastest vector processor used in the Cray X1 and ranks fourth right after IBM P690 1.3. The second fastest processor NEC SX-6 used in the Earth Simulator, which is no.1 in the TOP500 list, is just twice faster than Itanium 2. In summary, the advantage of the most expensive vector processors has decreased to a factor of 3.

Also, the technology in fast Ethernet network has been rapidly developing with fast data transfer rate, low latency, and high bandwidth. The improvement in microprocessors and the price/performance gained in the fast Ethernet network technology make HPCC affordable in the academic setting. An HPCC is a type of parallel or distributed processing system, which consists of a collection of interconnected stand-alone computers working together as a single, integrated computing resource. A computer node can be a single or multiprocessor system, such as PCs

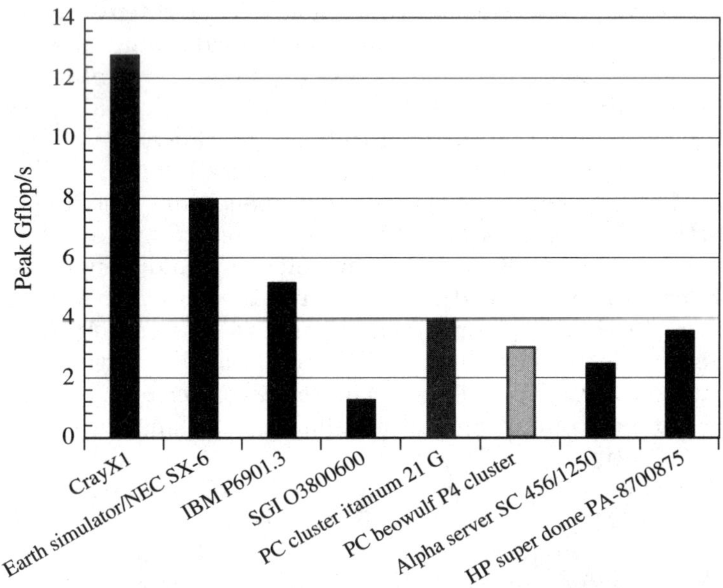

Figure 6.2.2 Peak performances of the fastest processors available in 2003.

or workstations with SMP. An HPCC generally refers to two or more computer nodes connected together. The nodes can exist in a single cabinet or be physically separated and connected via a local area network (LAN). An interconnected (LAN-based) cluster of computers appear as a single system to users and applications. Such a system can provide a cost-effective way to gain features and benefits (fast and reliable services) that have been found only on much more expensive, centered parallel machines such as nCube, CM5, Cray T3D, CRAY T3E.

6.2.3 Beowulf system

A Beowulf system is a type of HPCC, a massively parallel, distributed memory MIMD computer built primarily out of commodity hardware components such as PC, running a free-software operating system such as Linux or FreeBSD, interconnected by a high-speed stand-alone network. The primary advantage of such system is its high performance/price ratio in comparison with other dedicated MPP systems.

The topology of a typical Beowulf system is shown in Fig. 6.2.3. The network interface hardware acts as a communication processor and is responsible for transmitting and receiving packets of data between cluster nodes via a network/switch. Communication software offers a means of fast and reliable data communication among cluster nodes and to the outside world. Often, clusters with a special network/switch like Myrinet (1.28 Gbps) use communication protocols such as active messages for fast communication among its nodes. They potentially bypass the operating system and thus remove the critical communication overheads providing direct user-level access to the network interface. Through job-scheduling system, the cluster nodes can work collectively, as an integrated computing resource, or they can operate as individual computers. The cluster middleware is responsible for offering an illusion of a united system image (single system image) and availability out of a collection on independent but interconnected computers. Programming environments can offer portable, efficient, and easy-to-use tools for development of applications. They include message passing libraries, debuggers, and profilers, such as message passing interface (MPI), multiprocessing environment (MPE), and parallel virtual machine (PVM). Such a system could be used for the execution of sequential with high-throughput computing and parallel applications with high-performance computing.

The first Beowulf system came to birth at NASA Goddard Space Flight Center in 1994 (Sterling et al. 1999). It comprised 16, 66MHz Intel 80486 processors and cost less than $50K, which could compare to the comparable performance of Cray YMP with cost about 1 million dollars at that time, as shown in Fig. 6.2.1. After that, the cost efficiency

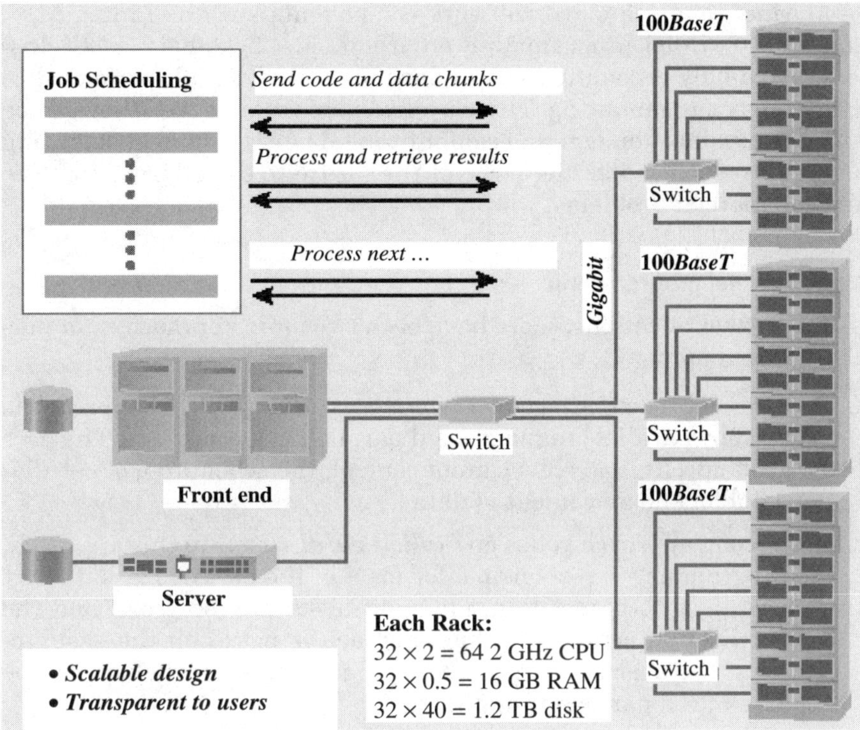

Figure 6.2.3 Topology of a typical Beowulf system.

quickly spread through NASA and into academic and research communities. In 1997 ASCI Red was assembled at Sandia National Laboratories as the first-stage Accelerated Strategic Computing Initiative (ASCI) plan to achieve a 100 TERAFLOP supercomputer system by 2004. ASCI Red comprises 9,216 PentiumPro processors and was born as the fastest system (1.8 Tflop/s) (Tomkins 2003). In 1997 and 1998, the systems and services to achieve 1 gigaflop cost $3K to $3.5K. The cost is decreasing dramatically. For example, the system shown in Fig. 6.2.3 has 194 Pentium Xeon 2GHz CPUs with 100BaseT network speed. The peak performance is about 384 gigaflop with total cost about $210K. That is, the cost to achieve 1 gigaflop has decreased to $550. Even if the efficiency is 40 percent considering the unbalance caused by nonprofessional setting of network, the actual price/performance is about $1.4K per gigaflop. And the price/performance will continue to drop in the future.

Because of the main attractiveness mentioned above, systems can be built using affordable, low-cost, commodity hardware, fast LAN such as Myrinet, and standard software components such as Linux, MPI, and PVM parallel programming environments. Scalable HPCC clusters are rapidly becoming the standard platforms of high-performance and large-scale computing. These systems are scalable, i.e., they can be tuned to available budget and computational needs. Cluster computing has been recognized as the wave of the future to solve large scientific and commercial problems.

6.2.4 Parallel programming

In practical applications, there have been two main approaches for parallel programming:

1. The first approach is based on *implicit parallelism*. This approach is followed by parallel languages and parallelizing compilers. The user does not specify, and thus cannot control, the scheduling of calculations and/or the placement of data.
2. The second approach relies on *explicit parallelism*. In this approach, the programmer is responsible for most of the parallelization effort such as task decomposition, mapping tasks to processors, and the communication structure. This approach is based on the assumption that the user is often the best judge of how parallelism can be exploited for a particular application.

Explicit parallelism is usually much more efficient than parallel languages or compilers that use implicit parallelism. Parallel languages, such as SISAL (Feo et al. 1990) and PCN (Foster and Tuecke 1991) have found little favor with programmers. Parallel languages and their functions are still very limited. Parallelizing compilers are still limited to applications that exhibit regular parallelism, such as computations in loops. Parallelizing compilers have been used for some applications on multiprocessors and vector processors with shared-memory, but are unproven for distributed-memory machines. The difficulties are caused by the nonuniform memory access (NUMA) in the latter systems. Therefore, explicit parallelism is currently the only way on cost effective HPCC.

In explicit parallelism, parallel computing paradigms fall into two broad classes: *functional decomposition* and *domain decomposition*, as shown in Fig. 6.2.4. These classes reflect how a problem is allocated to process. Functional decomposition divides a problem into several distinct tasks that may be executed in parallel; one field where this is popular today is that of Multidisciplinary Design Optimization. For CFD application, there are always many global variables inside each decomposed function. They will occupy a log of memory space. It is rarely used in CFD. On the other hand, domain decomposition distributes data

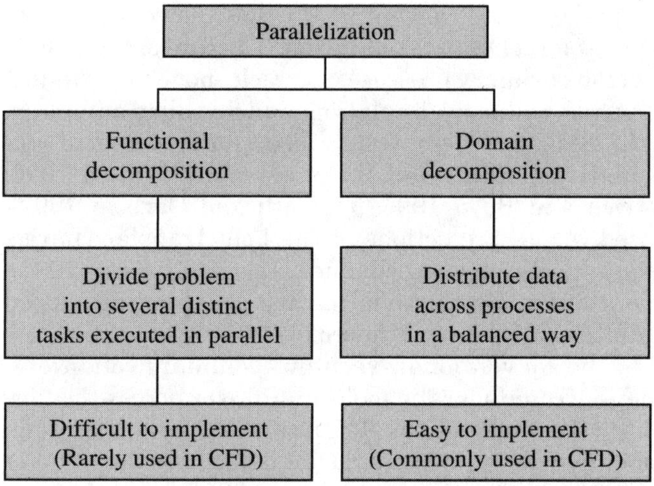

Figure 6.2.4 Parallel paradigms.

across processes, with each process performing more or less the same operations on the data. In the application of CFD, domain decomposition splits a large problem into small problems and is warmly welcomed by coding programmers.

On the system of HPCC, the parallel programming paradigms can be classified into the well known ones: *single-program multiple data* (SPMD), *task-farming* (or *master/slave*), *data pipelining*, *divide and conquer*, and *speculative parallelism*. SPMD is the most commonly used paradigm. Each process executes basically the same piece of code but on a different part of the data. This involves the splitting of application data among the available processors. SPMD applications can be very efficient if the data is well distributed by the processes and the system is homogeneous. If the processes present different workloads or capabilities, the paradigm requires the support of some load-balancing scheme able to adapt the data distribution layout during run-time execution. Therefore, there is still a lot of research work to do in the area of cost-effective high-performance parallel computing.

6.3 Parallel Implementation of DSMC

The conventional DSMC simulations describe the time-dependent evolution of molecules. The solution for a steady-state case is considered as an asymptotic limit of a corresponding unsteady flow. As the simulated molecules move, they collide with the other simulated molecules and interact with the physical boundaries. The locations and the velocities of these simulated molecules are determined and stored in sequences of time. Figure 5.1.1 represents the conventional DSMC flowchart, showing the main procedures for a DSMC application. The core of the DSMC algorithm consists of four primary processes: particle movement, indexing and cross-referencing, collision simulation, and flow field sampling.

The conventional DSMC has been widely used for hypersonic rarefied gas flow calculations (Rault and Woronowicz 1995; Bird 1990; LeBeau 1999; Dietrich and Boyd 1995; Wilmoth and Carlson 1992). DSMC has been used in the predictions of the heat transfer characteristics of supersonic flows in microchannel (Liou and Fang 2001). An implicit treatment of the flow boundaries has also been developed for the DSMC simulations of subsonic microfluid flow (Liou and Fang 2000). In contrast to the conventional "vacuum" boundary conditions, the implicit treatment accounts for the molecular fluxes across the flow boundaries due to local thermal motions. The local mean velocities, temperature, and number density at the boundaries determine the number of entering molecules, their velocities, and internal energies in the computational time interval. The implicit treatment has been successfully applied to the simulations of microCouette flows and microPoiseuille flows. The detailed implicit boundary treatment will be described in Chap. 9. A DSMC code with the implicit boundary treatment is parallelized with the library of message-passing interface (MPI) (Gropp et al. 1994) and is described in the following sessions.

The present parallelization of the DSMC method is based on the SPMD model. The physical domain is decomposed and each of its subdomains, along with its associate cells and molecules, is allocated to an individual processor (see Fig. 6.3.1). A complete DSMC code is loaded on to all processors and the simulation in subdomain proceeds independent of the other subdomains at each time step. Information of

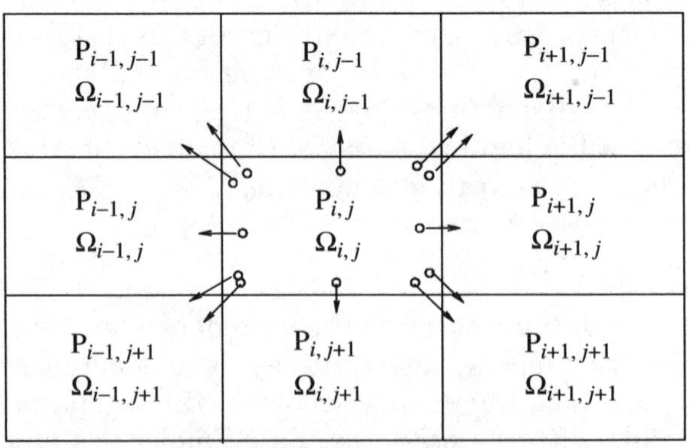

Figure 6.3.1 Sketch of domain decomposition.

the molecules that will cross the decomposition boundaries during the movement procedure of the DSMC algorithm is sent to the appropriate processor using MPI. Thus, the simulations are synchronized between the movement and collision routines at each time step. Since each processor contains a complete copy of the base simulation code, any addition to or modification of the base simulation algorithm to include, for example, a more complex geometry or collision physics can be readily implemented. In the following sessions, the data distribution (domain decomposition) and data communication model through MPI are described.

6.3.1 Data distribution

The physical region of interest is decomposed into a set of subdomains as shown in Fig. 6.3.1. The subdomains are mapped onto different processors. For example, Fig. 6.3.1 shows the subdomain $\Omega_{i,j}$ and all its neighboring subdomains in a 2-D space. The interface, which separates $\Omega_{i,j}$ from its neighbors, represents flow interface boundary. Molecules may move through the interfaces between $\Omega_{i,j}$ and the neighboring subdomain during a time step. It is important to parallel DSMC (PDSMC) that all the molecules that exit $\Omega_{i,j}$ through the interface to one of its neighboring subdomains should be registered and appropriately accounted for in their new "home" and be deleted from the original subdomain. In other words, there should be no loss/addition of molecules due to the decomposition of the flow domain.

To this end, sending buffers are used in the present work. Each sending buffer is assigned to one of the neighboring subdomains to store the information of the molecules that will travel to that particular subdomain. For example, in the case of a 2-D flow as shown in Fig. 6.3.1, the subdomain $\Omega_{i,j}$ has eight sending buffers. The outgoing molecules have, at most, three possible subdomains to go to through any flow interface boundary. At each time step, a molecule that is going to exit the current subdomain after a full time step is dealt with by a "remove" procedure. The remove procedure determines the new home of the exiting molecule based on its location after the current time step. Since the future subdomain where this molecule will reside has been found, the current information of the position, velocity, internal energy, rotational energy, gas species, and the remaining portion of the time step associated with this molecule should be saved in the corresponding sending buffer. The molecule is then taken out of its current subdomain.

There are also molecules entering this subdomain $\Omega_{i,j}$ from its neighbors. This is not shown in Fig. 6.3.1 for clarity. The information for the entering molecules is saved in respective receiving buffers after the data communication between the related processors has completed.

6.3.2 Data communication

After the DSMC procedure of movement for all simulated molecules and before the DSMC procedure of molecular collision during each time step, all the processors must make necessary data-exchange of the information stored in the sending buffers described in data distribution. The communication between the processors proceeds with information on the number of molecules in transfer and their coordinates, velocity, rotational energy, and time step remaining. For data communication, there is an additional startup time needed for the processors to be ready to transfer each data package. The startup time for data communication for PC clusters is much lengthier than that for other high-performance parallel architecture machines. Therefore, the number of data packages to transfer should not be large in order to improve the parallel efficiency of the parallel computation on distributed PC cluster. In the present work, a new data type *mesg_mpi_t* is derived and all the data associated with the molecule crossing the flow interface boundary is transferred as a whole using this newly defined data type.

Inside this new data type, all the values are continuously distributed logically on a segment of RAM for each processor, so that they can be transferred as a whole using MPI. Therefore, this new data type is used as a template to send and receive data for the transferred molecules. Before the sending of the molecules, all the information is encoded into this template. And after the receiving, all the information is decoded to a proper receiving buffer from the template.

While exchanging information between the processors, the processors must avoid sending data to or receiving data from each other at the same time. Otherwise, it could cause a dead lock among the running processors. It is also important to be sure that the targeted processor receives the data that have been sent by another processor. Asynchronous communication is one way to avoid this problem. But there is a risk of package loss. In this work, a synchronous communication model has been used for the *myid*th processor, which calculates the $(myid + 1)$th subdomain as shown in Fig. 6.3.2. Before communication, **MPI_Barrier** is employed to make sure that all the involved processors be ready to exchange data packages. In the process of communication, the relative value of processor identification is used to determine the order of data receiving (**MPI_Recv**) and data sending (**MPI_Ssend**). After communication, the sending buffer should be reset to be ready for the next time step.

```
MPI_Barrier make sure that all processors are ready to exchange data
To find the total number ii of its neighboring subdomains
do ith = 1, ii
    ip = the processor number of its ith neighboring subdomains;
    if (myid < ip) then
        copy data of the ith sending buffer to template of mesg_mpi_t;
        MPI_Ssend send mesg_mpi_t to processor ip;
    else
        MPI_Recv receive mesg_mpi_t from processor ip;
        copy the data of mesg_mpi_t to its receiving buffer;
    endif
enddo
do ith = 1, ii
    ip = the processor number of its ith neighboring subdomains;
    if (myid > ip) then
        copy data of the ith sending buffer to template of mesg_mpi_t;
        MPI_Ssend send mesg_mpi_t to processor ip;
    else
        MPI_Recv receive mesg_mpi_t from processor ip;
        copy the data of mesg_mpi_t to its receiving buffer;
    endif
enddo
do ith = 1, ii
    clear the ith sending buffer
enddo
```

Figure 6.3.2 MPI synchronous communication model.

The communication is followed by an additional procedure of "cross-move," where the new comers stored in the receiving buffers are allowed to finish their remaining time interval to reach new special locations. The newly arrived molecules can then be indexed into field cells. After this cross movement, the procedures of molecular collision and flow sampling proceed in the same manner as those in a conventional DSMC.

Figure 6.3.3 shows the flowchart of the PDSMC procedure (Fang and Liou, 2001). The additional steps, in comparison with the conventional DSMC given in Fig. 5.1.1, include decomposing geometry and mapping subdomains to processors, deriving data type for communication, moving outgoing molecules to sending data buffer, data communication, crossing moving, and subsonic downstream implicit boundary condition, as shown inside the dash box. These steps are dedicated to PDSMC simulation of microflows.

6.3.3 A parallel machine

All the parallel computations referred to in this book have been performed on the cost efficient parallel computing mast (CEPCOM) at the

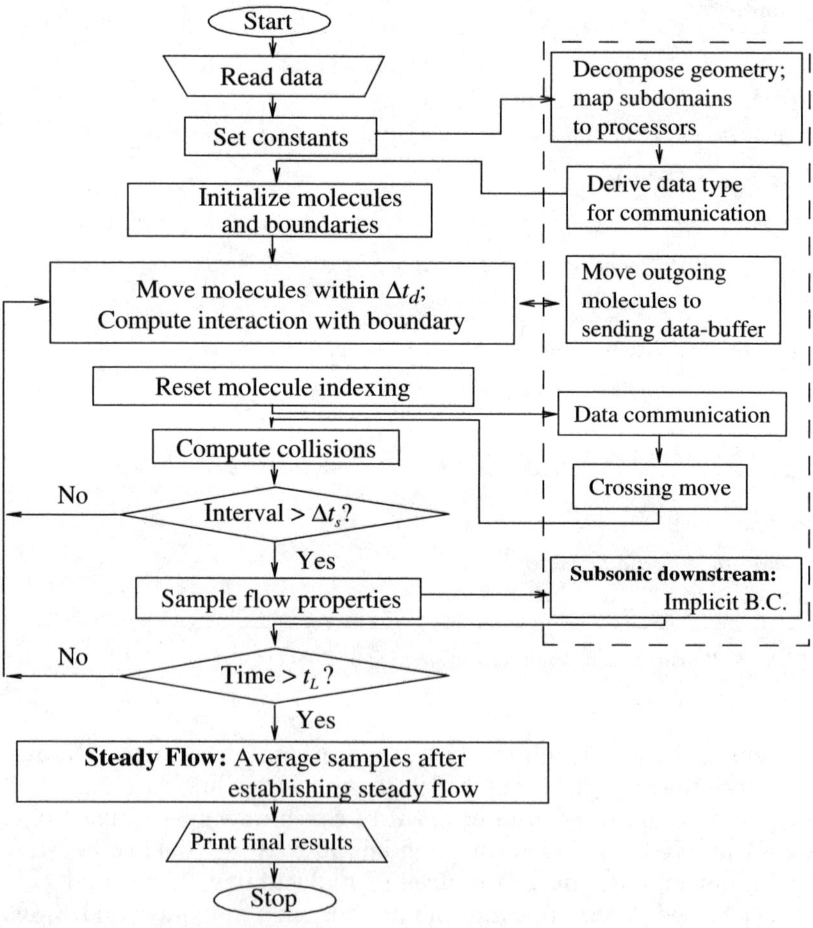

Figure 6.3.3 Parallel DSMC flowchart.

CFD laboratory in Western Michigan University. CEPCOM is a 23-node Beowulf-class cluster assembled in September 2001. Each node has two 1-GHz Pentium III processors and 512-MB RAM. The machines are connected via the fast Ethernet network, which can support up to 100-Mbps bandwidth for each node. An HP ProCurve Switch 4000 M with a backplane speed of 3.8 Gbps is used for the networking. All the processors are dedicated to run parallel jobs or high throughput jobs. The operating system is the Red Hat Linux 7.0. MPI library, *mpich-1.2.2*, is used for parallel programming in C/C++/Fortran77/Fortran90. The overall peak performance of CEPCOM is 46 Gflop/s, with total of 11.5-GB RAM and 302-GB hard drive space.

6.4 Parallel Performance of PDSMC

The performance and efficiency of the PDSMC method are examined using the microfluid flows in microchannels, which include a straight microchannel and a microchannel with a patterned surface structure.

A microPoiseuille flow has been calculated to benchmark the performance of the PDSMC code. The size of the microchannel is 5×1 μm, with the computational grid of 400×80 uniform rectangular cells. The total number of the simulated molecules is about 1.6 million. The pressure ratio is 4.0. The microchannel is divided into multiple numbers of subdomains of equal size and each subdomain is assigned to a processor. The parallel calculation achieves a satisfactory load balancing and the total computing time decreases as the number of processors increases. Figure 6.4.1 shows a sketch of the domain decomposition for 32 processors. Figure 6.4.2 shows the variation of the total computing time for 400 iterations for a range between 1 and 32 processors. For such a problem, runs on a single Pentium III processor take nearly 2,560 s, and for a 32-processor run, it is about 166 s. The speedup enhancement with the increasing number of processors is given in Fig. 6.4.3. The speedup is defined as the ratio of the total computing time on one processor to that on multiple processors. It deviates from the ideal linear variation at eight processors, beyond which the communication time and synchronization overhead increase relatively to the calculating time on each processor

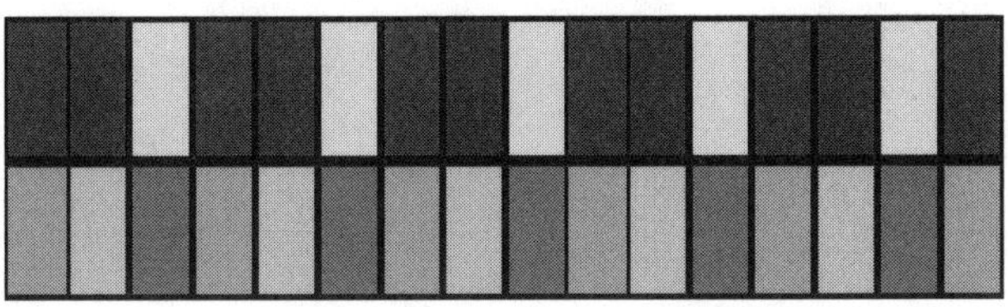

Figure 6.4.1 Domain decomposition for a 2-D microchannel flow.

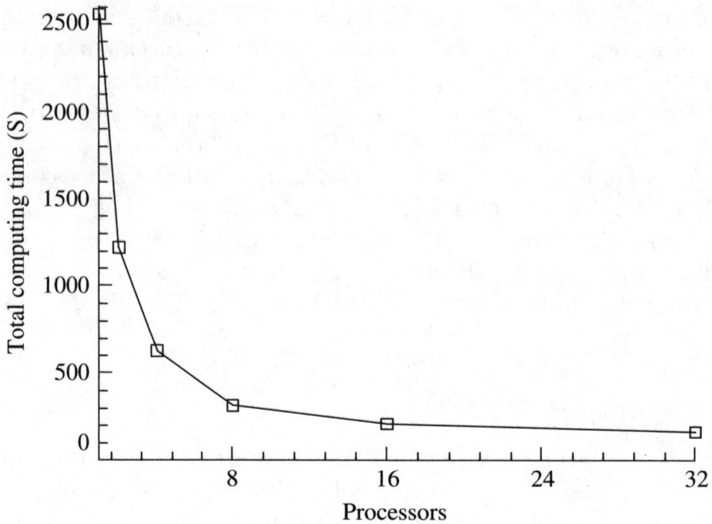

Figure 6.4.2 Total computing time for 2-D microchannel flow.

for this problem of fixed size. The speedup can be obtained up to 15.5 on 32 processors. Figure 6.4.4 shows the corresponding efficiency, which is defined as the ratio of the speedup to the corresponding number of processors used to run the problem. A nonlinear speedup is achieved

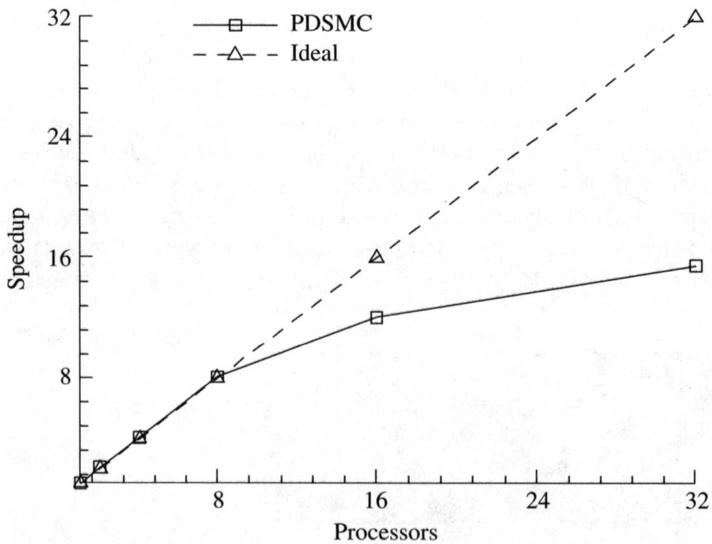

Figure 6.4.3 Speedup for a 2-D microchannel flow.

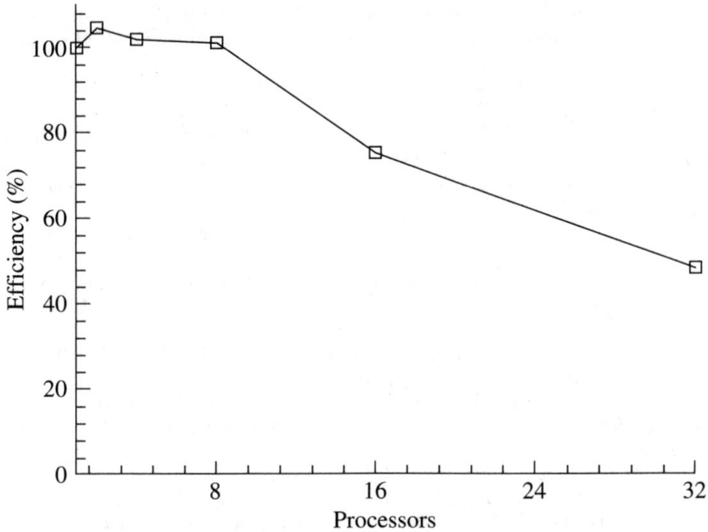

Figure 6.4.4 Efficiency for the 2-D microchannel flow.

within eight processors. Especially on two processors, the efficiency of parallel computing is about 105 percent. Running on 32 processors, the efficiency can be kept within 48 percent.

From this simple benchmark, it can be seen that the parallel processing can clearly provide significant reductions in total computing times for DSMC calculations. And DSMC can be parallelized very efficiently to run on PC clusters, such as CEPCOM. The PDSMC algorithm presented here can be implemented in the attached DSMC/IP1D computer program.

References

Bell, G., Scalable parallel computer: Alternatives, issues, and challenges, *Int. J. Parallel Programming*, Vol. 22, pp. 3–46, 1994.

Bird, G.A., Application of the direct simulation Monte Carlo method to the full shuttle geometry, AIAA paper 90-1692, 1990.

Boyd, I.D., Vectorization of a Monte Carlo scheme for nonequilibrium gas dynamics, *J. Comp. Phys.*, Vol. 96, pp. 411–427, 1991.

Cray Inc., En route to petaflop computing speed: Introducing the Cray X1TM supercomputer, Presentation at *http://www.cary.com/products/systems/*, 2003.

Dietrich, S. and Boyd, I.D., A scalar optimized parallel implementation of the DSMC method, AIAA paper 94-0355, Reno, NV, 1994.

Dietrich, S. and Boyd, I.D., Parallel implementation on the IBM SP-2 of the direct simulation Monte Carlo method, AIAA paper 95-2029, 1995.

Fang, Y. and Liou, W.W., Predictions of MEMS flow and heat transfer using DSMC with implicit boundary conditions, AIAA paper 2001-3074, 2001.

Feo, J., Cann, D., and Oldehoeft, R., A report on the SISAL language project, *J. Parallel Distrib. Comput.*, Vol. 10, pp. 349–366, 1990.

Flynn, M., Very high speed computing systems, *Proc. IEEE*, Vol. 54, pp. 1901–1909, 1996.
Foster, I. and Tuecke. S., Parallel programming with PCN, Technical report ANL-91/32, Argonne National Laboratory, Argonne, France, 1991.
Fox, G.C., Williams R.D., and Messina, P.C., *Parallel Computing Works*, Morgan Kaufmann, San Francisco, CA, pp. 85–101, 1998.
Gropp, W., Lusk, E., and Skjellum, A., *Using MPI-Portable Parallel Programming with the Message-Passing Interface*, The MIT Press, Cambridge, MA, 1994.
Intel, Intel® Itanium® 2 processor features, *http://www.intel.com/products/server/processors*, 2003.
LeBeau, G.J., A parallel implementation of the direct simulation Monte Carlo method, *Comput. Methods. Appl. Mech. Engrg.*, Vol. 174, pp. 319–337, 1999.
Liou, W.W. and Fang, Y., Implicit boundary conditions for direct simulation Monte Carlo method in MEMS flow predictions, *Comput. Model. Eng. Sci.*, Vol. 1, No. 4, pp. 119–128, 2000.
Liou, W.W. and Fang, Y., Heat transfer in microchannel devices using DSMC, *J. Microelectromech. Syst.*, Vol. 10, No. 2, pp. 274–279, 2001.
Moss, J.N., Mitcheltree, R.A., Dogra, V.K., and Wilmoth, R.G., Direct simulation Monte Carlo and Navier-Stokes simulations of blunt body wake flow, *AIAA J.*, Vol. 32, pp. 1399–1406, 1994.
Rault, A.G., and Woronowicz, M.S., Application of direct simulation Monte Carlo satelite contamination studies, *J. Spacecr. Rockets*, Vol. 32, pp. 392–397, 1995.
Sterling, T.L., Salmon, J., Becker, D.J., and Savarese, D.F., *How to Build a Beowulf: A Guide to the Implementation and Application of PC Clusters*, MIT Press, Cambridge, MA, 1999.
Tomkins, J.L., The ASCI red TFLOPS supercomputer, SAND96-2659C, *http://www.llnl.gov/asci/sc96fliers/snl/ASCIred.*, 2003.
TOP500, *http://www.top500.org/list/2003/11*
Wilmoth, R.G., DSMC application of a parallel direct simulation Monte Carlo method to hypersonic rarefied flows, *AIAA J.*, Vol. 30, pp. 2447–2452, 1992.
Wilmoth, R.G. and Carlson, A.B., DSMC analysis in a heterogeneous parallel computing environment, AIAA paper 92-2861, 1992.
Wilmoth, R.G., Carlson, A.B., and Bird. G.A., DSMC analysis in a heterogeneous parallel computing environment, AIAA paper 92-286, 1992.

Chapter

7

Gas–Surface Interface Mechanisms

7.1 Introduction

Gas–surface interactions involve many, such as physical and chemical, mechanisms that occur near the interface. The result of these interactions, however, can influence the operation characteristics of the microdevices. In the near surface region, behavior of the individual molecules cannot be ignored and the gas cannot be regarded as in equilibrium. For instance, the momentum and the energy transfer between the gas molecules and the surface need to be considered. Such regime or sublayer, which is estimated to be about a few mean free paths thick, is called the Knudsen layer. In the Knudsen layer, the gas behavior should be considered from gas kinetic theory. This will bring in the Boltzmann equation and the Liouville equations, which have been discussed earlier. In the cases where the local Knudsen number is small and the effect of the Knudsen layer is not significant, the continuum solution for bulk flow can be extrapolated satisfactorily to the surface, implying that the interactions of gas molecules and surface are the same as those between the gas molecules and a gas/surface in equilibrium. This then results in the average velocity and temperature of gas on the wall being continuous and equal to the velocity and the temperature of the wall. At high Knudsen number, the collision frequency can be low and the distribution function of the gas near the surface will not be an equilibrium distribution. The gas–surface interactions then need to be considered microscopically. The boundary conditions for the distribution function of a gas in contact with solid surfaces are also required in the solution of the Boltzmann equation, which contains spatial derivatives of the distribution function.

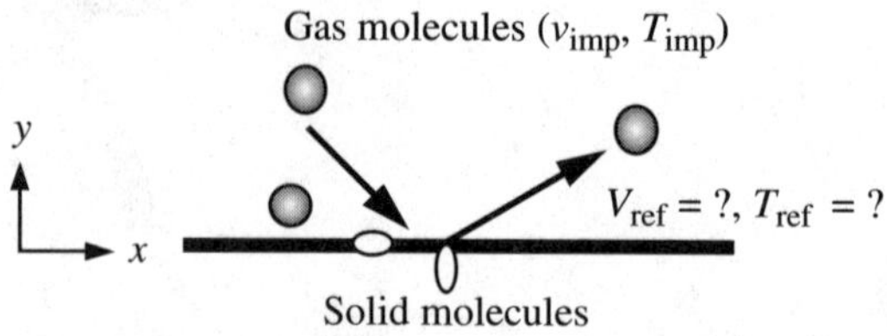

Figure 7.1.1 Simplified view of gas–surface interactions.

Interactions between the gas molecules and the solid molecules are complicated from a microscopic point of view. Figure 7.1.1 depicts such a view where $()_{imp}$ and $()_{ref}$ represent the impinging and the reflected molecular properties, respectively.

On the surface, the velocity distribution would be the sum of the distribution $f^{(-)}$ associated with the incident molecules with $c_y < 0$ and $f^{(+)}$ for the reflected molecules with $c_y > 0$. Or

$$f = f^{(+)} + f^{(-)} \tag{7.1.1}$$

If gas molecules strike and are bounced off the surface with no accumulation on the surface, then the number flux (per unit area per unit time) of the impinging molecules should equal to that of the reflected molecules. That is,

$$\int_{-\infty}^{\infty}\int_{-\infty}^{0}\int_{-\infty}^{\infty} c_y f^{(-)} dc_x\, dc_y\, dc_z = \int_{\infty}^{\infty}\int_{0}^{\infty}\int_{\infty}^{\infty} c_y f^{(+)} dc_x\, dc_y\, dc_z \tag{7.1.2}$$

The probability **P** of a surface reflection for a molecule with velocity c to leave with velocity c^* can be defined as

$$f^{(+)} = \int_{\infty}^{\infty}\int_{-\infty}^{0}\int_{\infty}^{\infty} \mathbf{P}\,(c \rightarrow c^*) f^{(-)} dc \tag{7.1.3}$$

If the reflection probability **P**, or scattering kernel is known, the above equation can be used to obtain an integral equation that relate $f^{(-)}$ to $f^{(+)}$. Except for simple cases, it is not easy to identify the reflection probability density function **P**. The incident molecules can be trapped by the weak van der Waals attraction and stay near the wall (physisorbed), which might cause a perturbation of the electronic structures of the substrate. Some of the adsorbates may move parallel to the surface and some might eventually be desorbed and be reemitted back into the gas.

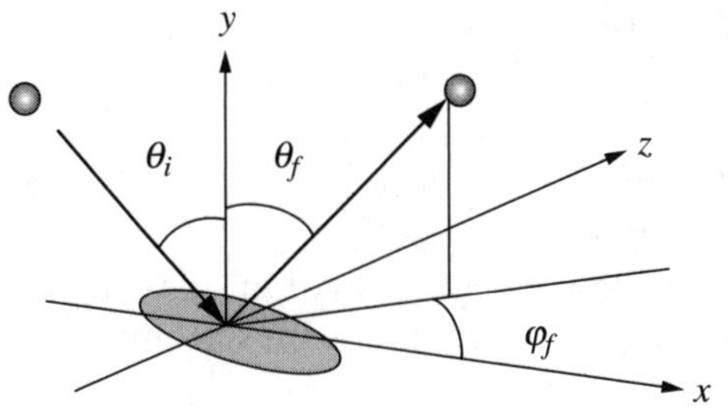

Figure 7.1.2 Three-dimensional scattering.

The trapping–desorbing process (Hurst et al. 1979) may result in out-of-plane scattering as shown in Fig. 7.1.2 where θ_i denotes the incident angle measured from the wall-normal direction, θ_f the reflected angle, and φ_f the out-of-plane scattering angle measured from the incident direction.

A dynamic equilibrium may be reached when the rate at which the number of molecules being adsorbed is equal to the rate of molecules being desorbed. The adsorbates may also be bounded by the chemical bounds (chemisorption) similar to those encountered in molecules that are stronger than the intermolecular van der Walls force. The lateral interactions between the species control the thermodynamic phase diagram of the combined adsorbates and substrate, and essentially form a new material. Concerns of such mechanisms would occur when considering thin film deposition and etching in the manufacturing of microdevices. It could also be a concern in developing micro gas sensors.

It has been proposed that the general requirement for equilibrium between a gas and surface at the molecular level is that the interaction satisfies the principles of reciprocity condition and normalization condition. The reciprocity principle is a relationship between the probability of a gas–surface interaction with a particular set of incident and reflected velocities and the probability of the inverse interaction. It may be written as (Cercignani 1969; Wenaas 1971; Kuščcer 1971)

$$c\mathbf{P}\,(c \rightarrow c^*)\exp[-E/(kT_w)] = -c^*\mathbf{P}\,(-c^* \rightarrow -c)\exp[-E^*/(kT_w)] \tag{7.1.4}$$

where E and E^* represent the energy of incident and reflected molecule, respectively. This condition is related to the detailed balance in energy. The normalization principle assumes that the interaction surface reflects all the incident molecules without adsorption. Then the scattering kernel must be a normalized probability function as

$$\int_{\infty}^{\infty}\int_{0}^{\infty}\int_{\infty}^{\infty} \mathbf{P}\,(c \rightarrow c^*)dc^* = 1 \qquad (7.1.5)$$

Macroscopically, the gas–surface interactions can be evaluated by using averaged parameters such as the tangential momentum accommodation coefficient σ_v.

$$\sigma_v = \frac{\tau_i - \tau_r}{\tau_i - \tau_w} \qquad (7.1.6)$$

where τ_i = tangential momentum of incoming molecules
τ_r = tangential momentum of reflected molecules
τ_w = tangential momentum of the wall (=0 for stationary surfaces)

The parameter represents a measure of the equilibrium of momentum of the reflected molecules with that of the wall. Thermal accommodation coefficient can be defined as

$$\sigma_T = \frac{dE_i - dE_r}{dE_i - dE_w} \qquad (7.1.7)$$

where dE_i = energy flux of incoming molecules
dE_r = energy flux of reflected molecules
dE_w = energy flux if all incoming molecules had been reemitted with energy flux corresponding to the wall temperature

The values of the coefficients could depend on the gas and surface properties, such as temperature and roughness, the local pressure, and even the local mean flow. It typically takes a molecule a few surface collisions to take on the average σ_v. More collisions are needed to obtain an energy level of the surface. Other parameters, such as chemical energy accommodation coefficient and recombination coefficient, were also used in characterizing macroscopically gas–surface interactions.

The interactions between gas particles and a solid surface are complex. It is unlikely that a general mathematical model can be rigorously developed that will be adequate for a quantitative description of the gas–surface interactions of different combinations of gases and surfaces at all conditions. Nevertheless, with the advances of *ab initio* simulations such as direct simulation Monte Carlo (DSMC) and molecular dynamics (MD) and new experimental methods, better understanding of the interaction phenomenon will be brought to light. This new knowledge can help develop empirical or phenomenological models that are more widely applicable.

7.2 Phenomenological Modeling

7.2.1 Specular and diffusive reflection models of Maxwell

In searching for boundary conditions for the distribution function of a gas that is in contact with solid surfaces, Maxwell (1879) developed two hypotheses. In the first, the surface was assumed to be elastic and smooth. The tangential molecular velocity component is then unchanged and the normal molecular velocity component is reversed as a result of the collision between the gas molecules and the surface. The incident angle θ_i, the angle between the wall normal and the incident molecular velocity is the same as the deflection angle θ_f, see Fig. 7.2.1.

This specular reflection model then results in no stress in the tangential direction and $\sigma_v = 0$, which is not generally observed. The reflection probability density function can then be written in the form of the delta function. That is,

$$\mathbf{P} = \delta(c_x - c_x^*)\delta(-c_y - c_y^*)\delta(c_z - c_z^*) \tag{7.2.1}$$

The second model, the diffuse reflection model (Fig. 7.2.2), hypothesizes that the gas molecules interact with lattices of surface molecules multiple times and, as they are desorbed and reemitted into the gas, they become in equilibrium with the solid surface, or are fully accommodated. Their velocity distributions assume the half-range Maxwellian distribution at the wall temperature.

For the diffuse model, $\tau_r = \tau_w$ and $\sigma_v = 1$. Experimental results show that for practical engineering surfaces at moderate temperature and gas velocities, the diffuse reflection model is a reasonably good approximation. For interactions with very clean surfaces or under vacuum condition, diffusive model becomes invalid. It can also be found (Koga 1970; Cercignani 1969) that

$$\mathbf{P} = c_y^* \left(\frac{2\pi m}{kT_w}\right)^{1/2} \left(\frac{m}{2\pi k T_w}\right)^{3/2} \exp\left[-\frac{m}{2kT_w}c^2\right] \tag{7.2.2}$$

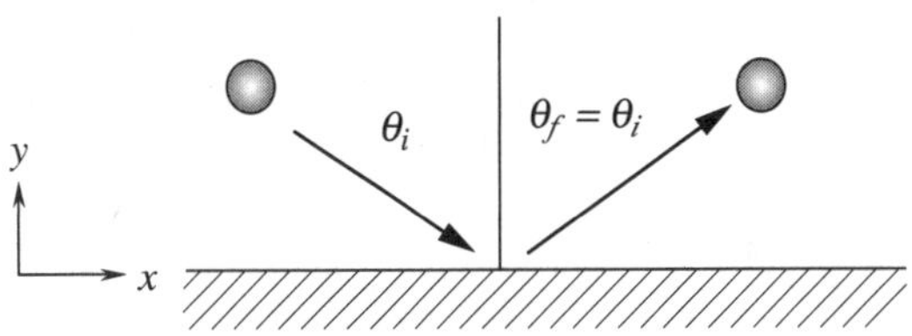

Figure 7.2.1 Specular reflection.

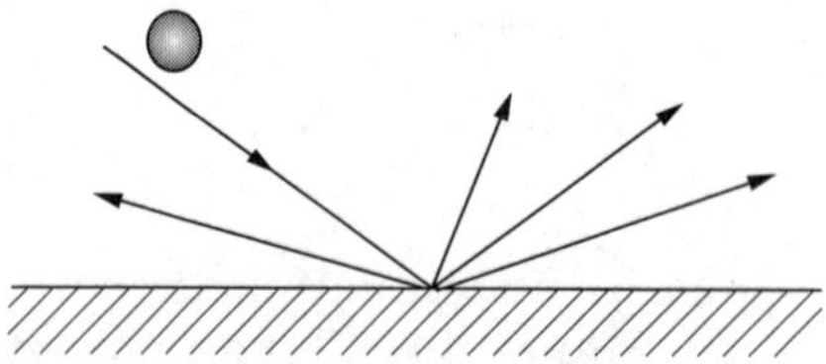

Figure 7.2.2 Diffuse reflection.

A model that describes the scattering as consisting of partly specular and partly diffuse was also considered by Maxwell. A portion of the surface is assumed to reflect incident molecules specularly and the remaining portion would trap and later desorb the gas molecules at the surface temperature. With σ_v representing the fraction of the molecules that are reflected diffusely and $(1-\sigma_v)$ the fraction that reflected specularly, the probability density function is the linear combination of that of the two models given above. That is,

$$\mathbf{P} = (1-\sigma_v)\delta(c_x - c_x^*)\delta(-c_y - c_y^*)\delta(c_z - c_z^*) + \sigma_v c_y^* \left(\frac{2\pi m}{kT_w}\right)^{1/2} \left(\frac{m}{2\pi kT_w}\right)^{3/2} \exp\left[-\frac{m}{2kT_w}c^2\right] \tag{7.2.3}$$

7.2.2 Cercignani, Lampis, and Lord model

The Maxwell models of compete and fractionally diffuse surfaces have been used widely. The models, however, do not produce the lobular distribution in the direction of the remitted molecules observed in experiments. Cercignani and Lampis (1971) proposed a phenomenological model that reproduced the lobular scattering distribution. The scattering of the normal and the tangential components are considered mutually independent in the model. Therefore, the scattering kernel consists of three parts, one for each of the velocity components. For the component tangential to the wall c_x, the term can be written as

$$\mathbf{P}\,(c_x \to c_x^*) = [\pi\sigma_v(2-\sigma_v)]^{-1/2} \exp\left[-\frac{\{c_x^* - (1-\sigma_v)c_x\}^2}{\sigma_v(2-\sigma_v)}\right] \tag{7.2.4}$$

For isotropic surfaces, the expression is the same for c_z. For the normal velocity component, the probability can be written as

$$\mathbf{P}\,(c_y \to c_y^*) = \frac{2c_y^*}{\alpha_n} I_0 \frac{2(1-\alpha_n)^{1/2} c_y c_y^*}{\alpha_n} \exp\left[-\frac{c_y^{*2} + (1-\alpha_n)c_y^2}{\alpha_n}\right] \tag{7.2.5}$$

Where α_n represents the accommodation coefficient for the kinetic energy of the normal component and I_0 the modified Bessel function of the first kind. Note that the molecular velocity components in Eqs. (7.2.4) and (7.2.5) are normalized by the most probable molecular speed at the surface temperature. The model satisfies the reciprocity condition. The model provides a continuous spectrum of behavior between the specular reflection on one end and the diffuse reflection on the other. The model contains two adjustable parameters, the normal energy accommodation coefficient α_n and the tangential momentum accommodation coefficient σ_ν. Several improvements of the Cercignani and Lampis model were reported by Lord (1991; 1995). In Lord (1991), the model was modified for diffuse scattering with partial energy accommodation and to include accommodation of vibrational energy of a diatomic molecule. The case with partially diffuse scattering was included by Lord (1995).

The Cercignani-Lampis-Lord (CLL) models are phenomenological and contain two adjustable parameters. These empirical constants may be determined by experimental data. A simple classical model for the scattering of gas atoms from a solid surface is the "hard-cube" model proposed by Logan and Stickney (1966). The model assumes that gas molecules are spherical, rigid, and elastic. The surface atoms are modeled as uniform cubes with one face parallel to the surface plane, which moves only in the direction normal to the surface with a one-dimensional Maxwellian distribution function. As a result, the tangential components of the gas molecules do not change during the interaction and each gas particle interacts with only one of these cubes. The change of the velocity component normal to the surface occurs according to the laws for the collision of rigid elastic bodies. The thermal motion of a surface atom then affects the scattering distribution. The hard-cube model contains no adjustable constant. The model can provide qualitative explanation of some experimental phenomena. These include the lobular scattering pattern and the observation that, with increasing surface temperature, the deviation from specular interaction also increases. However, the model cannot produce out-of-plane scattering due to the assumption of the conservation of tangential momentum.

With the advance of modern computers, first principle-based simulation methods, such as the molecular dynamics (MD) method and the DSMC methods can be effectively used to guide the development of models and the calibration of model constants. MD can offer detailed description of the collision process as affected by potential energy surface, lattice structure, and bond length. The simulations results obtained from these *ab initio* methods can also be used along side experimentals to provide further evaluation and assessment of these models.

A typical example of this approach is the development of the multistage gas–surface interaction model (Yamanishi and Matsumo 1999). Based on the MD results of surface interactions for three different gases, i.e., O_2, N_2, and Ar with clean perfect graphite surfaces, and comparisons with molecular beam experimental results, the gas–surface interaction was assumed to be made up of three stages. In the first stage, the translational and rotational energies are determined by model equations. The out-of-plane scattering direction is determined by the potential energy surface in the section stage. At stage 3, the molecules can either scatter, reenter the gas stream, or be trapped to the surface, depending on the translational energy. The model parameters are determined by the MD simulations. The multistage method produced quantitative agreement with molecular beam experiments for in-plane and out-of-plane scattering distributions. The model contains a number of parameters, and their values are not the same for different gases.

Ab initial simulations of the interaction between a platinum (Pt) solid surface with monatomic (Yamamoto 2002) and diatomic (Takeuchi et al. 2004) gases in a Couette flow have also been performed. In these studies, the gas–surface interactions are simulated by MD and the motion of the gas molecules are simulated by using DSMC. The surface is made up of a few sheets of Pt atoms and is assumed periodic in the directions horizontal to the surface. In Takeuchi et al. (2004), the N_2 molecules are assumed to be a rigid rotor with a fixed bound length. The vibration of the molecules is neglected and the rotational energy is assumed continuous. The Lennard-Jones potential was used for the interaction potential. The simulations provide direct calculations of the momentum accommodation coefficients and energy accommodation coefficients.

The microproperties of the surface can also affect the interactions between the gas and the surface. The lattice structures can produce molecular level roughness. Roughness, either at micro- or macro-scales, influence the momentum and energy exchange during interactions. The statistical model proposed by Sawada et al. (1996) considered conical surface roughness. New fractal function models for rough surfaces profiles, which derive the probability distribution of heights for general sample lengths, have also been proposed (Ling 1990; Blackmore and Zhou 1996).

7.3 DSMC Implementation

The Maxwellian specular and diffuse wall models are widely used in DSMC simulations. In this section, the implementation procedure for the Maxwellian models in a DSMC framework will be described. We will also examine the steps with which the CLL model can be realized in DSMC.

7.3.1 Specular reflection

In the DSMC simulations, the specular reflection model can be readily implemented. Only the velocity component normal to the surface changes its sign, other components do not change. For the remainder of the simulation time step, the reflected molecule continues to travel at the postreflection speed.

7.3.2 Diffusive reflection

The velocity distribution function for the Maxwellian diffuse reflection model is

$$f = \frac{\beta^3}{\pi^{3/2}} \exp[-\beta^2 c^{*2}] \tag{7.3.1}$$

where

$$\beta = \left(2\frac{k}{m}T_w\right)^{-1/2} \tag{7.3.2}$$

In diffusive reflection, the reflected velocity c^* can be written in three components in the form of (u, v, w), where v is the component normal to the surface, and u and w the components parallel to the surface. The distribution function for the velocity component normal to the surface is sampled from the number flux on the surface, and assumes the form of

$$f_v \propto v \exp(-\beta^2 v^2) \tag{7.3.3}$$

The proportionality coefficient can be obtained by applying the normalization condition to Eq. (7.3.3), which gives

$$f_v = 2\beta^2 v \exp(-\beta^2 v^2) \tag{7.3.4}$$

Therefore

$$f_v dv = \exp(-\beta^2 v^2) d(\beta^2 v^2) \tag{7.3.5}$$

That is, the distribution function for $\beta^2 v^2$ can be defined as

$$f_{\beta^2 v^2} = \exp(-\beta^2 v^2) \tag{7.3.6}$$

A typical value of v can be found by using the sampling of the accumulative distribution function in the App. 5A,

$$v = [-\ln(R_f)]^{1/2}/\beta \tag{7.3.7}$$

where R_f is a random fraction.

The distribution function for the velocity components parallel to the surface is the same as that for a velocity component in a stationary gas. That is,

$$f_u = \frac{\beta}{\sqrt{\pi}} \exp[-\beta^2 u^2] \tag{7.3.8}$$

$$f_w = \frac{\beta}{\sqrt{\pi}} \exp(-\beta^2 w^2) \tag{7.3.9}$$

The sampling pair of values for u and w is

$$f_u \, du \, f_w \, dw = \frac{\beta^2}{\pi} \exp[-\beta^2(u^2 + w^2)] du \, dw \tag{7.3.10}$$

One can rewrite Eq. (7.3.10) by using the following transformation

$$u = r\cos\theta \qquad w = r\sin\theta \qquad r = u^2 + w^2 \tag{7.3.11}$$

Or

$$f_u \, du \, f_w \, dw = \exp(-\beta^2 r^2) d(\beta^2 r^2) d(\theta/2\pi) \tag{7.3.12}$$

Equation (7.3.12) means that θ is uniformly distributed between 0 and 2π, and that the variable $\beta^2 r^2$ is distributed between 0 and ∞ with distribution function of

$$f_{\beta^2 r^2} = \exp(-\beta^2 r^2) \tag{7.3.13}$$

A typical value of θ is given as

$$\theta = 2\pi \cdot R_f \tag{7.3.14}$$

and Eq. (7.3.13) gives

$$r = [-\ln(R_f)]^{1/2}/\beta \tag{7.3.15}$$

that follows the same form as in Eq. (7.3.7).

Therefore, in the DSMC realization of Maxwellian model of diffusive reflection, the normal component of the reflected molecule velocity can be sampled by Eq. (7.3.7). A pair of values of r and θ may be sampled from Eqs. (7.3.14) and (7.3.15) using successive random fractions. The normally distributed values of u and w follow Eq. (7.3.11) and provide values for the parallel velocity components for the reflect molecule. In this model, the velocity of the reflected molecule depends on the surface temperature T_w only, and is not related to its incident velocity.

7.3.3 CLL model

Lord (1991; 1995) realized the CLL model in the DSMC method. Consider first the parallel components of the CLL reflection. The tangential momentum accommodation coefficient σ_v in Eq. (7.1.6) can be defined as

$$\sigma_v = \frac{c_x - c_x^*}{c_x} \tag{7.3.16}$$

and the tangential energy accommodation coefficient α_t

$$\alpha_t = \frac{c_x^2 - c_x^{*2}}{c_x^2} \tag{7.3.17}$$

Then relationship between these two accommodation coefficients can be obtained as

$$\alpha_t = \sigma_v(2 - \sigma_v) \tag{7.3.18}$$

Without the loss of the generality, the coordinate is chosen according to the velocity (combined parallel velocities) direction of the incident molecule. That is $c_x = u_i$, and $c_z = 0$. Using Eqs. (7.3.1) and (7.3.18), the probability that a molecule is reflected with tangential velocities of (u, w) from incident velocity of $(u_i, 0)$ can be obtained as

$$\begin{aligned} &P(u_i \to u)P(0 \to w)du\,dw \\ &\quad = \frac{1}{\pi\alpha_t}\exp\left[-\frac{\left\{u-(1-\alpha_t)^{1/2}u_i\right\}^2 + w^2}{\alpha_t}\right]du\,dw \end{aligned} \tag{7.3.19}$$

The coordinate $(u,\ w)$ is transformed to a polar coordinate system $(r,\ \theta)$, following the same rule with Eq. (7.3.11) with its origin at $Q((1-\alpha_t)^{1/2}u_i, 0)$, where

$$r = \sqrt{\{u-(1-\alpha_t)^{1/2}u_i\}^2 + w^2} \tag{7.3.20}$$

The probability is also transformed as

$$P(u_i \to u)P(0 \to w)du\,dw = \frac{r}{\pi\alpha_t}\exp\left(-\frac{r^2}{\alpha_t}\right)dr\,d\theta \tag{7.3.21}$$

The distribution function for r^2/α_t can be obtained by integrating Eq. (7.3.21) with θ from 0 to 2π

$$f(r^2/\alpha_t) = \exp(-r^2/\alpha_t) \tag{7.3.22}$$

The sampling for r and θ can be obtained independently as

$$\begin{cases} \theta = 2\pi \cdot R_f \\ r = \{-\alpha_t \ln(R_f)\}^{1/2} \end{cases} \tag{7.3.23}$$

and the parallel components for the reflected molecule are sampled as

$$u = (1 - \alpha_t)^{1/2} u_i + r \cos\theta \tag{7.3.24}$$

$$w = r \sin\theta \tag{7.3.25}$$

The normal velocity component v of the reflected molecule can be considered as the combined vector of the parallel components. With normal velocity v_i of the incident molecule and the same sampling for the parallel components of reflected velocity, the value of v is sampled as

$$\begin{cases} \theta = 2\pi \cdot R_f \\ r = \{-\alpha_n \ln(R_f)\}^{1/2} \end{cases} \tag{7.3.26}$$

$$v = \{r^2 + (1 - \alpha_n)v_i^2 + 2r(1 - \alpha_n)^{1/2} v_i \cos\theta\}^{1/2} \tag{7.3.27}$$

where α_n is the normal energy accommodation coefficient.

In the DSMC realization of the CLL model, the inward normal component of the incident velocity is v_i and the parallel components are u_i and w_i. These velocity components have been normalized by the most probable speed at the surface temperature. The new axes are chosen such that u_i lies in the interaction plane that contains the incident molecular path and the surface normal. w_i is therefore zero. The distributions of θ and r in Eq. (7.3.23) are applied to the parallel components of reflected molecular velocity in the interaction plane, u and w, in Eqs. (7.3.24) and (7.3.25). The distributions of θ and r in Eq. (7.3.26) are applied to the outward normal components of reflected velocity v in Eq. (7.3.27). Note that the accommodation coefficient in Eq. (7.3.23) is different from that in Eq. (7.3.26) and that different sets of random values for θ and r must be used for the different velocity components. The reflected components must be transformed back to the original coordinate system. The CLL model reduces to the specular reflection model when all the accommodation coefficients are set equal to zero and to diffuse reflection with values of one.

The physical meaning for the CLL model can be explained with the following. The average value u_0 of the tangential reflection velocity for reflected molecules can be obtained through integrating the reflection kernel equation (7.3.19) as

$$\begin{aligned} u_0 &= \int u\mathbf{P}(u_i \to u)du \\ &= \int_{-\infty}^{\infty} u\,[\pi\sigma_v(2 - \sigma_t)]^{-1/2} \exp\left[-\frac{[u - (1 - \sigma_t)u_i]^2}{\sigma_t(2 - \sigma_t)}\right] du \\ &= (1 - \alpha_t)^{1/2} u_i \end{aligned} \tag{7.3.28}$$

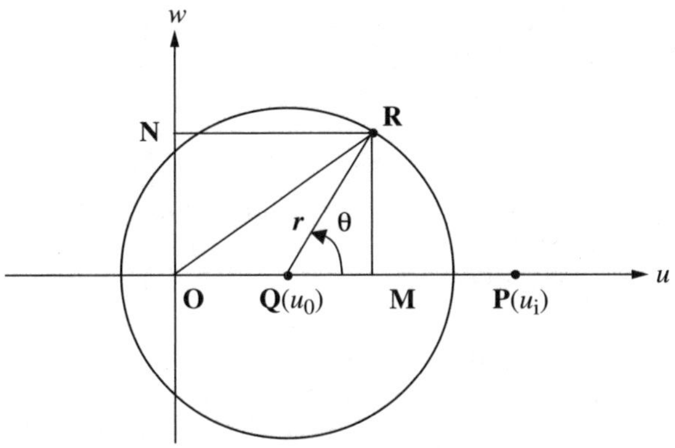

Figure 7.3.1 Graphic representation of CLL model.

Equation (7.3.28) gives the averaged state of reflected molecules by using the accommodation coefficients. Lord (1991) introduced the graphic shown in Fig. 7.3.1 to illustrate this general distribution function in the CLL model. The point P on the u-axis may represent the state of the incident molecule, and the distance OP can represent the magnitude of either the parallel velocity component u_i or the normal component v_i. The point Q represents the mean state u_0 of the reflected molecules under the condition that $OQ/OP = (1-\alpha)^{1/2}$, where α is the accommodation coefficient. Point R represents the actual state of the reflected molecule. The probability distribution of this state is given by a two-dimensional Gaussian distribution in Eq. (7.3.21) centered at Q, where r is the distance QR and θ is the angle $\angle$PQR. The distance OR represents either the reflected normal velocity v or the reflected combined parallel velocity $(u^2 + w^2)^{1/2}$, while the projections OM and ON of OR onto the axes represent either u or w. To summarize, the CLL model samples the reflection velocities to a mean state u_0 according to its accommodation coefficient and incident velocities and an actual state is sampled by a two-dimensional Gaussian distribution based on the mean state.

7.4 Wall-Slip Models for Continuum Approaches

The first- and the second-order approximation of the Chapman-Enskog expansion of the Boltzmann equation result in the Navier-Stokes equations and the various forms of the Burnett equations. These equations are valid in the region away from the wall. Since these equations involve spatial derivatives, boundary conditions are required in terms of the averaged properties of the flow, such as the velocity. For a specular wall, Eq. (7.2.1) suggests that

$$\overline{c_y} \equiv v_{\text{gas}} = 0$$

for a stationary surface. This applies to the diffuse reflection model as well. For the tangential velocity component u_s, the specular reflection model indicates that it is not changed by the gas–surface interaction. With diffuse reflection, the reflected molecules adopt the surface velocity that is independent of the incident velocity. An estimate of the average tangential velocity can be made by examining the momentum balance in the near-surface region. The average flux of particle impinging on the wall can be written as, for $Q = n[c \bullet (-\mathbf{n}_w)] = -nc_y$, where $\mathbf{n}_w$ denotes the unit vector normal to the wall, which points in the y-direction in Fig. 7.1.2,

$$-\int_{-\infty}^{\infty}\int_{-\infty}^{0}\int_{-\infty}^{\infty} nf c_y \, dc_x \, dc_y \, dc_z$$

Assuming a Maxwellian distribution function, we can write the above equation as

$$\frac{1}{4}n\overline{c'}$$

where $\overline{c'}$ denotes the average thermal speed. The tangential momentum each of the impinging molecules carries before the interaction, say at one mean free path away from the wall, can be written as

$$m\left(u_{\text{gas}} + \lambda \left.\frac{du}{dy}\right|_w\right)$$

by using a first-order approximation, where du/dy is evaluated on the wall. The averaged tangential momentum flux, or shear stress, would then be

$$\frac{1}{4}n\overline{c'}m\left(u_{\text{gas}} + \lambda\frac{du}{dy}\right) = \frac{1}{4}\overline{c'}\rho\left(u_{\text{gas}} + \lambda\frac{du}{dy}\right) \tag{7.4.1}$$

For the Navier-Stokes equations to be valid, this shear stress would be $\mu \frac{du}{dy}\big|_\lambda$. For monatomic gases,

$$\mu = \frac{1}{2}\rho\overline{c'}\lambda$$

Equating the shear stresses from the kinetic theory consideration and that from the Navier-Stokes approximation, we get

$$u_{\text{gas}} = \lambda \frac{du}{dy} \tag{7.4.2}$$

Equation (7.4.2) shows that the averaged tangential velocity component relative to the surface is not zero. The averaged flow shows a finite slip on the surface. The amount of the slippage is proportional to the local mean free path λ. When properly normalized, say, using the velocity and the length of the averaged flow, the equation shows that the slip velocity is proportional to the local Knudsen number. Therefore, for the first-order Chapman-Enskog expansion, the slip tangential velocity is small compared with the velocity scale of the averaged bulk flow. The "no-slip" boundary conditions that are normally used in the solution of the Navier-Stokes equations are then valid for flows that are in equilibrium. For microflows, the Knudsen numbers are large and the velocity slip may not be neglected.

For the partially specular and partially diffuse walls considered by Maxwell (1879), an expression for the slip velocity for an isothermal wall of temperature T_w is

$$u_{\text{gas}} - u_w = \frac{2-\sigma_v}{\sigma_v} \lambda \frac{\partial u}{\partial y} \tag{7.4.3}$$

For monatomic gases, the equation can be written as

$$u_{\text{gas}} - u_w = \frac{2-\sigma_v}{\sigma_v} \frac{\mu\sqrt{\pi}}{\rho\sqrt{2RT_w}} \frac{\partial u}{\partial y} \tag{7.4.4}$$

The contribution by the streamwise temperature gradient associated with the thermal creep phenomenon is added by von Smoluchowski (1898) and can be written as

$$u_{\text{gas}} - u_w = \frac{2-\sigma_v}{\sigma_v} \lambda \frac{\partial u}{\partial y} + \frac{3}{4} \frac{\mu}{\rho T_w} \frac{\partial T}{\partial s} \tag{7.4.5}$$

where $\partial/\partial s$ is taken tangential to the surface. According to Eq. (7.4.5), slip velocity can be caused by the velocity gradient normal to the wall and the temperature gradient tangent to the wall. Using a similar argument made for the velocity slip, von Smoluchowski also proposed a boundary condition for the temperature

$$T_{\text{gas}} - T_w = \frac{2-\sigma_T}{\sigma_T} \left[\frac{2\gamma}{\gamma+1}\right] \frac{\lambda}{\text{Pr}} \frac{\partial T}{\partial y} \tag{7.4.6}$$

The finite Knudsen number effects thus generate a difference of temperature between that of the surface and that of the gas, or a temperature jump. The nondimensional form of the velocity-slip and temperature-jump conditions can be written as

$$u_{\text{gas}} - u_w = \frac{2-\sigma_v}{\sigma_v}\text{Kn}\frac{\partial u}{\partial y} + \frac{3}{2\pi}\frac{\gamma-1}{\gamma}\frac{\text{Kn}^2\text{Re}}{\text{Ec}}\frac{\partial T}{\partial s} \tag{7.4.7}$$

$$T_{\text{gas}} - T_w = \frac{2-\sigma_T}{\sigma_T}\left[\frac{2\gamma}{\gamma+1}\right]\frac{\text{Kn}}{\text{Pr}}\frac{\partial T}{\partial y} \tag{7.4.8}$$

The Eckert number Ec is defined as,

$$\text{Ec} = \frac{u_r^2}{C_p \Delta T_r}$$

where u_r and ΔT_r represent the reference velocity and the reference temperature difference, respectively.

Various second-order slip boundary conditions have also been examined (Schamberg 1947; Cercignani and Daneri 1963; Deissler 1964; Hsia and Domoto 1983). Beskok and Karniadakis (1994, 1999, 2002) proposed a modified form of the slip conditions,

$$u_{\text{gas}} - u_w = \frac{2-\sigma_v}{\sigma_v}\frac{\text{Kn}}{1-b\text{Kn}}\frac{\partial u}{\partial y} + \frac{3}{2\pi}\frac{\gamma-1}{\gamma}\frac{\text{Kn}^2\text{Re}}{\text{Ec}}\frac{\partial T}{\partial s} \tag{7.4.9}$$

The slip coefficient b is written as

$$b = \frac{1}{2}\left(\frac{u''}{u'}\right)_w$$

where the prime denotes derivatives of the tangential velocity field, obtained with no-slip conditions, along the normal direction to the surface. The procedure of Beskok and Karniadakis can be extended to high orders for both the velocity-slip and the temperature-jump conditions.

For liquids, there is no advanced first-principle based theory as it is for dilute gases. The surface interaction modeling is mostly phenomenological. Thompson and Troian (1997) performed MD simulations to quantify the dependence of the slip-flow boundary condition on shear rate in a Couette flow arrangement. Similar to the slip modeling for gases, a slip length L_s was defined,

$$u_{\text{liquid}} - u_w = L_s\frac{du}{dy} \tag{7.4.10}$$

They found that L_s begins to diverge well below the shear rate where the linear constitutive relation would become invalid. The divergence was also observed to be highly nonlinear and quick, indicating that a small change of shear rate or a change of surface properties may result in large variation of velocity slip.

Based on their MD data, they suggested that

$$L_s = L_s^0 \left(1 - \frac{\dot{\gamma}}{\dot{\gamma}_c}\right)^{-1/2} \tag{7.4.11}$$

where L_s^0 denotes the slip length at small shear rates, $\dot{\gamma}_c$ the critical shear rate.

The boundary conditions can be more complex when effects such as the chemical, physical, and electrical properties are included. For instance, more drugs today are protein based. The interaction between the pharmaceutical products with glass containers, such as adsorption, can affect the packaging and the delivery of the solutions. At microscale, the surface mechanisms play an important role in the behavior of the flow and there are many phenomena that are not fully understood and modeled.

References

Beskok, A. and Karniadakis, G.E., Simulation of heat and momentum transfer in complex micro-geometries, *J. Thermophys. Heat Transf.*, Vol. 8, p. 355, 1994.

Beskok, A. and Karniadakis, G.E., A model for flows in channels, pipes, and ducts at micro and nano scales, *Microscale Thermophysics Eng.*, Vol. 3, p. 43, 1999.

Beskok, A. and Karniadakis, G.E., *Microflows, Fundamentals and Simulation*, Springer, New York, 2002.

Blackmore, D. and Zhou, J.G., A general fractal distribution function for rough surface profiles, *SIAM J. Appl. Math.*, Vol. 56, p. 1694, 1996.

Cercignani, C., *Mathematical Methods in Kinetic Theory*, Plenum Press, New York, 1969.

Cercignani, C. and Daneri, A., Flow of a rarefied gas between two parallel plates, *J. Appl. Phys.*, Vol. 34, p. 3509, 1963.

Cercignani, C. and Lampis, M., Kinetic models for gas–surface interactions, *Transp. Theory Stat. Phys.*, Vol. 1, p. 101, 1971.

Deissler, R., An analysis of second-order slip flow and temperature jump boundary conditions for rarefied gases, *Int. J. Heat Mass Transf.*, Vol. 7, p. 681, 1964.

Hurst, J.E., Becker, C.A., Cowin, J.P., Janda, K.C., and Wharton, L., Observation of direct inelastic scattering in the presence of trapping-desorption scattering: Xe on Pt (111), *Phys. Rev. Lett.*, Vol. 43, p. 1175, 1979.

Hsia, Y. and Domoto, G., An experimental investigation of molecular rarefied effects in gas lubricated bearings at ultra low clearances, *J. Lubrication Technol.*, Vol. 105, p. 120, 1983.

Koga, T., *Introduction to Kinetic Theory Stochastic Process in Gaseous Systems*, Pergamon Press, Oxford, 1970.

Kuščcer, J., Reciprocity in scattering of gas molecules by surfaces, *Surface Sci.*, Vol. 25, p. 225, 1971.

Ling, F., Fractals, engineering surfaces and tribology, *Were*, Vol. 136, p. 141, 1990.

Logan, R.M. and Stickney, R.E., Simple classical model for the scattering of gas atoms from a solid surface, *J. Chemical Phys.*, Vol. 44, p. 195, 1966.

Lord, R.G., Some extensions to the Cercignani-Lampis gas–surface scattering kernel, *Phys. Fluids*, Vol. 3, p. 706, 1991.

Lord, R.G., Some further extensions of the Cercignani-Lampis gas–surface interaction model, *Phys. Fluids*, Vol. 7, p. 1159, 1995.

Maxwell, J.C., On stresses in rarefied gases arising from inequalities of termperature, *Phi. Trans. R. Soc.*, Part 1, Vol. 170, pp. 231–256, 1879.

Schamberg, R., The fundamental differential equations and the boundary conditions for high speed slip-flow, and their application to several specific problems, Ph.D. Thesis, California Institute of Technology, CA, 1947.

Sawada, T., Horie, B.Y., and Sugiyama, W., Diffuse scattering of gas molecules from conical surface roughness, *Vacuum*, Vol. 47, pp. 795—797, 1996.

Takeuchi, H., Yamamoto, K., and Hyakutake, T., Behavior of the reflected molecules of a diatomic gas at a solid surface, 24th *Rarefied Gas Dynamics Symposium*, Bari, Italy, 2004.

Thompson, P.A. and Troian, S.M., A general boundary condition for liquid flow at solid surfaces, *Phys. Rev. Lett.*, Vol. 63, p. 766, 1997.

von Smoluchowski, M., Ueber Warmeleitung in verdunnten gasen, *Annalen der Physik und Chemi*, Vol 64, p. 101, 1898.

Wenaas, E.P., Scattering at gas–surface interface, *J. Chemical Phys.*, Vol. 54, p. 476, 1971.

Yamamoto, K., Slip Flow over a Smooth Platinum Surface, *JSME Int. J.*, Series B, Vol. 45, p. 788, 2002.

Yamanishi, N. and Matsumoto, Y., Multistage gas–surface interactionmodel for the direction simulation Monte Carlo method, *Phys. Fluids*, Vol. 11, pp. 35–40, 1999.

Chapter

8

Development of Hybrid Continuum/Particle Method

8.1 Overview

The Chapman-Enskog expansion of the velocity distribution in terms of the power series of the Knudsen number gives rise to various forms of approximation for the Boltzmann equation. As was shown in Chap. 4, for the thermal equilibrium condition, the Euler equations are obtained. As the departure from equilibrium increases, measured by appropriately defined Knudsen numbers, higher order terms need to be included. The first-order expansion results in the Navier-Stokes equations and the Burnet equations have also been derived as the second-order approximation form of the Boltzmann equation. For micro flows of small Knudsen number, say, less than 0.01, the Navier-Stokes equations have generally been found satisfactory. At higher Knudsen number, the Burnett equations become more appropriate. The application of the slip boundary condition for the velocity and the temperature jump condition often enables both the Navier-Stokes equations and the Burnett equations to provide solutions at high Knudsen numbers. At higher Knudsen number, it becomes necessary to use discrete based approaches, such as molecular dynamics (MD), direct simulation Monte Carlo (DSMC), or lattice Boltzmann method, for numerical simulations of practical flows. These first principle methods are physically sound and valid for flows at all Knudsen numbers. They, however, demand more computational resources than the differential equation models for flows of low speed and Knudsen number.

For microfluidic devices, the operational value of the Knudsen number can spread over quite a large range in the same system. A single-scale approach based on a continuum equation model is then not uniformly valid in the entire system. On the other hand, the costs of using discrete methods, although physically accurate regardless of the local Knudsen number, are high. In these situations with physics that are of multiple scales in nature, a hybrid approach becomes an attractive option. Such an approach would involve the use of both the discrete as well as the continuum-based methods for the solution of the appropriate flow domain in the same flow field. For the sake of computational efficiency, a differential equation model of the continuum approach, such as the Euler and the Navier-Stokes equations, is normally applied to the largest domain possible. The discrete method is then used in regions where a breakdown of the continuum assumption is anticipated.

The hybrid approaches have been explored in many studies of hypersonic rarefied gas flows. The flight trajectory of an aerospace vehicle can cover the entire atmosphere of a planet, ranging, for instance, from high-altitude rarefied gas regime to low-altitude continuum flow regime. There are also flows, such as the blunt body wakes and control system plumes that are in a mixed continuum and rarefied condition. Aerospace engineers have had quite some successes in applying hybrid methods to these flows. There are also recent studies with focuses on the hybrid approaches for microflows. In this chapter, these hybrid approaches are briefly discussed.

8.2 Breakdown Parameters

One of the key issues in the development of the hybrid approaches is the determination of the location of the interface boundary between the domains, or patches, where the continuum assumption can be applied and the domain where a discrete method should be used. The decomposition of the domain is normally accomplished through the use of parameters that measure the breakdown of the continuum approach. Although the Navier-Stokes equations can be derived from the Boltzmann equation with the assumption that the velocity distribution is a small perturbation of the equilibrium or Maxwellian function, the Chapman-Enskog theory did not provide explicitly a limit when such perturbation can be considered small and the Navier-Stokes equations are valid. Several breakdown parameters have been proposed in the literature. Bird (1970) proposed a breakdown parameter

$$P = \frac{1}{\nu}\left|\frac{D(\ln\rho)}{Dt}\right| \tag{8.2.1}$$

The materials derivative is evaluated along the flow streamlines. Bird's study of the expanding flows showed that the breakdown of the continuum approach correlated with a value of P of approximately 0.02

(Bird 1994). Boyd et al. (1995) considered a gradient-length local Knudsen number defined by

$$\mathrm{Kn}_{\mathrm{GLL}} = \frac{\lambda}{Q}\left|\frac{dQ}{dL}\right| \tag{8.2.2}$$

Boyd et al. (1995) concluded in their study of one-dimensional normal shock waves and two-dimensional bow shocks that the continuum approach broke down wherever the value of the gradient-length local Knudsen number exceeded 0.05. Garcia and Alder (1998) proposed a breakdown parameter B.

$$B = \max\left\{\left|\tau_{ij}^*\right|, \left|q_i^*\right|\right\} \tag{8.2.3}$$

where

$$\tau_{ij}^* = \frac{\mu}{p}\left(\frac{\partial V_i}{\partial x_j} + \frac{\partial V_j}{\partial x_i} - \frac{2}{3}\frac{\partial V_k}{\partial x_k}\delta_{ij}\right) \qquad q_i^* = -\frac{\kappa}{p}\left(\frac{2m}{kT}\right)^{1/2}\frac{\partial T}{\partial x_i} \tag{8.2.4}$$

are the normalized stress tensor and heat flux vector. p denotes pressure. The continuum breakdown parameter includes both the viscous stress and heat transfer coefficients in the first-order Chapman-Enskog expansion. The validity of the Chapman-Enskog distribution was found questionable for $B > 0.2$. This parameter has also been used by Garcia et al. (1999). Sun et al. (2004) also used the breakdown parameter B as the continuum/particle interface indicator in the implementation of a hybrid approach of the Navier-Stokes equations and information preservation (IP) method. A Knudsen number type of breakdown parameter was proposed by Wang and Boyd (2003).

$$\mathrm{Kn}_{\max} = \max[\mathrm{Kn}_D, \mathrm{Kn}_V, \mathrm{Kn}_T] \tag{8.2.5}$$

where the Knudsen numbers Kn_D, Kn_V, Kn_T were calculated by using the formulation of Boyd et al. (1995) and the subscripts D, V, and T denoted density, velocity magnitude, and translational temperature. Tiwari (1998) proposed a criterion for local equilibrium by using the norm of the first order Chapman-Enskog expansion function Φ_1 defined by Eq. (4.3.2) in Chap. 4.

$$\|\Phi_1\| = \frac{1}{\rho RT}\left[\frac{2}{5}\frac{|q_i|^2}{RT} + \frac{1}{2}\|\tau\|_E^{\,2}\right]^{1/2} \tag{8.2.6}$$

where $\|\tau\|_E$ denotes the Euclidean norm of the stress tensor matrix τ_{ij}. The parameter was used to automatically determine the domain decomposition between the regions where the Boltzmann and the Euler solvers would be applicable. Both equations were solved by particle methods.

8.3 Hybrid Approaches for Microfluid Flow

Aktas and Aluru (2002) used the Stokes equations as the continuum fluid model and the DSMC as the discrete model. For the microfluidic filters considered, the computational domains for the two methods are overlapped and Dirichelet-Dirichelet type boundary conditions are used. There was no need for interface flux evaluation. In the overlapped domain, the solution was interpolated from one subdomain to the other by using scattered point interpolation. The Stokes equations were solved by using the scattered point finite cloud method. It was argued that in comparison with a mesh-based method, such as the finite element methods, the scattered point method allowed arbitrary treatment of the interface between the DSMC and the continuum domains. The multiscale method was applied to the steady-state analyses of microfluidic filters. In these simulations, the continuum and the discrete regions were assumed and remained unchanged during the simulations.

Sun et al. (2004) presented a coupling of the information preservation (IP) method described in Chap. 5 with a continuum model using the Navier-Stokes equations. Reservoir cells and buffer cells were used to generate particles from the continuum region. The interface location was determined by using the breakdown parameter proposed by Garcia et al. (1999). The IP method preserves information at the macroscopic level, which allows a direct feed of information to the continuum approach. For low-speed flows, the IP method exhibits very small statistical scatter compared with the DSMC method. The application of the IP method can be quite significant in the future development of hybrid approaches for low-speed microflows.

The hybrid of continuum Navier-Stokes equations and atomics MD is implemented to simulate unsteady sudden-start Couette flow and channel flow with nano-scale rough walls by Nie et al. (2004). The spatial coupling between the continuum equations and the MD method was achieved through constrained dynamics in an overlap region. Continuity of fluxes was imposed at the boundaries of the overlap region. The results were validated with analytical solutions and full molecular dynamics simulations.

8.4 Development of Additional Hybrid Approaches

A zonally decoupled hybrid approach was used by Wilmoth et al. (1994) to examine the wake closure behind a hypersonic blunt body. At the simulated conditions of Mach 20, the forebody flow was regarded as continuum flow. The wake flow was solved by using DSMC with the exit-plane Navier-Stokes solution as the inflow condition. Overlapping cells were used for coupling and sampling from Maxwellian distribution function. There was no feedback from the DSMC solution to the Navier-Stokes forebody solution.

Gatsonis et al. (1999) studied the nozzle and plume flows at the firing of a small cold-gas thruster onboard a suborbital spacecraft. Three-dimensional Navier-Stokes simulations were performed first until the breakdown surface of the continuum flow was established according to Bird's breakdown parameter (Bird 1970). Three-dimensional DSMC simulations using the DAC code (Wilmoth et al. 1996) were subsequently performed inside the breakdown surfaces. The input data necessary for the DSMC runs were taken from the Navier-Stokes simulations using a linear interpolation scheme of visualization software.

Glass and Gnoffo (2000) proposed a coupled 3-D CFD-DSMC method for highly blunt bodies using the structured grid.

The work of Roveda et al. (1998; 2000) on hybrid approach was motivated by the need to study thrust plume impingement on spacecraft solar panels. In this paper, the Euler equations were used for the equilibrium region and the DSMC method was applied in the nonequilibrium patches. The Euler equations were solved by using adaptive discrete velocity (ADV) method of Gadiga (1995), which was a quasi-particle approach developed from kinetic theory. This was a strongly coupled approach, in that the DSMC and the ADV solutions were coupled to exchange information within every calculation cycle. The translational nonequilibrium patches were adaptively identified based on Bird's breakdown parameter (Bird 1970), which involved the density gradients in determining the breakdown of the continuum approach. At the interface, a buffer layer of two DSMC cells was treated as an extension of the ADV domain and the macroscopic DSMC properties were used to evaluate the half-fluxes. To reduce the statistical noises of DSMC at the interface, ghost cells were used near the interface to locally increase the number of particles.

In the hybrid method developed by Wadsworth and Erwin (1990; 1992), the flux into the DSMC region was calculated by interpolation to the interface of the cell-centered Navier-Stokes solution. The boundary conditions to the Navier-Stokes simulations were provided by the cumulative sampled value of the DSMC cell centered conservation quantities adjacent to the interface. The solutions therefore improved as they approached a steady-state condition.

In an experiment by Golse (1989), the coupling was accomplished through a half-flux method borrowed from radiative heat transfer problem. The half-flux of mass, momentum, and energy of the particles crossing the interface based on the DSMC and the Navier-Stokes equations were considered. The method was developed for application in space vehicle reentry problems. The computational domains were not overlapped. Hash and Hassan (1996) considered various interface coupling methods in terms of their physical and numerical advantages by using rarefied Couette flows and the half-flux method was found to offer the best performance. In a later paper, Hash and Hassan (1997) found that the Marshak condition of half fluxes was unacceptable for low-speed flow applications due to the large statistical scatter of DSMC in evaluating the fluxes.

Garcia et al. (1999) used the adaptive mesh refinement (AMR) method to solve the Navier-Stokes equations. DSMC is used at the finest level of the AMR hierarchy. The breakdown of the Navier-Stokes model is determined by a breakdown parameter for the Chapman-Enskog distribution (Garcia and Alder 1998). The adaptive mesh and algorithm refinement (AMAR) coupling between the particle and the continuum regions was found (Garcia et al. 1999) to conserve mass, momentum, and energy within round-off error.

Lian et al. (2005) developed a parallel 3-D hybrid DSMC-NS method using an unstructured grid topology and the breakdown parameter of Wang and Boyd (2003). Alexander et al. (2002) constructed a hybrid particle/continuum algorithm for linear diffusion in the fluctuating hydrodynamic limit with the adaptive mesh and algorithm refinement.

LeTallec and Mallinger (1997) used the Boltzmann equation in the discrete domain of a space vehicle reentry problem. The Navier-Stokes model was used in the continuum part of the flow. The coupling was achieved by matching half-fluxes at the interface of the two nonoverlapping domains.

8.5 Remarks

Microfluidic applications can have physics of vastly different time scales in the different spatial areas. The hybrid approaches are best suited for analyses of problems that involve physics of disparate scaling. Hybrid approaches can provide physically realistic results at greater computation efficiency than a single-scale approach. Such a multiscale, hybrid approach may include the continuum theories, the statistical modeling of DSMC, and the deterministic method of MD. As the functions of the microfluidic devices expand, hybrid approaches become more useful in the modeling of multiscale, multiphysic phenomena. Much work is still needed in areas such as the coupling of the different methods, the proper identification of the interfaces, and the interface conditions.

References

Aktas, O. and Aluru, N.R., A combined continuum/DSMC technique for multiscale analysis of microfluidic filters, *J. Comp. Phys.*, Vol. 178, p. 342, 2002.

Alexander, F.J., Garcia, A.L., and Tartakovsky, D.M., Algorithm refinement for stochastic partial differential equations, *J. Comp. Phys.*, Vol. 182, p. 47, 2002.
Bird, G.A., Breakdown of transitional and rotational equilibrium in gaseous expansion, *AIAA J.*, Vol. 8, p. 1998, 1970.
Bird, G.A., *Molecular Gas Dynamics and the Direct Simulation of Gas Flows*, Oxford University Press, New York, 1994.
Boyd, I.D., Chen, G., and Chandler, G.V., Predicting failure of the continuum fluid equations in transitional hypersonic flows, *Phys. Fluids*, Vol. 7, p. 210, 1995.
Gadiga, T.T., An Euler solver based on locally adaptive discrete velocities, *J. Stat. Phys.*, Vol. 81, p. 129, 1995.
Garcia, A.L. and Alder, B.J., Generation of the Chapman-Enskog distribution, *J. Comp. Phys.*, Vol. 140, p. 66, 1998.
Garcia, A.L., Bell, J.B., Crutchfield, W.Y., and Alder, B.J., Adaptive mesh and algorithm refinement using direct simulation Monte Carlo, *J. Comp. Phys.*, Vol. 154, pp. 134–155, 1999.
Gatsonis, N.A., Nanson, R.A., and LeBeau, G.J., Navier-Stokes/DSMC simulations of cold-gas nozzle/plume flows and flight data comparisons, AIAA paper 99-3456, 1999.
Glass, C.E. and Gnoffo, P.A., A 3-D coupled CFD-DSMC solution method with application to the Mars sample return orbiter, NASA report TM-2000-210322, 2000.
Golse, F., Applications of the Boltzmann equation within the context of upper atmosphere vehicle aerodynamics, *Comput. Methods Appl. Mech. Eng.*, Vol. 75, p. 299, 1989.
Hash, D.B. and Hassan, H.A., Assessment of schemes for coupling Monte Carlo and Navier-Stokes solution methods, *J. Thermophys. Heat Trans.*, Vol. 10, p. 242, 1996.
Hash, D.B. and Hassan, H.A., Two-dimensional coupling issues of hybrid DSMC/Navier-Stokes solvers, AIAA paper 97-2507, 1997.
LeTallec, P. and Mallinger, F., Coupling Boltzmann and Navier-Stokes equations by half fluxes, *J. Comp. Phys.*, Vol. 136, p. 51, 1997.
Lian, Y.-Y., Wu, J.-S., Cheng, G., and Koomullil, R., Development of a parallel hybrid method for the DSMC and NS solver, AIAA paper 2005-0435, 2005.
Nie, X.B., Chen, S.Y., E, W.N., and Robbins, M.O., A continuum and molecular dynamics hybrid method for micro- and nano-fluid flow, *J. Fluid Mech.*, Vol. 500, p. 55, 2004.
Roveda, R., Goldstein, D.B., and Varghese, P.L., Hybrid Euler/particle approach for continuum/rarefied flows, *J. Spacecr. Rocket*, Vol. 35, p. 8, 1998.
Roveda, R., Goldstein, D.B., and Varghese, P.L., Hybrid Euler/direct simulation Monte Carlo calculation of unsteady slit flow, *J. Spacecr. Rocket*, Vol. 37, p. 753, 2000.
Sun, Q., Boyd, I.D., and Candler, G.V., A hybrid continuum/particle approach for modeling subsonic, rarefied gas flows, *J. Comp. Phys.*, Vol. 194, p. 256, 2004.
Tiwari, S., Coupling the Boltzmann and Euler equations with automatic domain decomposition, *J. Comp. Phys.*, Vol. 144, p. 710, 1998.
Wadsworth, D.C. and Erwin, D.A., One-dimensional hybrid continuum/particle simulation approach for rarefied hypersonic flows, AIAA paper 90-1690, 1990.
Wadsworth, D.C. and Erwin, D.A., Two-dimensional hybrid continuum/particle approach for rarefied hypersonic flows, AIAA paper 92-2975, 1992.
Wang, W.L. and Boyd, I.D., Predicting continuum breakdown in hypersonic viscous flows, *Phys. Fluids*, Vol. 15, p. 91, 2003.
Wilmoth, R.G., LeBeau, G.J., and Carlson, A.B., DSMC grid methodologies for computing low density hypersonic flows about reusable launch vehicles, AIAA paper 96-1812, 1996.
Wilmoth, R.G., Mitcheltree, R.A., Moss, J.N., and Dogra, V.K., Zonally decoupled direct simulation Monte Carlo solutions of hypersonic blunt-body wake flows, *J. Spacecr. Rockets*, Vol. 31, p. 971, 1994.

Chapter

9

Low-Speed Microflows

9.1 Introduction

Numerical solutions of a number of microfluid flow cases are presented in this chapter. Results obtained by using both the continuum as well as the discrete approaches will be presented, discussed, and, when possible, compared. The classical problems of the Couette and Poiseuille flows will be used. This allows us to prime the reader with some of the fundamental differences between low-speed microflows and those at macroscales by simple analyses before the numerical solutions are introduced.

9.2 Analytical Flow Solutions

9.2.1 MicroCouette flows

The geometry of the microCouette flow is simple. The flow develops between two infinite parallel walls. The top wall moves at a constant speed and the lower surface is stationary. Without considering other physical forces, such as electrical and magnetic forces, the wall shear provides the driving mechanism. The flow is a classical problem in continuum fluid mechanics. In microscale devices, the flow is also representative of many flows seen in microfluidic devices, such as micropumps, microbearings, and micromotors. In this section, we will examine its analytical solution in the slip-flow regime and discuss the resulting corrections to the continuum solution due to the finite Knudsen number.

The flow geometry is shown in Fig. 9.2.1. The flow is assumed homogeneous in the z-direction and, therefore, two-dimensional. As in the continuum problem, solutions will be sought when the flow has become steady and fully developed in its velocity profiles. Without losing

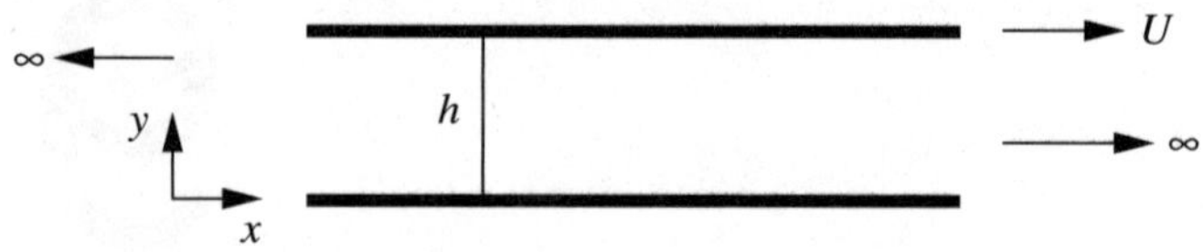

Figure 9.2.1 Couette flow.

generality, the fluid will be assumed incompressible. The Navier-Stokes equations become

$$\frac{\partial v}{\partial y} = 0 \tag{9.2.1}$$

$$\rho v \frac{\partial u}{\partial y} = \mu \frac{\partial^2 u}{\partial y^2} \tag{9.2.2}$$

$$\rho v \frac{\partial v}{\partial y} = -\rho g - \frac{\partial p}{\partial y} + \mu \left(\frac{\partial^2 v}{\partial y^2} \right) \tag{9.2.3}$$

With the no-slip boundary condition, $v_w = 0$, equation (9.2.1) gives $v = 0$ and Eq. (9.2.2) gives

$$u = C_1 y + C_2 \tag{9.2.4}$$

With the no-slip boundary condition, $u_w = 0$ the stream wise velocity profile of the continuum flow solution become

$$\frac{u}{U} = \frac{y}{h}$$

Or, in a dimensionless form,

$$u = y \tag{9.2.5}$$

where u and y now represent nondimensional quantities and $\in (0,1)$. For the slip flow solutions, Eq. (9.2.4) still holds but subject to the velocity slip condition. The first-order surface-slip condition can be written as

$$u_s - U_w = \frac{2 - \sigma_v}{\sigma_v} \mathrm{Kn} \frac{\partial u_s}{\partial n} = \alpha \frac{\partial u_s}{\partial n} \tag{9.2.6}$$

where

$$\alpha = \frac{2 - \sigma_v}{\sigma_v} \mathrm{Kn}$$

On the lower wall, $U_w = 0$ and $\partial u_s/\partial n = \partial u_s/\partial y \cdot U_w = 1$ and $\partial u_s/\partial n = -\partial u_s/\partial y$ on the upper wall. It can then be obtained

$$C_1 = \frac{1}{1 + 2\alpha} \qquad C_2 = \frac{\alpha}{1 + 2\alpha}$$

Equation (9.2.4) thus becomes

$$u = \frac{1}{1+2\alpha}y + \frac{\alpha}{1+2\alpha} \tag{9.2.7}$$

Or,

$$u = \frac{y + \frac{2-\sigma_v}{\sigma_v}\mathrm{Kn}}{1 + 2\frac{2-\sigma_v}{\sigma_v}\mathrm{Kn}}$$

The slip flow velocity solutions, compared with the continuum flow solution Eq. (9.2.5), show not only a velocity slip on the wall, but also a correction to the slope of the profile. The corrections to both the wall gas velocity and the slope depend on σ_v the tangential momentum accommodation coefficient and the flow Knudsen number.

Example 9.1 For $\sigma_v = 0.8, \mathrm{Kn} = 0.05$, Eqs. (9.2.6) and (9.2.7) give

$$\alpha = 0.075 \qquad u = \frac{y + 0.075}{1.15} = 0.87y + 0.065$$

Therefore the slip velocity at the lower wall ($y = 0$)

$$u_s = 0.065$$

or 6.5 percent of the top wall velocity U. At the upper wall ($y = 1$), the gas flow velocity is $0.935U$ and the slip velocity relative to the wall is

$$U_s - U_w = U_s - 1 = -0.065$$

The slip flow solutions of the streamwise gas velocities on the walls thus show a 6.5 percent correction to that of the continuum solutions. Similarly, the nondimensional velocity slope is now 0.87, compared to one based on the continuum solutions. Figure 9.2.2 shows a comparison of the two solutions.

A few other flow characteristics of interest for this flow can now be examined. The nondimensional volume flow rate per unit span in the z-direction is

$$Q = \int_0^1 u\,dy = 0.5 \tag{9.2.8}$$

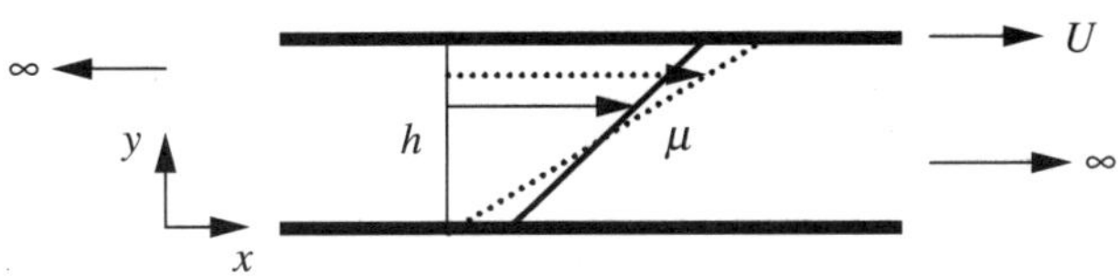

Figure 9.2.2 Comparison of the continuum and the slip Couette flow solutions: - - - - continuum, —slip flow.

Therefore, the slip flow solution gives the same nondimensional volume flow rate as that by the continuum flow solution. The volume flow rate is independent of Kn. The skin friction coefficient

$$C_f = \frac{1}{\text{Re}}\frac{\partial u}{\partial y}$$

where $\text{Re} = \rho U h/\mu$. Or

$$C_f = \frac{1}{\text{Re}}\frac{1}{1+2\alpha} \tag{9.2.9}$$

For the continuum solution, the skin friction coefficient is 1/Re. If we revisit Example 9.1, the slip flow solution of the skin friction coefficient represents a 13 percent decrease of skin friction coefficient compared to the continuum flow value for a microCouette flow with $\sigma_v = 0.8$ and $\text{Kn} = 0.05$.

Since the velocity profile is linear, there is no correction to the streamwise velocity profile if the second-order velocity slip condition of Karniadakis and Beskok (2002) is used.

9.2.2 MicroPoiseuille flows

The simpler form of the Poiseuille flow considered here comprises a flow confined between two infinite parallel stationary plates. The flow is driven by pressure gradient in the mainstream direction of the flow. Pressure-driven flows mostly occur in microchannels where flows are pressurized to pass through microdevices for measurement, sensing, or pumping purposes. A sketch of the flow geometry is shown in Fig. 9.2.3.

We consider a steady, two-dimensional flow that is fully developed in its velocity distributions. The fluid is assumed incompressible. The Navier-Stokes equations can be written as

$$\frac{\partial v}{\partial y} = 0$$

$$\rho v\frac{\partial u}{\partial y} = -\frac{\partial p}{\partial x} + \mu\frac{\partial^2 u}{\partial y^2}$$

$$\rho v\frac{\partial v}{\partial y} = -\rho g - \frac{\partial p}{\partial y} + \mu\frac{\partial^2 v}{\partial y^2}$$

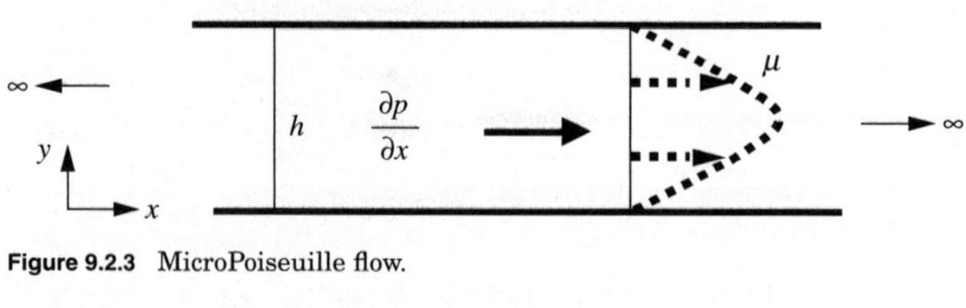

Figure 9.2.3 MicroPoiseuille flow.

The nondimensional forms of the solutions for the velocity components are

$$u = \frac{\mathrm{Re}}{2}\frac{dP}{dx}y^2 + C_1 y + C_2 \tag{9.2.10}$$

where u and y have been nondimensionalized by a certain velocity scale U and the height of the channel h, respectively. Therefore, $y \in (0,1)$. With the no-slip boundary conditions, the continuum solution becomes.

$$u = \frac{\mathrm{Re}}{2}\frac{dP}{dx}(y^2 - y) \tag{9.2.11}$$

The parabolic streamwise velocity profile is indicated in Fig. 9.2.3. The slip flow solution for the microPoiseuille flow can be obtained by applying the velocity slip boundary conditions. Applying the first-order velocity slip boundary condition Eq. (9.2.6), we can find the integration constants C_1 and C_2. The slip flow solutions can then be written as

$$u = \frac{\mathrm{Re}}{2}\frac{dP}{dx}(y^2 - y - \alpha) \tag{9.2.12}$$

Or

$$u = \frac{\mathrm{Re}}{2}\frac{dP}{dx}\left(y^2 - y - \frac{2-\sigma_v}{\sigma_v}\mathrm{Kn}\right) \tag{9.2.13}$$

Equation (9.2.13) is the slip flow solution to the microPoiseuille flow. The distribution of the streamwise component of the velocity is parabolic, similar to Eq. (9.2.11). There is a correction in the amount of

$$-\frac{\mathrm{Re}}{2}\frac{dP}{dx}\alpha$$

to the gas velocity at the wall and to the entire profile of u. This is a positive quantity, indicating that for the same pressure gradient and Reynolds number, the first-order slip flow solution gives an increase of flow speed. The maximum flow speed $u_{\max}$ occurs at the centerline of the channel and can be written as

$$u_{\max} = -\frac{\mathrm{Re}}{2}\frac{dP}{dx}\left(\frac{1}{4} + \alpha\right)$$

For $u_{\max}$, the ratio of the slip flow correction to that of the continuum flow value is

$$\frac{-\dfrac{\mathrm{Re}}{2}\dfrac{dP}{dx}\alpha}{-\dfrac{\mathrm{Re}}{2}\dfrac{dP}{dx}\dfrac{1}{4}} = 4\alpha \tag{9.2.14}$$

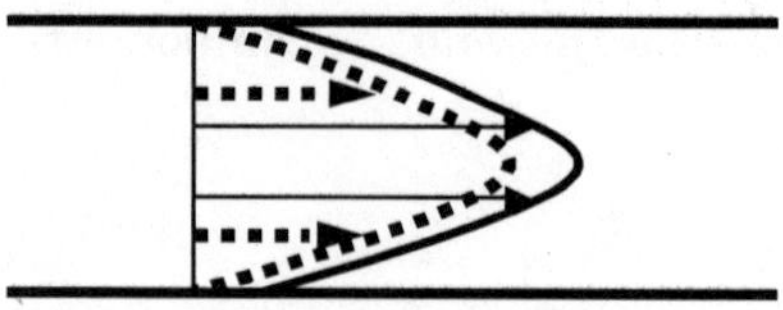

Figure 9.2.4 Qualitative comparison of the continuum and the slip Poiseuille flow solutions: - - - - continuum, —slip flow.

Figure 9.2.4 shows a qualitative comparison of the continuum and the slip flow solutions.

The volume flow rate per unit width can be obtained as

$$Q = \int_0^1 u\,dy = \int_0^1 \frac{\mathrm{Re}}{2}\frac{dP}{dx}(y^2 - y - \alpha)\,dy$$

which becomes

$$Q = -\frac{\mathrm{Re}}{2}\frac{dP}{dx}\left(\frac{1}{6} + \alpha\right)$$

The correction to the continuum solution is therefore

$$-\frac{\mathrm{Re}}{2}\frac{dP}{dx}\alpha$$

and the ratio of the correction to that of the continuum solution is

$$\frac{-\dfrac{\mathrm{Re}}{2}\dfrac{dP}{dx}\alpha}{-\dfrac{\mathrm{Re}}{2}\dfrac{dP}{dx}\dfrac{1}{6}} = 6\alpha \tag{9.2.15}$$

Example 9.2 For $\sigma_v = 0.9$ and $\mathrm{Kn} = 0.05$,

$$\alpha = \frac{2 - \sigma_v}{\sigma_v}\mathrm{Kn}$$
$$= 0.06$$

The increase of the volume flow rate and the maximum streamwise velocity, according to Eqs. (9.2.14) and (9.2.15) are 36 and 24 percents, respectively.

Using the second-order velocity slip boundary conditions of Karniadakis and Beskok (2002), the velocity profile with the second-order correction can be obtained.

$$u = \frac{\mathrm{Re}}{2}\frac{dP}{dx}\left(y^2 - y - \frac{2 - \sigma_v}{\sigma_v}\frac{\mathrm{Kn}}{1 + \mathrm{Kn}}\right) \tag{9.2.16}$$

The slip velocity correction of the second order is then

$$-\frac{\mathrm{Re}}{2}\frac{dP}{dx}\frac{\alpha}{1+\mathrm{Kn}} \tag{9.2.17}$$

Similarly, one can find the corrections for other important flow parameters.

The above results show that for microflows in the slip flow regime, say with $0.01 < \mathrm{Kn} < 0.1$, the velocity slip effects are important. The fact that gas molecules slip over solid surface accounts significantly for large deviation from continuum solutions and the effects of velocity slips need to be considered in microfluidic microelectromechanical systems (MEMS) design. The slip flow corrections remain important when the effect of compressibility and thermal conditions are considered (Harley et al. 1995). The compressible analyses are involved and not described here. Further examination of the low-speed microflows will be shown in the later section by comparisons with DSMC and Burnett equations solutions.

9.3 Numerical Flow Simulations

9.3.1 Subsonic flow boundary conditions

Microflows often operate with a given pressure (gradient) at the inlet and the outlet boundaries. In this section, an implicit treatment for low-speed inlet and exit boundaries for the DSMC of microflows in such operating conditions is briefly described for completeness. A detailed derivation can be found in Liou and Fang (2000).

From the microscopic point of view, gas molecules translate, in addition to a mean molecular velocity, by the thermal or random velocity. For flows at high speeds, such as hypersonic flows, the thermal velocity can be smaller in magnitude compared with the mean velocity. For a DSMC simulation of high-speed flow, a conventional approach is to impose a "vacuum" condition at the exit boundary, where no molecules enter the computational domain from the region external to the flow domain. For the low-speed flows in fluidic MEMS, the thermal motion can be of the same order of magnitude as the mean molecular motion. It then becomes inappropriate to neglect the mass influxes due to thermal motion at, for example, an exit flow boundary. A boundary treatment has been proposed for such MEMS flow simulations in Liou and Fang (2000). In this method, the number of molecules entering the computational domain and their corresponding internal energy and velocity components are determined in an implicit manner by the local mean flow velocity, temperature, and number density. The number flux of the molecules entering the computational domain can be described by using the Maxwellian distribution function,

$$F_j = \frac{n_j}{2\sqrt{\pi}\beta_j}\{\exp(-s_j^2\cos^2\vartheta) + \sqrt{\pi}s_j\cos\vartheta[1 + \mathrm{erf}(s_j\cos\vartheta)]\} \tag{9.3.1}$$

where

$$s_j = U_j\beta_j \qquad \text{and} \qquad \beta_j = 1/\sqrt{2RT} \tag{9.3.2}$$

F_j represents the number flux through a cell face of the boundary cell j. "erf" denotes the error function, R the gas constant, and n_j the number density of molecules in cell j. T_j and U_j denote the local temperature and the streamwise mean velocity component, respectively. The value of ϑ is zero for the upstream boundary and π for the downstream exit boundary. The velocity components of the entering molecule can be determined by using the acceptance–rejection method of Bird (1994) and the Maxwellian distribution function.

At the upstream inlet boundary, the streamwise velocity c_x, the cross-stream velocities c_y, and c_z of the molecules entering the computational domain through the cell face of a boundary cell j can be written as,

$$\begin{aligned} c_x &= (U_j + 3c'_{\mathrm{mp}})R_f \\ c_y &= A\cos\varphi \\ c_z &= A\sin\varphi \end{aligned} \tag{9.3.3}$$

where

$$A = \sqrt{-\ln(R_f)c'_{\mathrm{mp}}} \qquad \text{and} \qquad \varphi = 2\pi R_f \tag{9.3.4}$$

and R_f represents a random fraction number and c'_{mp} the local most probable thermal speed of molecules.

At the downstream, the velocity components for the molecule entering the computational domain through the exit flow boundary are

$$\begin{aligned} c_x &= (U_j - 3c'_{\mathrm{mp}})R_f \\ c_y &= V_j + A\cos\varphi \\ c_z &= A\sin\varphi \end{aligned} \tag{9.3.5}$$

where V_j denotes the local transverse mean velocity.

With the vibrational energy neglected, the internal energy of the entering equilibrium gases of diatomic molecule consists of translational energy E_{tr} and rotational energy E_{rot},

$$E_{\mathrm{tr}} = \frac{m\,|\mathbf{c}|^2}{2} \tag{9.3.6}$$

$$E_{\mathrm{rot}} = -\ln(R_f)kT \tag{9.3.7}$$

where $\mathbf{c}$ is the velocity of an entering molecule, m the mass of the simulated gas, and k the Boltzmann constant.

To implement the conditions shown in Eqs. (9.3.1–7), the number density, temperature, and the mean velocity at the flow boundaries are needed. At the upstream boundary, the pressure p_{in} and density ρ_{in} are normally given. The number density n_{in} and temperature T_{in} can then be obtained according to the conservation of mass and the equation of state. That is,

$$n_{\mathrm{in}} = \frac{\rho}{m} \qquad \text{and} \qquad T_{\mathrm{in}} = \frac{p_{\mathrm{in}}}{\rho_{\mathrm{in}} R} \tag{9.3.8}$$

The transverse mean velocity $(V_{\mathrm{in}})_j$ is set zero. A first-order extrapolation is used to determine the streamwise mean velocity $(U_{\mathrm{in}})_j$ from that of the computed for cell j. That is,

$$(U_{\mathrm{in}})_j = U_j \tag{9.3.9}$$

At the downstream boundary, the only given flow parameter is normally the exit pressure p_e. The other mean properties of the flow are to be determined implicitly as the calculation proceeds. In the present method, the flow variables are first computed by the following characteristics theory-based equations:

$$(n_e)_j^k = n_j^k + \frac{p_e - p_j^k}{m\left(a_j^k\right)^2} \tag{9.3.10}$$

$$(u_e)_j^k = u_j^k + \frac{p_j^k - p_e}{m n_j^k a_j^k} \tag{9.3.11}$$

$$(v_e)_j^k = v_j^k \tag{9.3.12}$$

$$(T_e)_j^k = \frac{p_e}{m(n_e)_j^k R} \tag{9.3.13}$$

The subscript e denotes the exit boundary, superscript k the computed quantities at the kth step, and the a_j^k local exit speed of sound. The exit mean flow velocities can then be obtained through a sampling of the following form:

$$(U_e)_j = \frac{1}{N_j} \sum_{i=1}^{N_j} c_{xe_i} \tag{9.3.14}$$

$$(V_e)_j = \frac{1}{N_j} \sum_{i=1}^{N_j} c_{ye_i} \tag{9.3.15}$$

Nance et al. (1998) first proposed an application of the characteristics theory that has been used extensively in continuum computational fluid dynamics (CFD) (Whitfield and Janus 1984) in DSMC simulations. The current implementation of the characteristic equations involves an updating procedure for the mean flow properties at the exit boundary as the DSMC calculation proceeds. This information, in turn, is used in Eqs. (9.3.1) and (9.3.3) to calculate the properties of the molecules entering the computational domain. It is shown in the following results that the implicit boundary treatment successfully drives a microchannel flow to a stationary state with matching exit pressure and overall mass balance, which are the fundamental criteria for validating internal flow computation. Because of the statistical scatter of the DSMC method, this technique becomes inappropriate for flows of extremely low speeds. The IP method described earlier has been shown to work well in such conditions.

In the following, the results of the simulations for three types of microchannel flows are presented. The flows include microCouette flows and microPoiseuille flows with various values for the ratio of wall temperature to the inlet flow temperature. Sketches of the simulated geometries are shown in Figs. 9.2.1 and 9.2.3. The variable hard sphere (VSH) model (Bird 1994) and diffuse wall boundary condition are used for the collision process. Some of the computational results are presented and, where appropriate, compared with the Navier-Stokes equations with slip boundary conditions.

9.3.2 MicroCouette flows

For the simulated microCouette flow of nitrogen, the upper wall moves with a speed U of 100 m/s. The pressure at the inlet and the outlet boundaries are both set at 0.83 atm ($p_{\text{in}} = p_e = 0.83$ atm). The inlet flow temperature T_{in} and the wall temperature T_w are set equal to 300K. The channel length L is 4.0 μm. Results will be shown for two channel heights h of 0.4 and 0.8 μm. The Knudsen numbers, based on the channel height and the inlet conditions, are 0.163 and 0.08, respectively. The physical parameters are set such that low-speed microflows are simulated. The computational grid consists of 100 × 60 uniform rectangular cells. The number of the simulated molecules in the DSMC simulations is about 320,000. Runs of 200,000 time steps and sampling after 2000 time steps of development on SGI Octane typically take roughly 48 hours of CPU time with a single processor.

Figures 9.3.1*a* and 9.3.1*b* show the predicted velocity profiles for Kn = 0.08 and 0.163, respectively. The corresponding linear profiles given in Eq. (9.2.7) are also shown for comparison. Overall, there is a good agreement between the continuum solutions and the DSMC result for Kn = 0.08, which suggests that the implicit flow boundary condition described earlier is suitable for DSMC simulations of low-speed microflows. In the center portion of the channel, the DSMC solution agrees well with the continuum-based linear analytical solution. There appears to be a slight difference between the two solutions away from the center portion. For the case with a higher Kn (0.163), a nonlinear velocity profile was obtained by the DSMC simulation. Compared with the linear continuum solution, the DSMC method has predicted a velocity profile with a lower value of slope in the center portion of the channel and with slight curvatures approaching the wall. Figure 9.3.2 shows the velocity differences between the DSMC results and the continuum analytical solution across the channel for both cases. The nonlinear, wavy behavior of the DSMC solution is evident, especially for the higher Kn case. The difference in the wall slip velocity between the DSMC and the analytical solution is small. For example, for Kn = 0.163, the calculated slip velocity on the lower wall, 11.87 m/s, is roughly 3 percent

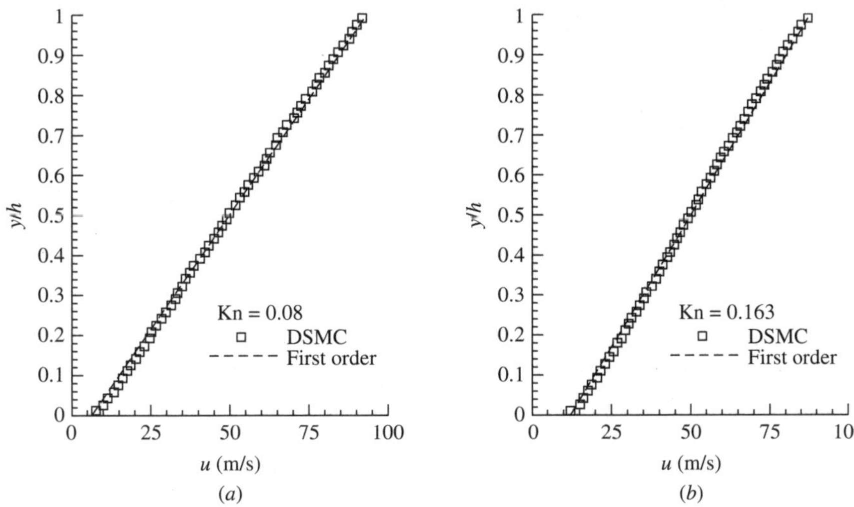

Figure 9.3.1 Comparison of the velocity profiles for microCouette flows: (*a*) Kn = 0.08; (*b*) 0.163. (*Liou and Fang 2000*.)

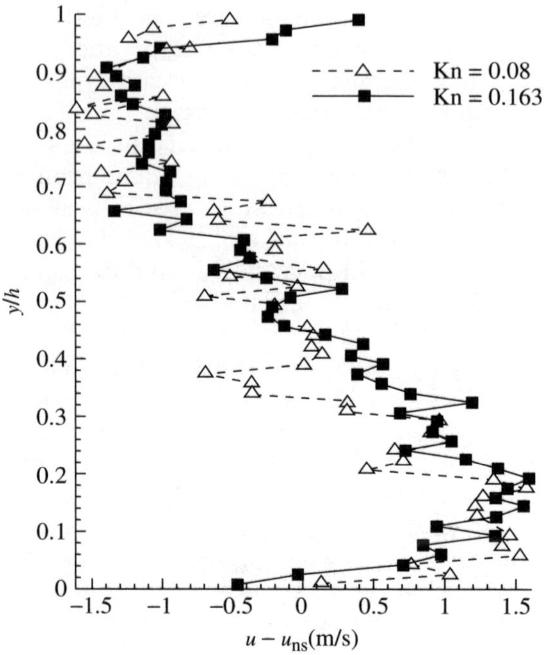

Figure 9.3.2 Distribution of the velocity difference between the DSMC and the continuum flow solutions. (*Liou and Fang 2000.*)

lower than that of given by Eq. (9.2.7). It suggests that the difference between the velocity profiles predicted by the DSMC method and the analytical method based on the continuum flow assumption is not necessarily due entirely to the low-order accuracy of the slip flow boundary condition.

The distribution of temperature in the wall-normal direction, at the station $x/L = 0.5$, is shown in Fig. 9.3.3 for the microCouette flow. Also included is an analytical approximation, which is derived from a simplified form of the Navier-Stokes equations using the first-order wall velocity-slip and the first-order wall temperature-jump condition of Maxwell. The analytical approximation can be written as

$$T = T_w + \frac{1}{2}\frac{\mu U^2}{K}\left(\frac{1}{2\alpha + 1}\right)^2\left(-\left(\frac{y}{h}\right)^2 + \frac{y}{h} + \frac{2\gamma}{\gamma + 1}\frac{\alpha}{\mathrm{Pr}}\right) \qquad (9.3.16)$$

where α is defined in Eq. (9.2.6). K represents the thermal conductivity and Pr the Prandtl number.

The values used for μ, σ_v, K, and Pr for nitrogen are 1.656×10^{-5} kg/(ms^2), 1.0, 0.023 J/(msK), and 0.72, respectively. The analytical form

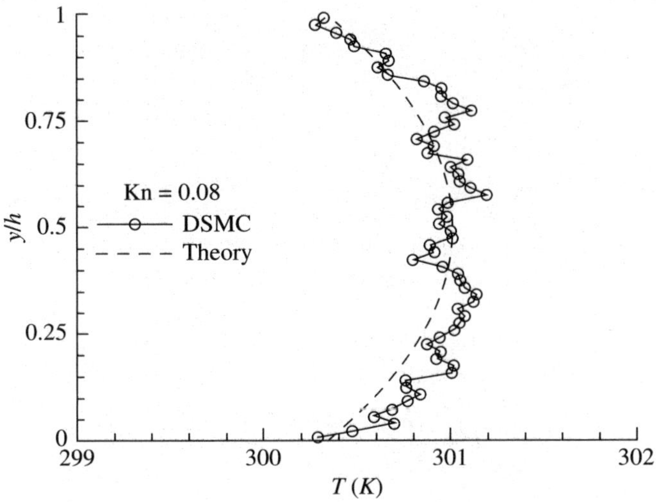

Figure 9.3.3 Comparison of the temperature profile for micro-Couette flows with Kn = 0.08. (*Fang and Liou 2002a.*)

gives a distribution that, in general, agrees well with the present DSMC results. The wall temperature-jump, about 0.3K, is predicted by both methods. The relative scattering error is less than 0.1 percent. Results at other locations are similar because of the one-dimensional nature of the microCouette flow. For the flow velocity, a distribution that shows a slight deviation from a linear variation is observed.

9.3.3 MicroPoiseuille flows

Two microPoiseuille flows were calculated. The height of the microchannel h is 0.4 μm for both cases and the channel length L is 2.0 μm. The inlet temperature is 300K and the wall temperature is 323K. The inlet pressures are 2.5 and 0.72 atm for Case 1 and Case 2, respectively. The pressure ratios are 2.5 for Case 1 and 4.54 for Case 2. A constant velocity of 100 m/s is used in the computational region, including the flow boundaries, to initiate the simulations. The operating conditions resulted in the local Kn to vary from 0.055 at the inlet to 0.123 at the exit for Case 1. For Case 2, they change from 0.12 to 0.72. The computational grid contains 100×60 rectangular cells. The simulated number of molecules is about 180,000. Runs of 200,000 time steps, with sampling after 2000 DSMC time steps, on SGI Octane typically took nearly 36 h of CPU time with a single processor. Significant speedup can be and has been achieved with a parallel version of the DSMC code (Fang and Liou 2002*b*). A multifold speedup is normally obtained. The results shown in the following for Case 2 were obtained using a dual-processor SGI Octane.

Figure 9.3.4*a* shows the predicted pressure distribution along the centerline of the channel for Case 1. A continuum-based analytical form (Piekos and Breuer 1995) was also included for comparison. Figure 9.3.4*b* shows the relative difference of the pressure distributions between the DSMC result and the analytical solution. The maximum difference is roughly 2 percent, indicating that the current flow boundary treatment is appropriate to use with the DSMC techniques. Both methods predicted pressure variations that are nonlinearly distributed along the channel, which has been observed experimentally for flows in microchannels. Figure 9.3.4*c* compares the deviations of the DSMC and of the analytical pressure drops from the linear distribution p_l. The nonlinearity exhibited by the DSMC solution is 0.5 percent lower than that of the continuum flow solution.

Figure 9.3.5 shows a comparison of the DSMC velocity profiles with the continuum-based analytical solution of Eqs. (9.2.13) and (9.2.16) at

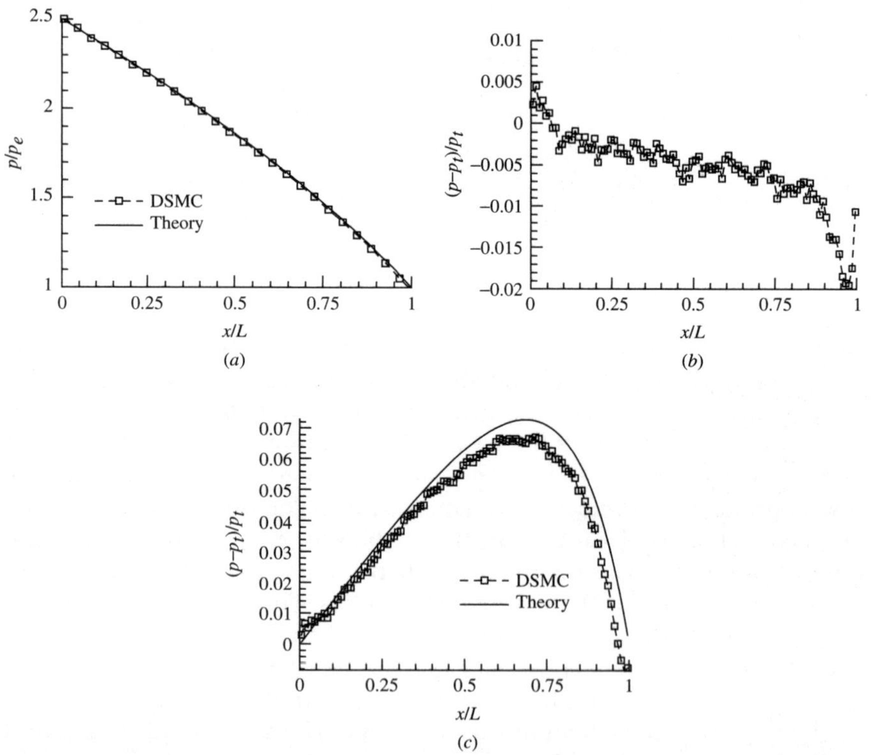

Figure 9.3.4 Comparison of pressures of microPoiseuille flow. Case 1: (*a*) Pressure distributions; (*b*) Relative errors; (*c*) Deviations from linear pressure drop. (*Liou and Fang 2000.*)

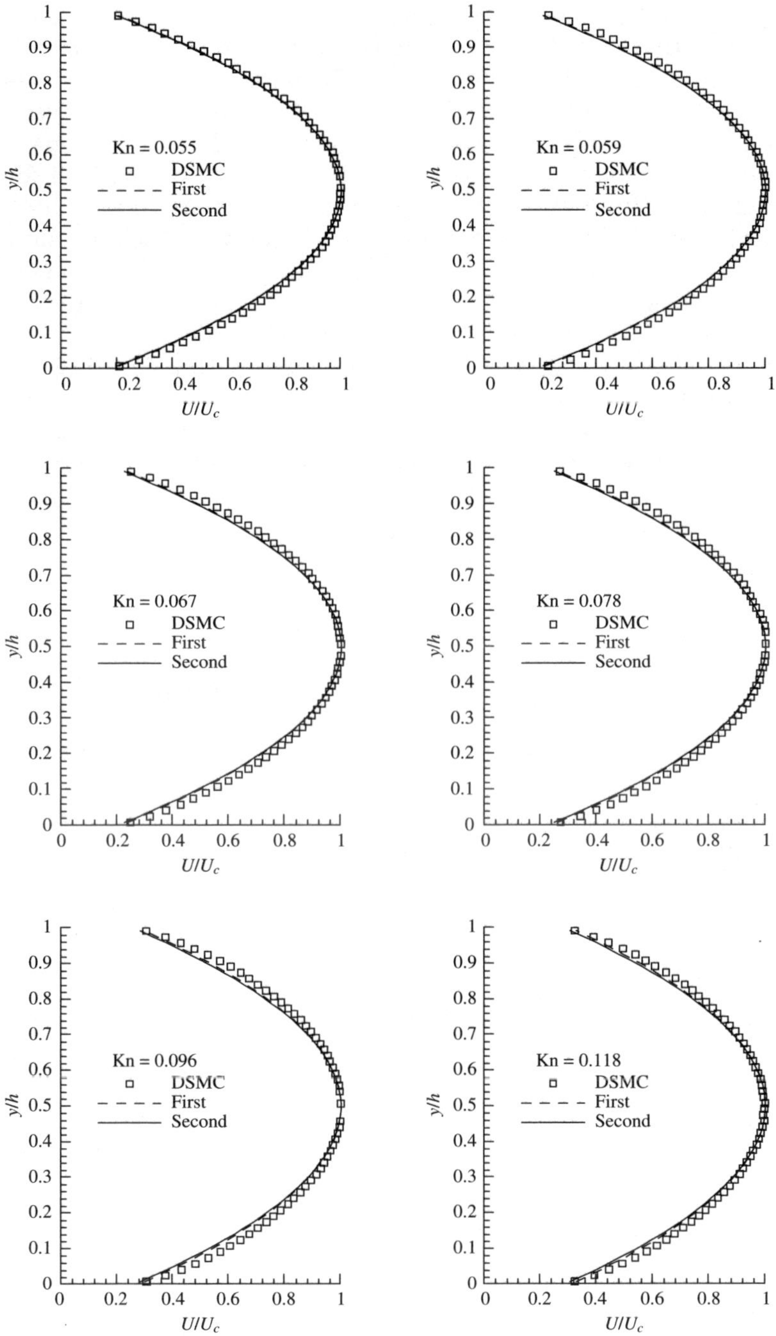

Figure 9.3.5 Comparison of the velocity profiles of microPoiseuille flow. Case 1. (*Liou and Fang 2000.*)

six different locations along the channel. The velocity profiles have been normalized by their respective maximum value, which occurs at the centerline of the channel, denoted by U_c. The value of the Kn changes from 0.055 to 0.118. For the small Kn, there is little difference between the first- and second-order accurate continuum-based velocity profiles. The calculated profiles agree well with the continuum solutions for all the stations compared. In this low Kn range, the continuum analytical solution does provide an approximated solution to the microflow considered. The results show that the present implicit flow boundary treatment is consistent the DSMC procedure and has produced accurate numerical predictions to the microflow considered.

The pressure ratio for Case 2 is 4.54, compared with 2.5 for Case 1. The local Kn is higher than that of Case 1. Figure 9.3.6*a* shows the pressure distribution along the centerline of the channel. Figure 9.3.6*b* shows the difference of pressure distribution between the DSMC

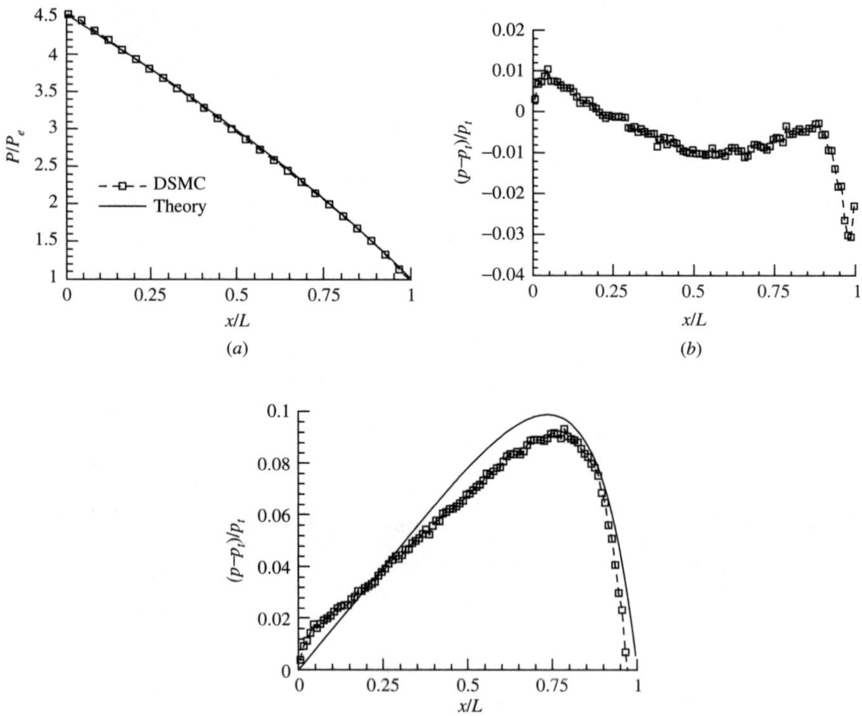

Figure 9.3.6 Comparison of the pressures of microPoiseuille flow. Case 2: (*a*) Pressure distributions; (*b*) Relative errors; (*c*) Deviations from linear pressure drop. (*Liou and Fang 2000.*)

solutions and the continuum-based analytical form, with a maximum difference of 3 percent. Compared with Case 1, a stronger nonlinearity is predicted by both methods. Figure 9.3.6*c* shows that the deviation from the linear distribution is now 9 percent for the DSMC solution and 10% for the continuum-based analytical solution.

Figure 9.3.7 shows the calculated mean velocity profiles at six different stations along the channel. Continuum-based solutions using the first- and second-order slip wall conditions were also included for comparison. The difference between the first-order profiles and those of the second order for this case are more significant than those for Case 1. While the second-order profiles give reasonable approximations to the DSMC results for Kn up to 0.641, the first-order profiles move away from the DSMC results as the value of Kn increases, with the largest difference occurring at the channel wall.

Figure 9.3.8 shows the evolution of the calculated pressure at the downstream boundary, nondimensionalized by the imposed exit pressure p_e as the solution progresses for Case 2. In each print cycle, there are ten time steps of sampling. The calculated pressure converges to the imposed value, e.g. $p/p_e = 1$, after the transient variation from the uniform initial conditions has subsided.

Figure 9.3.9 shows the approach to a steady-state solution using the same print cycles as in Fig. 9.3.8 for Case 2 with two different initial streamwise velocities. As was described earlier, an initial uniform velocity field (U_{initial}) is required to start the calculations and $U_{\text{initial}} =$ 100 m/s has been used. For a pressure-driven internal flow, this computational initial condition set at the interior flow domain and the flow boundaries where the boundary treatment is used should have no bearing on the final DSMC numerical solution. For validation, Fig. 9.3.9 also includes results for a second $U_{\text{initial}}(=15$ m/s). Figure 9.3.9*a* shows the variation of the mass fluxes at the upstream and downstream boundaries. The mass flux is defined as

$$\dot{m} = \frac{1}{h}\int_0^h \rho u \, dy \tag{9.3.17}$$

The corresponding initial values are denoted by the solid symbols in Fig. 9.3.9. The initial mass flux is about 12 kg/ms for the case with $U_{\text{initial}} = 15$ m/s and 80 kg/ms for $U_{\text{initial}} = 100$ m/s. For both cases Fig. 9.3.9*a* shows that, after a transient from the uniform initial conditions, the mass fluxes at the upstream and the downstream converge to the same constant value and an overall mass balance in the microchannel is established. Figure 9.3.9*a* also shows that the converged mass fluxes for the two different initial fields agree well, which is about 39.4 kg/ms with a 0.25 percent variation. Figure 9.3.9*b* shows

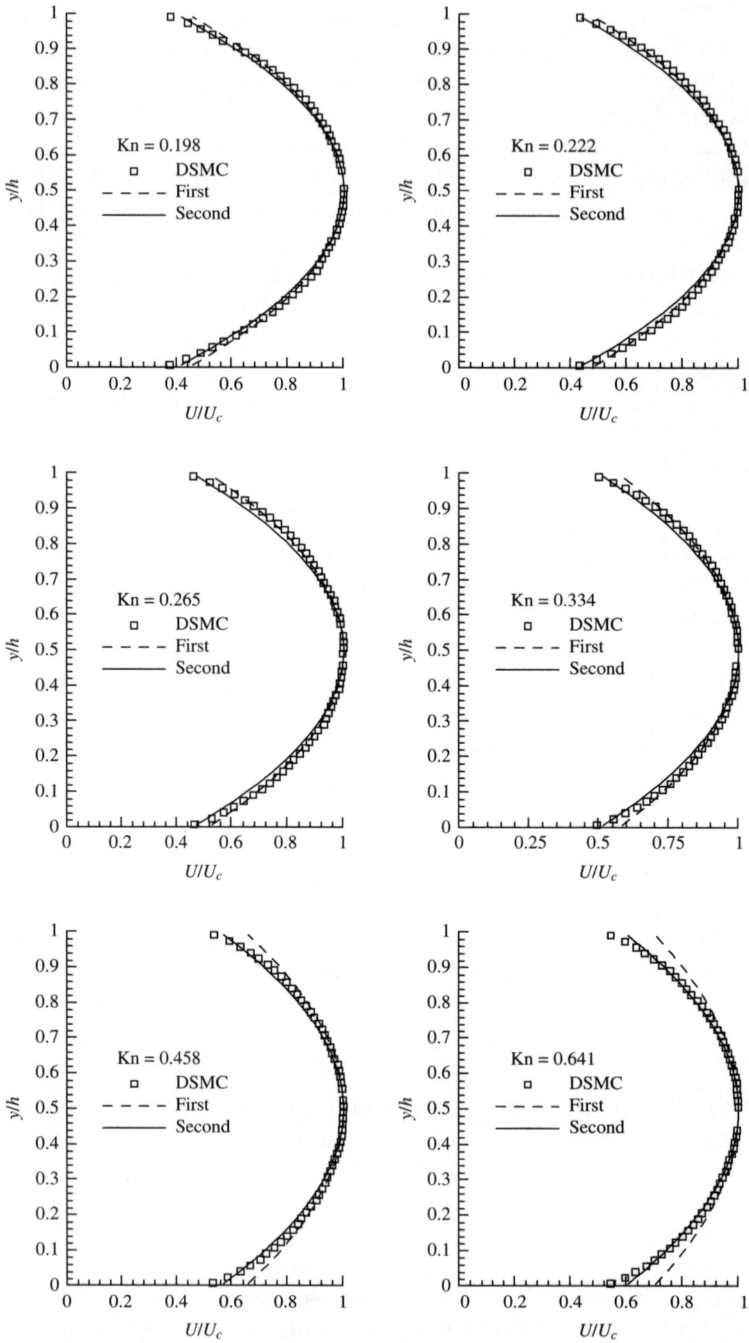

Figure 9.3.7 Comparison of the velocity profiles of microPoiseuille flow. Case 2. (*Liou and Fang 2000.*)

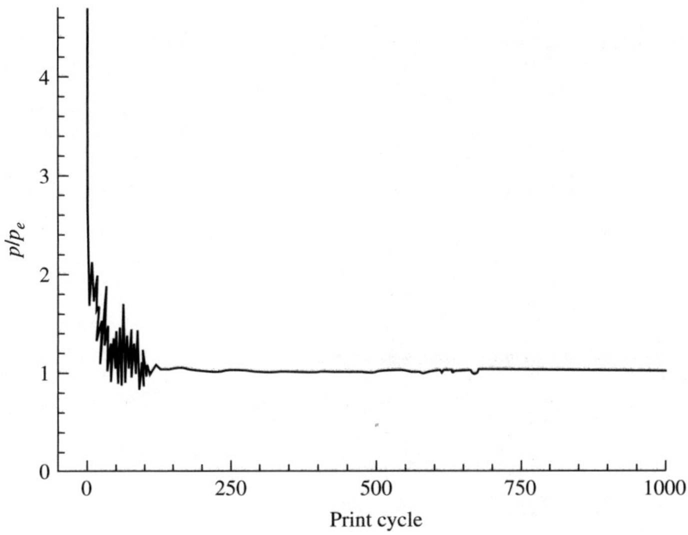

Figure 9.3.8 Evolution of the downstream pressure. Case 2. (*Fang and Liou 2002a.*)

the variation of the average streamwise velocity component for an inlet cell and an outlet cell located at $(0, h/2)$ and $(L, h/2)$, respectively. For both initial conditions, the streamwise velocity component for the inlet cell converges to 58.78 ($\pm$0.54%) m/s. For the outlet cell, the velocity converges to 232.05 ($\pm$0.32%) m/s. These results agree well with Bird's calculation using extended buffer domains and vacuum boundary conditions (Bird 2002). The results shown in Fig. 9.3.8 and 9.3.9 indicate

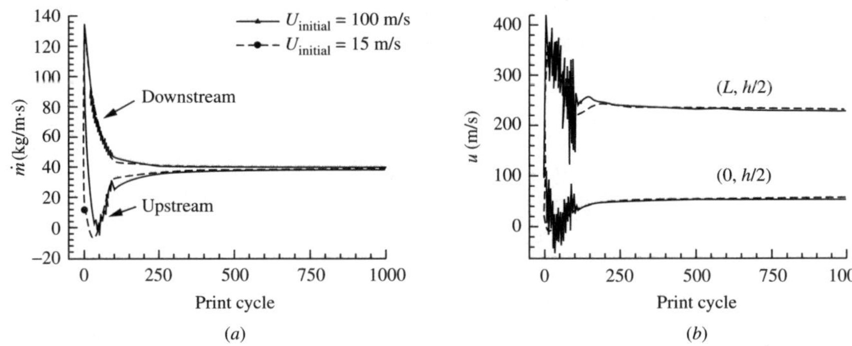

Figure 9.3.9 (*a*) Evolution of mass fluxes; (*b*) Evolution of the streamwise velocities. Case 2. (*Fang and Liou 2002a.*)

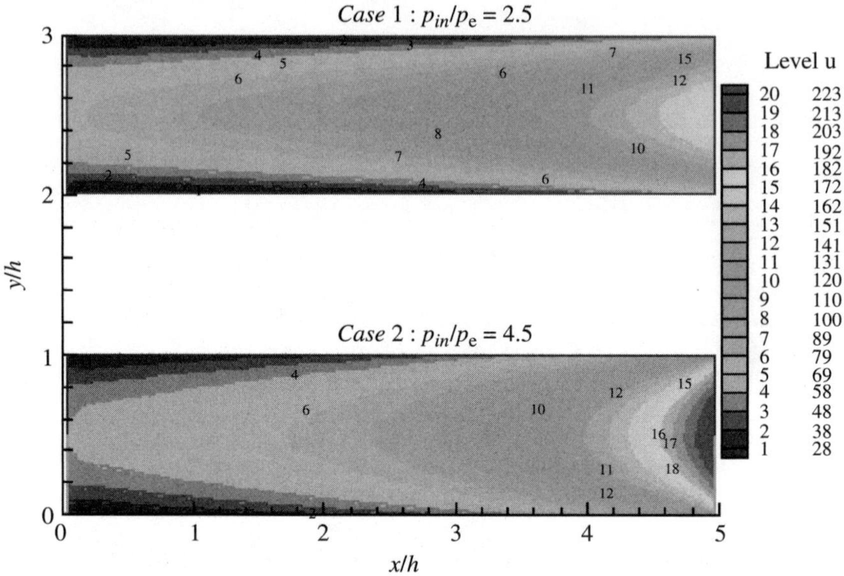

Figure 9.3.10 Streamwise velocity magnitude contours. (*Fang and Liou 2002a.*)

that the present implicit treatment of the low-speed flow boundaries supports a stable and efficient solution process for the DSMC of the internal microflows studied.

Figure 9.3.10 shows the contours of the streamwise mean velocity magnitude for Cases 1 and 2. The flow velocity increases as it develops downstream. Partly because of the high-pressure gradient, the flow speed at the exit of Case 2 is higher than that of Case 1. There is only small statistical scattering in the computed velocity magnitude for both cases.

Figure 9.3.11 shows the variations of the gas slip-velocity along the wall for both cases. The slip-velocity is defined as

$$u_s = u_g - u_w \tag{9.3.18}$$

where u_g represents the gas velocity on a wall and u_w the wall velocity. It can be seen that the slip-velocity increases as the flow develops downstream. For the high Kn case, Case 2, the increase is more significant. At the exit of the microchannel, the slip-velocity for Case 2 is nearly twice as high as that of Case 1.

Analytical forms of the velocity distributions can be obtained using the Navier-Stokes equations with slip-wall conditions described in

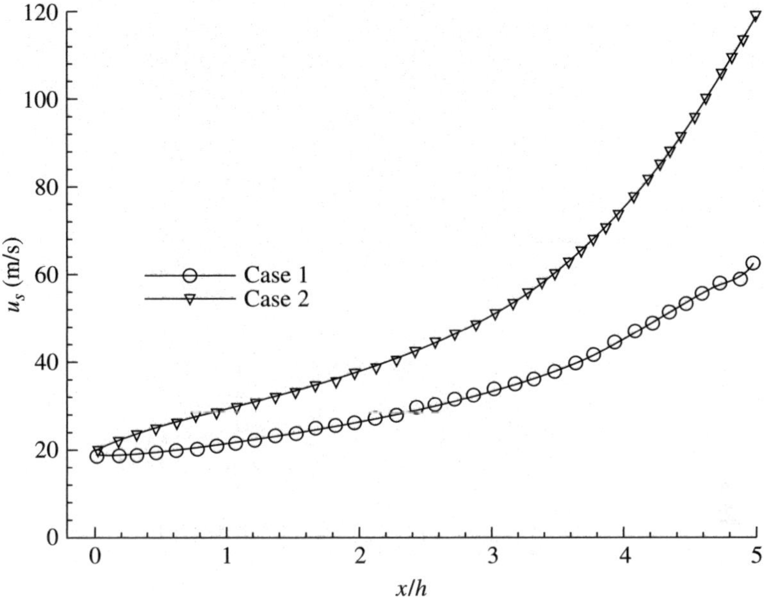

Figure 9.3.11 Slip velocity distributions on the wall. (*Fang and Liou 2002a.*)

Eqs. (9.2.14) and (9.2.17). The resulting wall slip velocity ratio can be written as

$$\frac{u_s}{u_c} = \mathrm{Kn}\bigg/\left(\frac{1}{4} + \mathrm{Kn}\right) \tag{9.3.19}$$

and

$$\frac{u_s}{u_c} = \left(\frac{\mathrm{Kn}}{1+\mathrm{Kn}}\right)\bigg/\left(\frac{1}{4} + \frac{\mathrm{Kn}}{1+\mathrm{Kn}}\right) \tag{9.3.20}$$

where Kn is the local Knudsen number evaluated on the wall, which varies with the streamwise flow development. Equation (9.3.19) is obtained by using the first-order slip-wall condition proposed by Maxwell and Eq. (9.3.20) by using Karniadakis and Beskok (2002) second-order condition. In Fig. 9.3.12, the wall slip velocity given by Eqs. (9.3.19) and (9.3.20) are compared with the present DSMC predictions. For Case 1 with the lower Kn, the continuum-based analytical values agree well with the DSMC results in the entire microchannel. As Kn is low, it is not surprising to observe such a good agreement between the continuum-based solutions and the current DSMC results. This agreement also further validates the DSMC solver used in the present study.

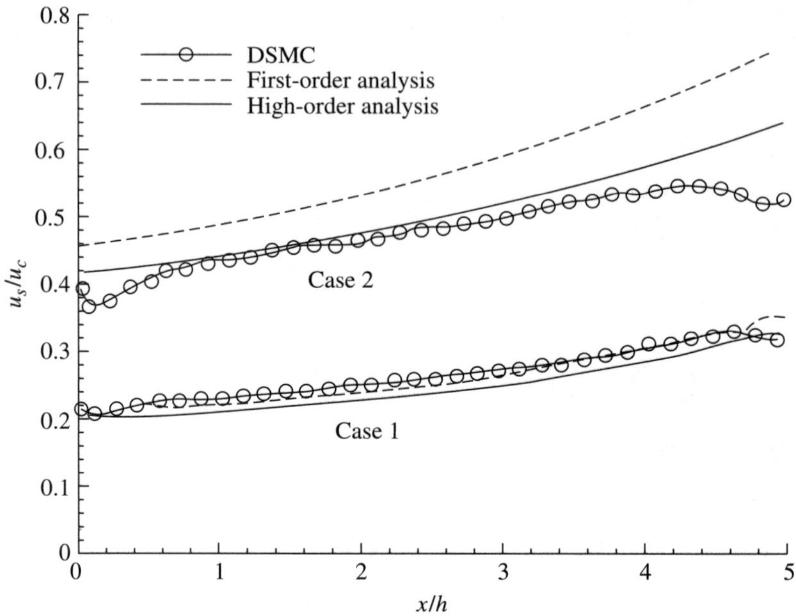

Figure 9.3.12 Comparison of normalized slip velocity distributions on the wall. (*Fang and Liou 2002a.*)

The difference between the first-order and the high-order approximations becomes more apparent for Case 2, where the Kn is large (>0.19). In fact, Fig. 9.3.12 shows that the first-order approximation is no longer valid for Case 2, producing a velocity slip significantly higher than both the DSMC result and the high-order analysis.

The difference between the wall temperature and the gas temperature near the wall, or the temperature-jump, for these two cases are given in Fig. 9.3.13. For both cases, there is a gradual decrease of temperature-jump in the first half of the channel, from about 10K to nearly the same, suggesting a corresponding gradual decrease of the heat transfer from the wall to the microflow. In the second half of the channel, the wall temperature-jump for Case 2 shows a significant increase as the flow develops toward the exit than that for Case 1.

The calculated distributions of temperature are shown in Fig. 9.3.14. The flows are seen to have an increase of temperature as the flow develops in the first half of the channel. This is likely to be caused by the temperature difference between the wall and the flow. The contours, however, show a subsequent decrease of temperature in the downstream half of the microchannel for both cases, with a more pronounced reduction in Case 2.

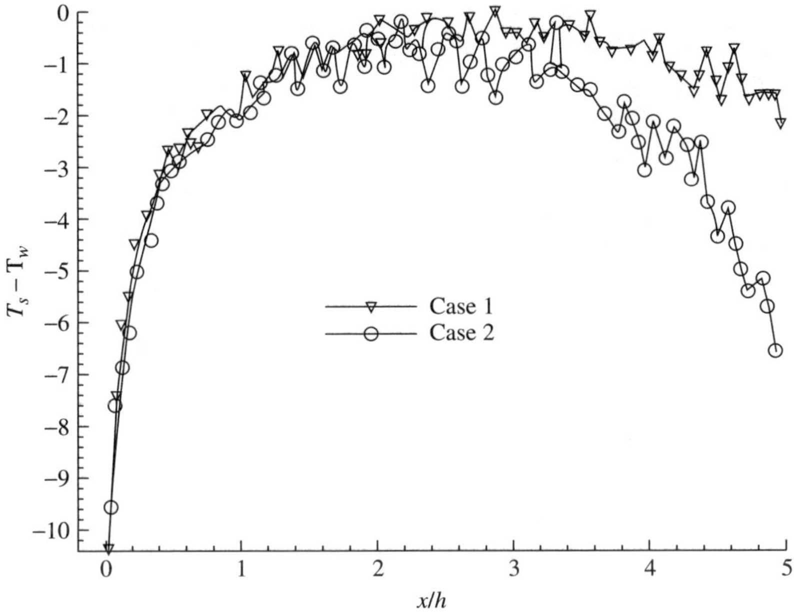

Figure 9.3.13 Comparison of wall temperature jump. (*Fang and Liou 2002a.*)

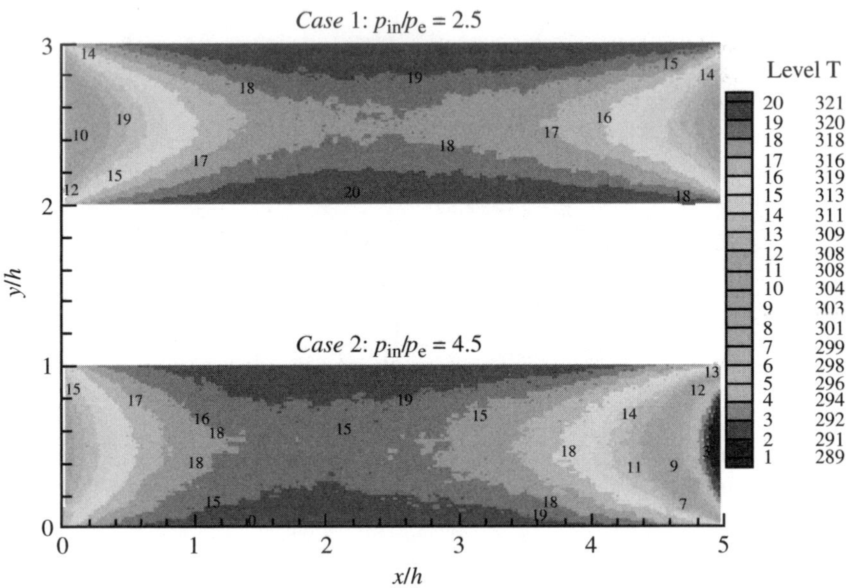

Figure 9.3.14 Temperature contours. Cases 1 and 2. (*Fang and Liou 2002a.*)

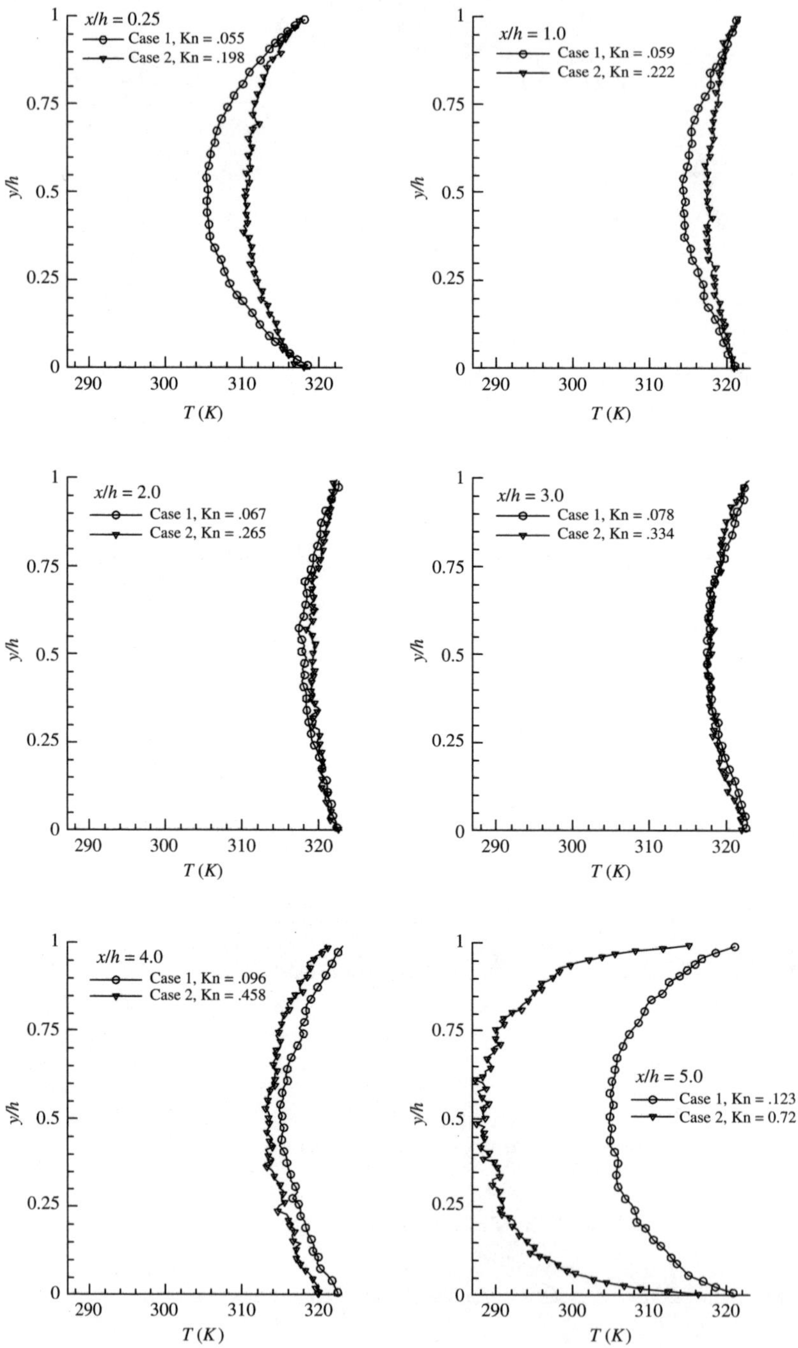

Figure 9.3.15 Comparison of temperature distributions. (*Fang and Liou 2002a.*)

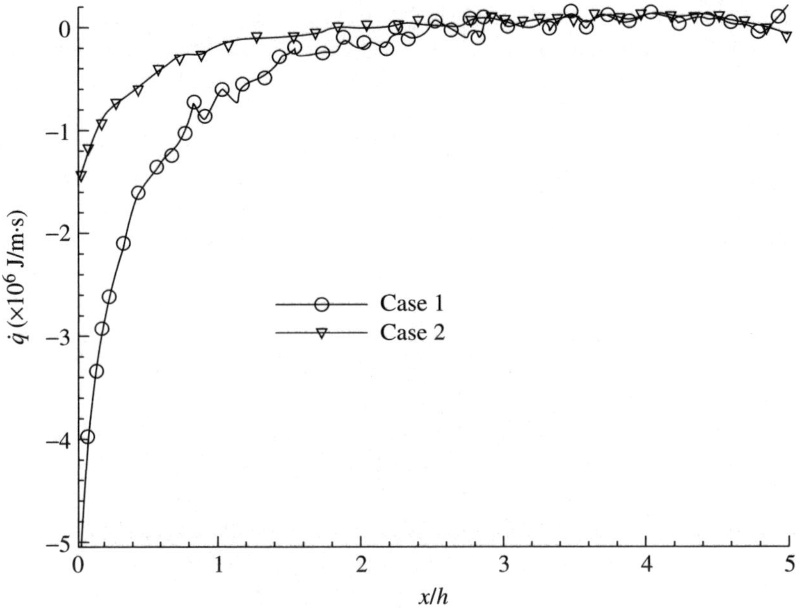

Figure 9.3.16 Wall heat transfer. (*Fang and Liou 2002a.*)

A different view of the temperature distribution is given in Fig. 9.3.15. The temperature distributions across the channel for these two cases develop in a similar manner before $x/h = 3.0$. Further downstream, the temperature decreases, with a more significant reduction for Case 2 with high Kn. Near the exit, the temperature in the centerline region of Case 2 is about 15K lower than that predicted for Case 1.

Figure 9.3.16 shows the wall heat transfer distributions. The net heat flux on a wall element with length $x\Delta$ can be evaluated as

$$\dot{q} = \frac{\left\{\left[\sum_{i=1}^{n}(E_{\text{tr}} + E_{\text{rot}})_i\right]_{\text{inc}} - \left[\sum_{i=1}^{n}(E_{\text{tr}} + E_{\text{rot}})_i\right]_{\text{ref}}\right\} \cdot N_0}{t_s(1 \cdot \Delta x)} \qquad (9.3.21)$$

where n denotes the total number of simulated molecules that strike the wall element during sampling, N_0 the number of gaseous molecules associated with a simulation cell, and t_s the time period of the sampling. The subscript "inc" and "ref" denote values before and after the molecule impacts the wall, respectively. Near the entrance region, there is a significant transfer of heat from the wall for both cases due to the high wall temperature. The level is higher for Case 1 than for Case 2 with higher Kn. As the flow develops downstream, the wall heat transfer diminishes and the difference between Case 1 and Case 2 falls within statistical error.

The results suggest that the flow expansion observed in Fig. 9.3.10 in the second half of the channel is nearly adiabatic for both cases, despite the finite temperature jump at the wall. This is particularly true for Case 2. Recall that the gas temperature near the exit for Case 2 is lower than that for Case 1 by about 15K and than the wall temperature by about 40K. It suggests that, for the high Kn case (Case 2), the large temperature difference between the wall and the flow is not accompanied by any significant amount of wall heat transfer. Note that the heat transfer characteristics of supersonic flows with three different values of Kn in a microchannel is shown in Chap. 10. For the case with the highest Kn (0.186), the results showed a significant increase of wall heat transfer with large temperature difference between the wall and the flow. On the other hand, the temperature along the streamwise direction develops in a manner similar to the current low-speed cases. That is, the temperature for the flow with large Kn is higher than that with low Kn at the upstream and, as the flow develops downstream, a more substantial drop of the flow temperature is found for cases with high Kn. These computational results appear to indicate that the heat transfer characteristics of high Kn microfluid flows at low speeds can be different from that at high speeds.

9.3.4 Patterned microchannel flow

Depending upon the specific applications, the internal flow passage in MEMS may contain various forms of partial blockages and cavities. These patterns can be micromachined with surface etching that selectively removes materials using, for example, imaged photoresist as a masking template. Figure 9.3.17 shows a typical cross-sectional view of such patterns. Compared with the microCouette and microPoiseuille flows described in the sections above, the flow inside the geometry shown in Fig. 9.3.17 is highly two-dimensional. To examine the thermal

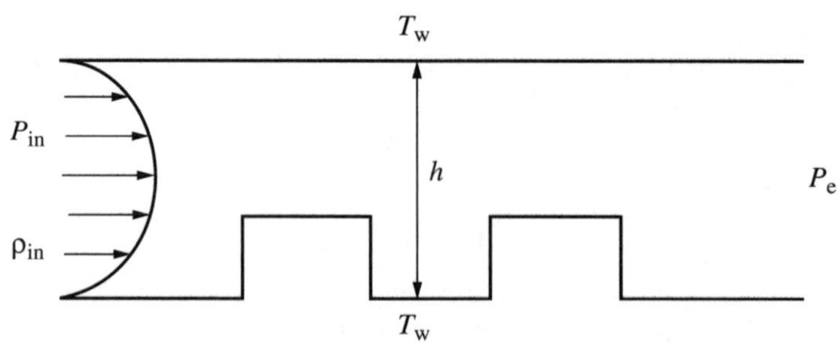

Figure 9.3.17 Sketch of a patterned microchannel. (*Liou and Fang 2002a.*)

and microflow phenomena in a more realistic geometry, the present DSMC procedure is applied to the microchannel shown in Fig. 9.3.17. The inlet pressure is 0.73 atm and temperature is 300K. For comparison, two cases with different exit pressure have been computed. The corresponding pressure ratios are 2.5 (Case 3) and 4.0 (Case 4), which are comparable to those of Case 1 and Case 2, respectively. The channel height is 0.9 μm with an aspect ratio of 6.7 and the block height is 0.3 μm. The wall temperature is 323K, except for the top of the blocks with a temperature of 523K. The Knudsen number, based on the inlet conditions, is about 0.08. The simulations have been performed in the parallel mode by domain decomposition (Fang and Liou 2002*b*). The number of the simulated molecules is about 1.62 million in each case, with 16,000 uniform rectangular cells. Runs of 200,000 time steps, with sampling after 2000 time steps of development, on ten processors of Pentium III-800 take roughly 50 h.

Figure 9.3.18 shows the temperature contours for both cases. Mushroom-like regions of high temperature can be observed surrounding the blocks, due to the high temperature at the top of the blocks. These high-temperature regions for the two cases are geometrically similar.

Figure 9.3.19 shows the streamwise variation of temperature at two height levels for both cases; one at the top of the blocks and the second at midway between the top of the blocks and the top wall. The temperature

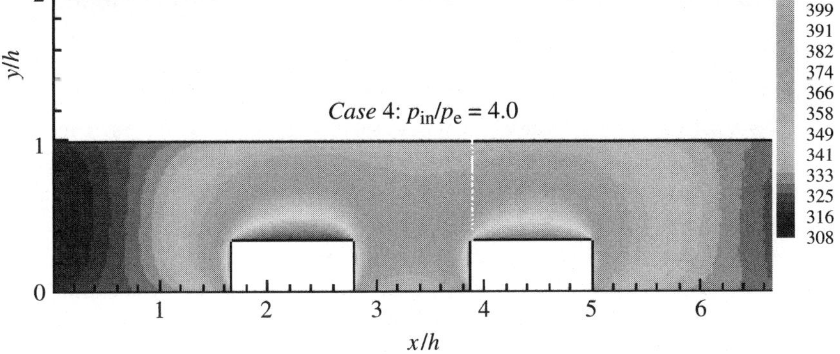

Figure 9.3.18 Temperature contours. Cases 3 and 4. (*Fang and Liou 2002a.*)

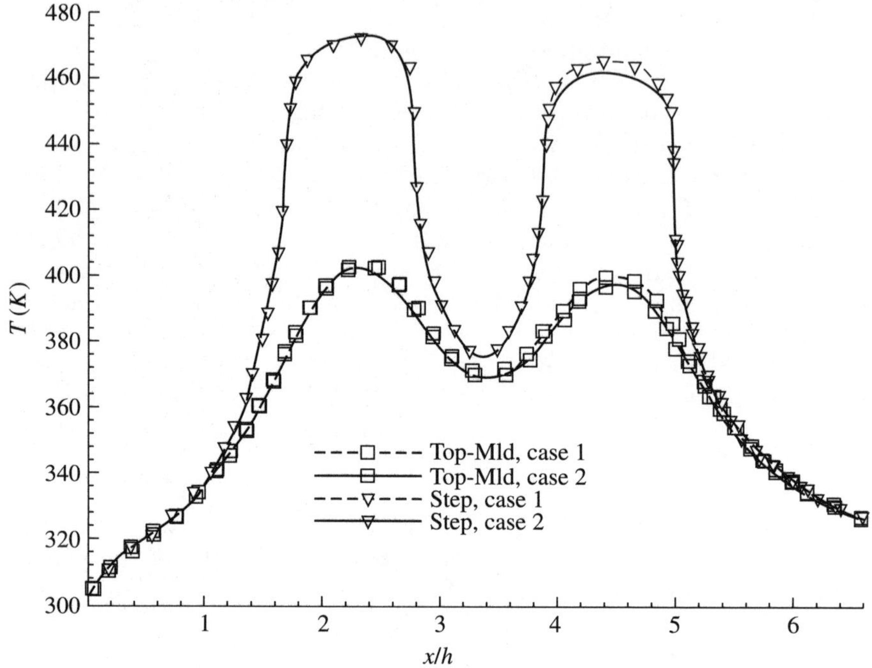

Figure 9.3.19 Temperature distributions. (*Fang and Liou 2002a.*)

in the regions above the blocks is significantly higher than the regions upstream and downstream of the blocks. The highest temperature change is roughly 60K between these two levels, which are 0.3 μm apart. Except for the region above the second block, there is a minimal difference in the streamwise temperature distributions between the two cases.

Figure 9.3.20 shows the temperature-jump distributions along the upper and the horizontal part of the lower walls. The temperature-jumps along both the upper and the lower walls are high in the regions around the blocks. The absolute value of the temperature-jump is higher on the top surface of the block than on the upper wall. For instance, on the top surface of the first block the temperature-jump is nearly 80K, while that for the upper surface is about 20K. Overall, the temperature-jump distributions for Case 3 are similar to those of Case 4.

A comparison of the heat transfer at the top channel wall for the two cases is given in Fig. 9.3.21. The peak values roughly correspond to the locations of the blocks. The heat transfer for the low pressure-ratio case (low Kn) is slightly higher than that for the higher value. This agrees well with the results for the microPoiseuille flows presented

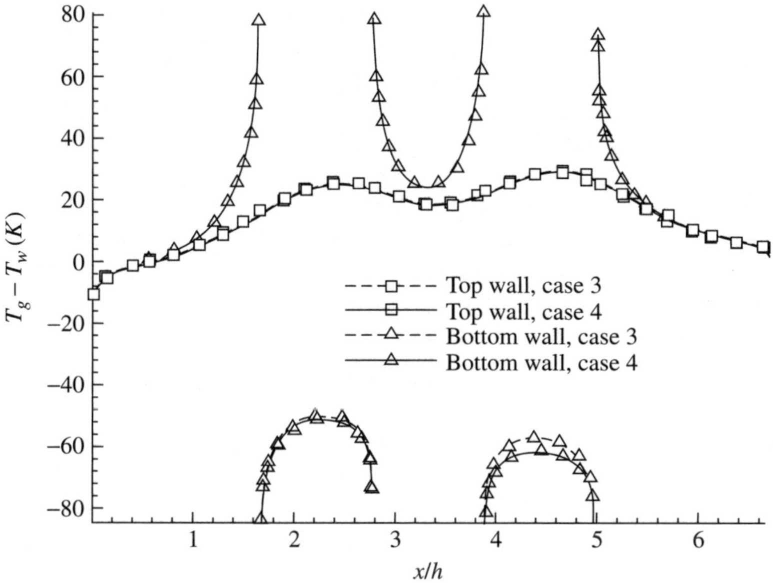

Figure 9.3.20 Temperature jump distributions on the upper and the lower wall. (*Fang and Liou 2002a.*)

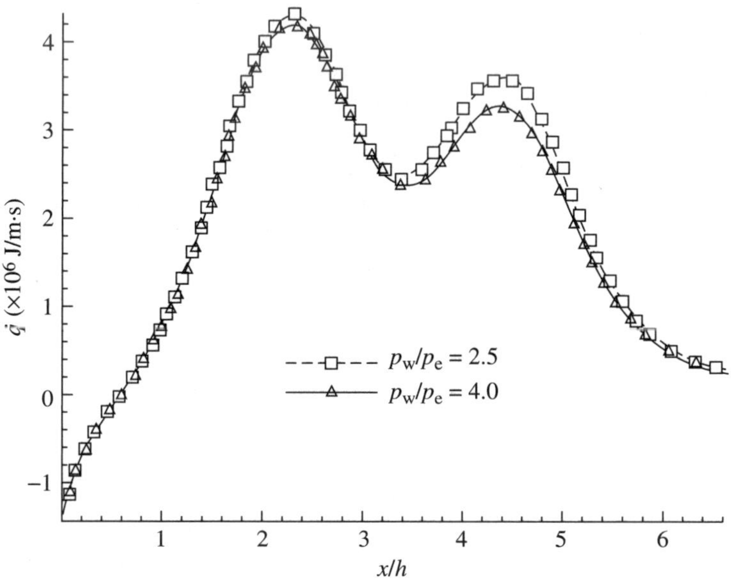

Figure 9.3.21 Comparison of heat transfer on the upper wall. (*Fang and Liou 2002.*)

above where the flow with high Kn shows less heat transfer compared to that with lower Kn.

Three types of subsonic microchannel flows have been computed using the DSMC method and an implicit treatment of the flow boundaries developed specifically for the simulation of low-speed microfluid flows. The heat transfer and the fluid dynamics of the computed flows are examined. Where applicable, the DSMC calculations were found to agree well with the analytical results. The wall heat transfer in the calculated subsonic microfluid flow decreases with an increase of Knudsen number. The present results show that the Knudsen number and the geometric complexity of the channel have significant effects on the fluid dynamic and the thermodynamic behavior of the microfluid flows studied.

9.3.5 Microchannels with surface roughness

There are experiments (Mala and Li, 1999) of microchannel flows that have reported that deviation of, for example, friction factor from the Moody chart, can depend on the material of the microchannel and on the flow transition to turbulence. The different materials produced different roughness on the wall. The reported mean roughness height was about 1.75 μm with a roughness/diameter value of about 0.035. According to the data, the roughness Reynolds number, a ratio of the roughness height and the near-wall viscous length scale, can be estimated to be about three. The roughness Reynolds number is below the critical Reynolds numbers for the roughness to affect the skin friction coefficient of a macroturbulent boundary layer (Schlichting and Gersten 2000). It appears that the effect of wall roughness on the microchannel flow can be more significant than that on macro scale flows. In this section, the effects of roughness are discussed based on the results obtained by using a phenomenological model (Liou and Fang, 2003) and a continuum approach. The model is built on simple scaling arguments and can best be used for qualitative, rather than quantitative, characterization of roughness effects. The model can be straightforwardly incorporated in the NB2D code attached in the back of this book.

The disturbance momentum transport generated by the roughness is assumed diffusive and the gradient assumption is used. The effective diffusivity associated with the roughness μ_R can then be written as

$$\mu_R \sim \rho u l \tag{9.3.22}$$

To proceed, it is assumed that the roughness has the same equivalent roughness height k and is uniformly distributed on the wall of the microchannel. The characteristic length scale l can be taken as the roughness height and the velocity scale u can be taken as the product of the mean rate of strain and the roughness height. For a microchannel, it can be written as

$$u \sim \left(\frac{dU}{dy}\right)_{y=k} k \tag{9.3.23}$$

Then

$$\mu_R \sim \rho \left(\frac{dU}{dy}\right)_{y=k} k^2 \tag{9.3.24}$$

The resulting effective diffusivity can then be high, say, a maximum value, near the wall and decreases as the distance from the wall increases. Or

$$\mu_R = c f_\mu \rho \left(\frac{dU}{dy}\right)_{y=k} k^2 \tag{9.3.25}$$

where c represents a model constant and f_μ a damping function for the assumed decreasing effects of roughness with the distance to the pipe wall. The model formulation is similar to what has been commonly used in turbulence modeling with the turbulent eddy viscosity evaluated by using the Prandtl's mixing length model. The total viscosity then is the sum of the molecular viscosity and the effective roughness viscosity.

$$\mu = \mu_m + \mu_R \tag{9.3.26}$$

A simple functional form that can be used for f_μ is

$$f_\mu = 1 - \exp\left[-\frac{k\,|y - h/2|}{(y - \beta)^2}\right] \qquad \beta = \begin{cases} 0, & y \le h/2 \\ h, & y > h/2 \end{cases} \tag{9.3.27}$$

The value of the damping function f_μ is unity near the wall and zero near the centerline. The value for the model constant c, determined currently by numerical runs, is set at 0.5.

Results of calculations of two air microchannels with roughness using the Navier-Stokes equations as well as the Burnett equations are presented. The slip wall boundary conditions were used in all the calculations. The first channel has a length L of 800 μm and a height of 50 μm. The outlet pressure is fixed at 0.5 atm. The wall temperature is set at 300K, which is also applied at the inlet of the channel. The Knudsen number, based on the channel height, is 0.0026. For this small Knudsen number case, the flow calculations were performed by solving the Navier-Stokes equations. Three different grids were used to evaluate grid independence. The grid sizes are 121×61, 181×31, and 241×41, respectively, in the streamwise and the vertical directions.

Figure 9.3.22 shows the variation of the mass flow rate with pressure gradient. The roughness heights calculated are 1.0 and 1.5 μm.

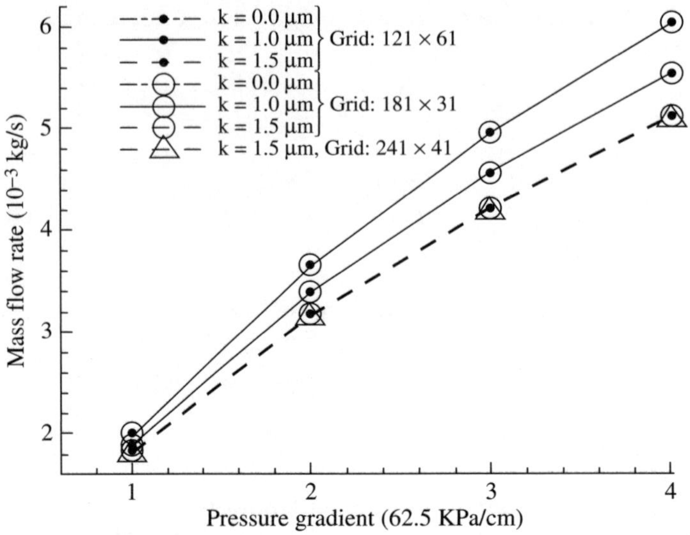

Figure 9.3.22 Changes of mass flow rate with pressure gradient.

The smooth-wall results ($k = 0.0\ \mu$m) are also included for comparison. The modeled roughness effects caused a reduction of the mass flow rate with the increase of the roughness height at the same pressure gradient. This effect of the roughness on the mass flow rate agrees with general experimental observations (Peiyi and Little 1983). As the pressure gradient increases, the differences among the smooth and rough wall

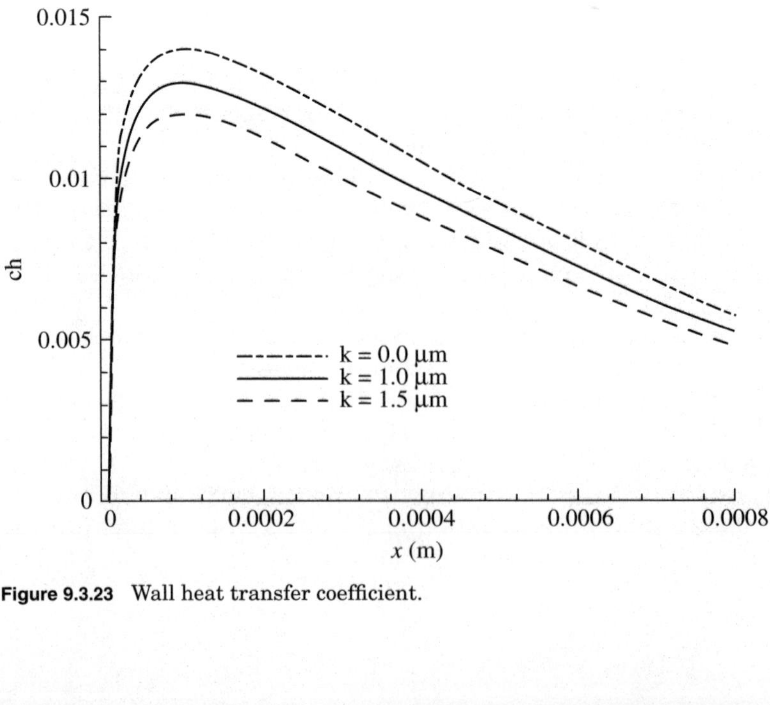

Figure 9.3.23 Wall heat transfer coefficient.

mass flow rates also increase, again, agreeing with experimental observations.

Figure 9.3.23 shows the distributions of the wall heat transfer coefficient with the pressure ratio of 1.2. With roughness, the results show a slight decrease of wall heat transfer as k increases. Their values remain positive in the entire length of the channel.

Figure 9.3.24 shows the variations of the wall slip velocity, normalized by the average velocity, along the microchannel. The average velocities u_{av} at $x/L = 0.5$ are 112, 104, and 96 m/s for the $k = 0.01$, 1.0, and 1.5 μm, respectively. The slip velocity is less than 0.1 percent of the average velocity and appears to increase with the roughness height. The Navier-Stokes equations calculations are then appropriate for this case.

The second channel calculated has a height of 4 μm. The length of the microchannel is 20 μm. The roughness height is 0.15 μm. The outlet pressure is 0.4 atm. The wall temperature and the inlet flow temperature are both set at 300K. The Knudsen number is about 0.04. The grid used for this case is 121 × 31. Because of the large Knudsen number, in addition to the Navier-Stokes equations, the Burnett equations have also been applied. Similar to the observations made previously for the 50 μm channel, the wall slip velocity for rough walls is higher than the smooth wall data (Fig. 9.3.25). The average velocities u_{av} at $x/L = 0.75$ are 130, 130, and 127 m/s for the Navier-Stokes, the Burnett and the Burnett with roughness calculations, respectively. The velocity slip is about 10 percent of the average velocity and is significantly higher than that in the previous small Knudsen case.

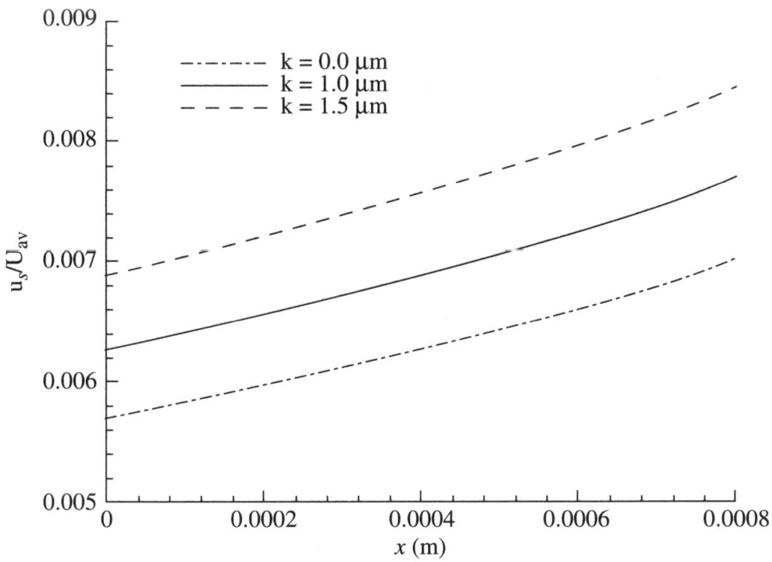

Figure 9.3.24 Comparison of slip velocities: Kn = 0.0026.

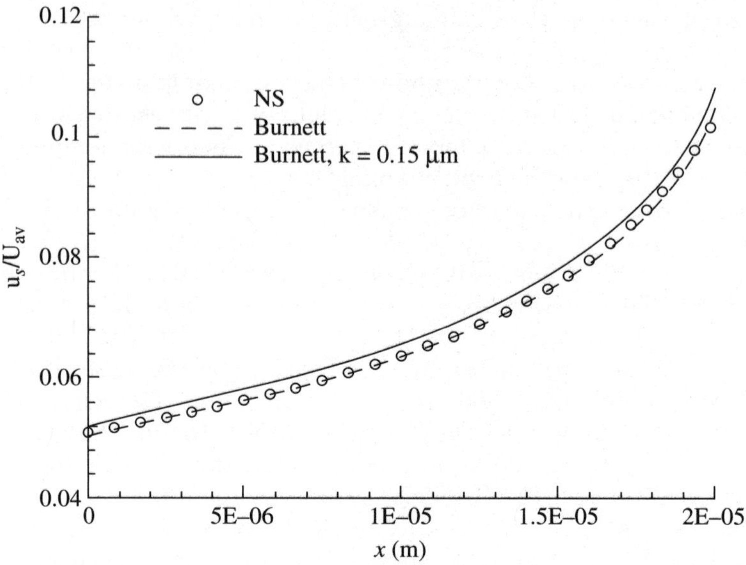

Figure 9.3.25 Comparison of slip velocities: Kn = 0.04.

Figure 9.3.26 shows the wall heat transfer distributions along the microchannel. There are some differences between the results with and without roughness. It is interesting to note that the wall heat transfer coefficient is negative at the second half of the channel, which is not

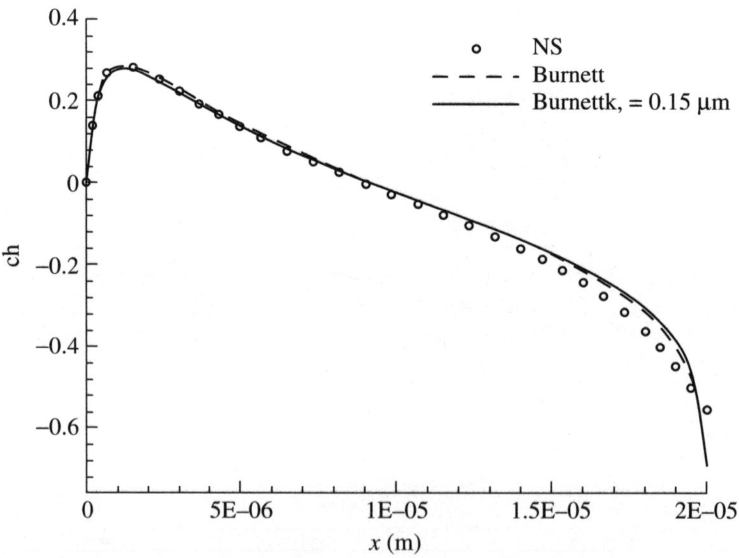

Figure 9.3.26 Wall heat transfer coefficients: Kn = 0.04.

present for the 50 μm case. This indicates a heat transfer into the flow at the second half of the channel. Similar heat transfer characteristics have also been observed in DSMC simulations of high Knudsen number smooth-wall microchannel flow in the previous section. The sign change of wall heat transfer results from the effects of rarefaction and compressibility. According to the computational results, the effect of roughness, although small, seems to decrease the heat transfer rate.

The model used here has been developed based on the simple gradient assumption that is often invoked in the macroscale modeling of diffusion effects. Quantitative experimental evidence at the microscale needed to support this assumption is scarce at present. For microchannel flows at high Knudsen number, the nonequilibrium effects in Knudsen layer near the wall become significant. Gas-surface interactions are complex and there is presently no general and comprehensive theory. In the mean time, for engineering computational fluid dynamics (CFD) calculations of microfluid flows in microdevices with roughness, phenomenological models can be implemented straightforwardly in the existing CFD framework.

References

Bird, G.A., *Molecular Gas Dynamics and the Direct Simulation of Gas Flows*, Oxford University Press, New York, 1994.

Bird, G.A., Private communication, 2002.

Fang, Y.C. and Liou, W.W., Computations of the flow and heat transfer in microdevices using DSMC with implicit boundary conditions, *J. Heat Transf.*, Vol. 124, pp. 338–345, 2002*a*.

Fang, Y.C. and Liou, W.W., Microfluid flow computations using a parallel DSMC code. AIAA paper 2002-1057, 2002*b*.

Harley, J.C., Huang, Y., Bau, H.N. and Zemel, J.N., Gas flow in microchannels, *J. Fluid Mech.*, Vol. 284, pp. 257–274, 1995.

Karniadakis, G.E. and Beskok, A., *Micro Flows: Fundamentals and Simulation*, Springer, New York, 2002.

Liou, W.W. and Fang, Y.C., Implicit boundary conditions for direct simulation Monte Carlo method in MEMS flow predictions, *Computer Modeling in Engineering & Science*, Vol. 1, No. 4, pp. 119–128, 2000.

Liou, W.W. and Fang, Y.C., Modeling of surface roughness effects for microchannel flows, AIAA paper 3586-2003, 2003.

Mala, G.M. and Li, D., Flow characteristics of water in microtubes, *Int. J. Heat Fluid Flow*, Vol. 20, pp. 142–148, 1999.

Nance, R.P., Hash, D.B., and Hassan, H.A., Role of boundary conditions in Monte Carlo simulation of MEMS Devices, *J. Thermophys. Heat Transf.*, Vol. 12, No. 3, pp. 447–449, 1998.

Piekos, E. and Breuer, K., DSMC modeling of micromechanical devices, AIAA paper 95-2089, 1995.

Peiyi, W. and Little, W.A., Measurement of friction factors for the flow of gases in very fine channels used for microminiture Joule-Thomson refrigerators, *Cryogenics*, Vol. 23, pp. 273–277, 1983.

Schlichting, H. and Gersten, K., *Boundary Layer Theory*, (revised and enlarged) 8th ed., Springer, Berlin, 2000.

Whitfield, D.L. and Janus, J.M., Three-dimensional unsteady Euler equations solution using flux vector splitting, AIAA paper 84-1552, 1984.

Chapter

10

High-Speed Microflows

10.1 Introduction

Maxwell (1879) studied the near-wall behavior of fluid flow of large Kn and proposed that there might be a finite slip of velocity and a jump of temperature for gaseous fluid when the mean free path is large compared to the flow dimensions. The large gradients of temperature and velocity may affect the transport of heat and momentum in a manner that is different from those observed in larger systems. As the number of industrial and scientific devices using microelectromechanical systems (MEMS) increases, a detailed understanding of the heat transfer in microchannel flows is becoming increasingly important for an accurate prediction of their performance and for a better design. In Chap. 9, low-speed microflows were described. In this chapter, the heat transfer characteristics of two-dimensional microchannels of high-speed inflows at atmospheric conditions, as opposed to vacuum conditions (Yasuhara et al. 1989), are examined. Particularly, the effects of Kn on the wall heat flux are investigated in detail. The value of Kn is changed by varying the channel height, while keeping the aspect ratio of channel constant. The bow shock structure, temperature distribution, and net heat flux on the wall for a range of Kn are examined. Detailed studies of shocks in and around microscale devices can be found in Dlott (2000), Ohashi et al. (2001), and Brouillette (2003).

10.2 High-Speed Channel Flows

The DSMC method is used to simulate the heat transfer in two-dimensional microchannel flows near the atmospheric condition. The two-dimensional simulations allow a better use of the computer resources for a detailed study of the heat transfer mechanisms in microchannels, which may be significantly different from that in three-dimensional cases (Sobek et al. 1994). In fact, there are many examples of two-dimensional high-speed microchannel flow in MEMS, for example, the flow in the gap between rotor and stator in micromotors (Guckel et al. 1993) and microengines (Janson et al. 1999). Each computational cell has been divided into two subcells in each direction in the present DSMC simulations. The time step has been chosen such that a typical molecule moves about one-fourth of the cell dimension in one computational time step. Nitrogen gas is used here and the variable hard sphere (VHS) model (Bird 1994) has been applied in all the simulations. For the standard atmospheric condition, the number density is high and the mean collision time is in the order of 10^{-10} sec. The time step used in the DSMC method, Δt_d should be less than the mean collision time, so that particle movement and collision may be uncoupled, i.e.,

$$\Delta t_d < \frac{\Delta x_d}{\overline{c'}} \tag{10.2.1}$$

where $\overline{c'}$ is the most probable molecular speed. Borrowing from the traditional computational fluid dynamics (CFD), this constraint, Eq. (10.2.1), may be expressed as a Courant-Friedrichs-Lewy (CFL) number. So

$$\mathrm{CFL} = \frac{\overline{c'}\Delta t_d}{\Delta x_d} < 1 \tag{10.2.2}$$

The cell size Δx_d is taken as less than a half of the mean free path in the current simulations, i.e.,

$$\mathrm{Kn}_c = \frac{\lambda}{\Delta x_d} \geq 2 \tag{10.2.3}$$

where Kn_c is defined as a "cell Knudsen number." Physically, this condition requires the particle to reside in the same cell for at least a few time steps to allow it to interact with other particles. This ensures that its information can be distributed properly throughout the domain of computation. Therefore, for DSMC, the CFL number is not a stability constraint, but rather a physical requirement. Violation of this condition may produce solutions that are not physically realistic. In the simulations that will be presented, the value of CFL is set at about 0.2.

The temperature at the upper and the lower walls are set equal. When the simulated particle collides with the wall, the diffuse reflection model is used to determine the reflection. In this model, the emission of the impinging molecules is not correlated with the preimpingement state of the molecules. The outgoing velocity is randomly assigned according to a half-range Maxwellian distribution determined by the wall temperature. This is also known as the full thermal and momentum accommodation.

For equilibrium gases of diatomic molecules, by neglecting the vibrational energy, the temperature can be calculated as

$$T = \frac{(3T_{\mathrm{tr}} + \xi_r T_{\mathrm{rot}})}{3 + \xi_r} \qquad (10.2.4)$$

where T_{tr} denotes the translational temperature and T_{rot} the rotational temperature. They can be written as

$$T_{\mathrm{tr}} = \frac{2}{3R}(\overline{mc^2} - m\bar{c}^2) \qquad \text{and} \qquad T_{\mathrm{rot}} = \frac{2}{\xi_r k}\hat{\varepsilon}_{\mathrm{rot}} \qquad (10.2.5)$$

For nitrogen, $\xi_r = 2$, and the averaged molecular rotational energy can be written as

$$\hat{\varepsilon}_{\mathrm{rot}} = -\ln(R_f)kT \qquad (10.2.6)$$

The net molecular energy lost per unit width resulting from a wall collision or the wall heat transfer can be calculated by

$$q = \frac{\left[\left(\sum_{N_w} \hat{e}_{\mathrm{tr}} - \sum_{N_w} \hat{e}_{\mathrm{rot}}\right)_{\mathrm{inc}} - \left(\sum_{N_w} \hat{e}_{\mathrm{tr}} - \sum_{N_w} \hat{e}_{\mathrm{rot}}\right)_{\mathrm{ref}}\right] N_0}{\Delta t \bullet \Delta x} \qquad (10.2.7)$$

where N_w represents the number of molecules that have collided with the wall during the sampling period of Δt and N_0 the number of gas molecules assigned to a computational molecules in DSMC.

Three example simulations, referring to as Case 1, Case 2, and Case 3, of microchannel flow will be discussed. The Knudsen numbers are 0.031, 0.062, and 0.186 for Case 1, Case 2, and Case 3, respectively. The flow conditions and the channel sizes are given in Table 10.2.1. The aspect

TABLE 10.2.1 Conditions of the Simulations

	Case 1	Case 2	Case 3
L(μm)	12	6	2
H(μm)	2.4	1.2	0.4
Free stream distance (μm)	1.2	0.6	0.2
Kn	0.031	0.062	0.186
Total number of particles	12,017	11,854	8,041
M_∞	4.15		
T_∞(K)	300		
T_w(K)	323		
Number density (m^{-3})	1.75×10^{25}		
Free stream mean free path	7.4×10^{-8}		

ratio of the channels is set at 5 for all cases and rectangular cells (100 × 60) were used. In the figures presented, the height of the channel, H, has been used for nondimensionalization. The microchannel is embedded in a high-speed stream of Mach number 4.15. The distance between the upstream computational boundary and the inlet of the channel is referred to as the free stream distance. The inlet is considered to have sharp leading edges. Such high-speed microchannel flows have been studied by Oh et al. (1997), Liou and Fang (2001), and Raju and Roy (2003). The following results are based on those of Liou and Fang (2001).

Figures 10.2.1*a*–*c* show the shock structures for the three cases. Figure 10.2.1*a* shows that in the immediate upstream of the leading edges, detached bow shocks form because of the viscous boundary Layers. For the case with Kn = 0.031 (Case 1), the two bow shocks

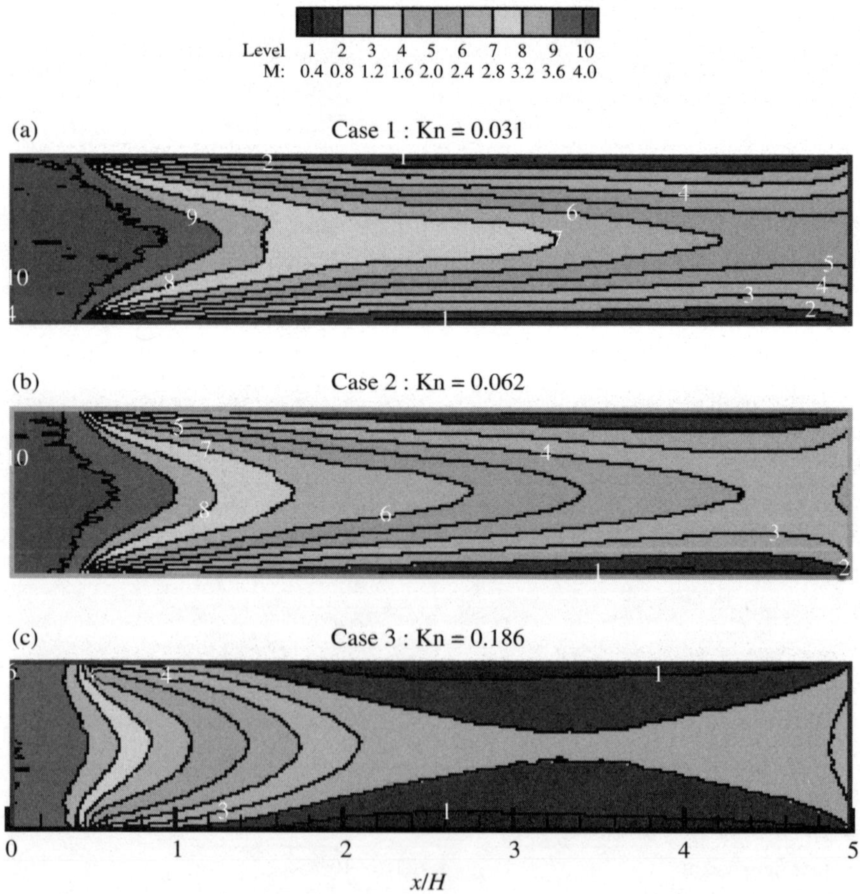

Figure 10.2.1 Mach number contours. (*Liou and Fang 2001.*)

emitted from the upper and the lower walls intersect near $x/H = 1.35$. The waves resulting from the intersection of the bow shocks diffuse as they separate and extend toward the wall due to the weakening incident shocks. As the Kn increases, the leading edge shocks thicken and the angle between the shock and the wall increases. As such, the intersection of the bow shocks moves upstream. For Case 3 with $\mathrm{Kn} = 0.186$, the intersect cannot be identified because of the relatively large shock thickness. The flow initially decelerates to subsonic speed, followed by acceleration. As will be seen in the following, the deceleration of the flow in the highest Kn case, Case 3, is accompanied by a significant increase of the wall heat transfer.

Figures 10.2.2*a–c* show the temperature contours. These figures show the incident bow shocks and the strong temperature gradients across

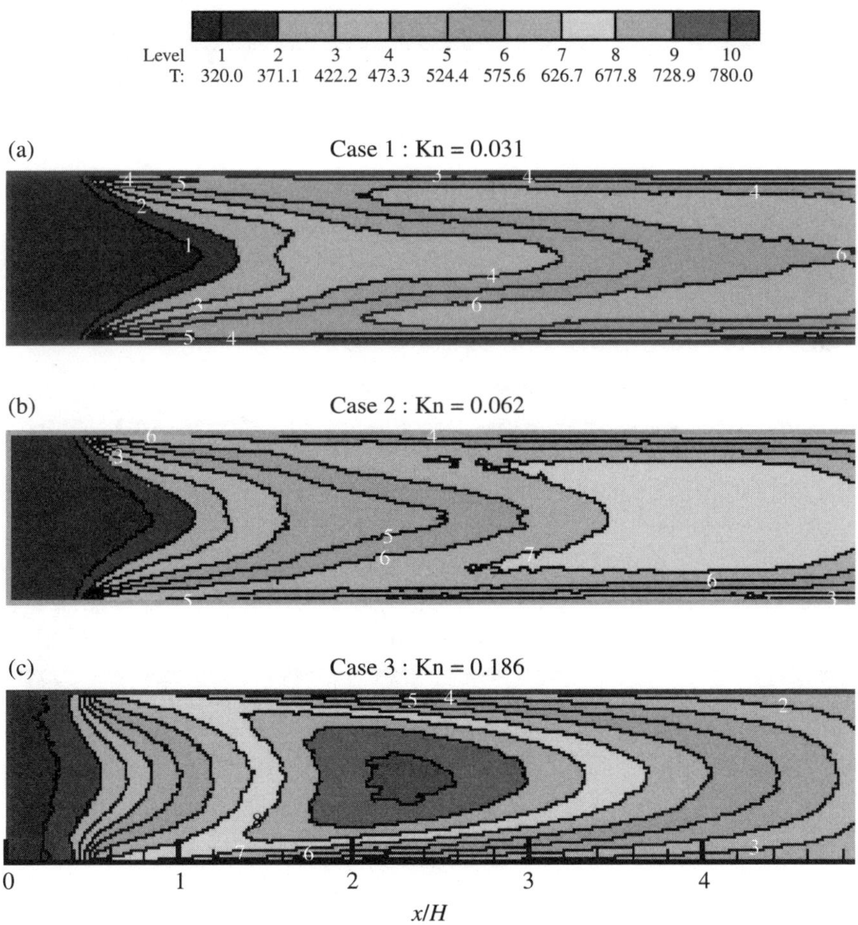

Figure 10.2.2 Temperature contours. (*Liou and Fang 2001.*)

the bow shocks near the leading edges. As the Kn increases, both the temperature and its gradient increase sharply. It is also apparent that an island of high temperature is formed as the Kn increases and the location of the high-temperature island moves upstream as the Kn increases. The formation of the high-temperature island is, as was mentioned above, accompanied by the deceleration of the flow. Also, the thickness of the thermal boundary layers in all the cases can be observed to increase with Kn.

Figures 10.2.3*a–c* show the density contours. For Kn = 0.031 and 0.062, the two shocks emanating from the leading edges of the microchannel intersect, and then extend to the wall. This causes the formation of islands of low density near the walls in the first half of the microchannel. Behind the intersection, a region of high density appears.

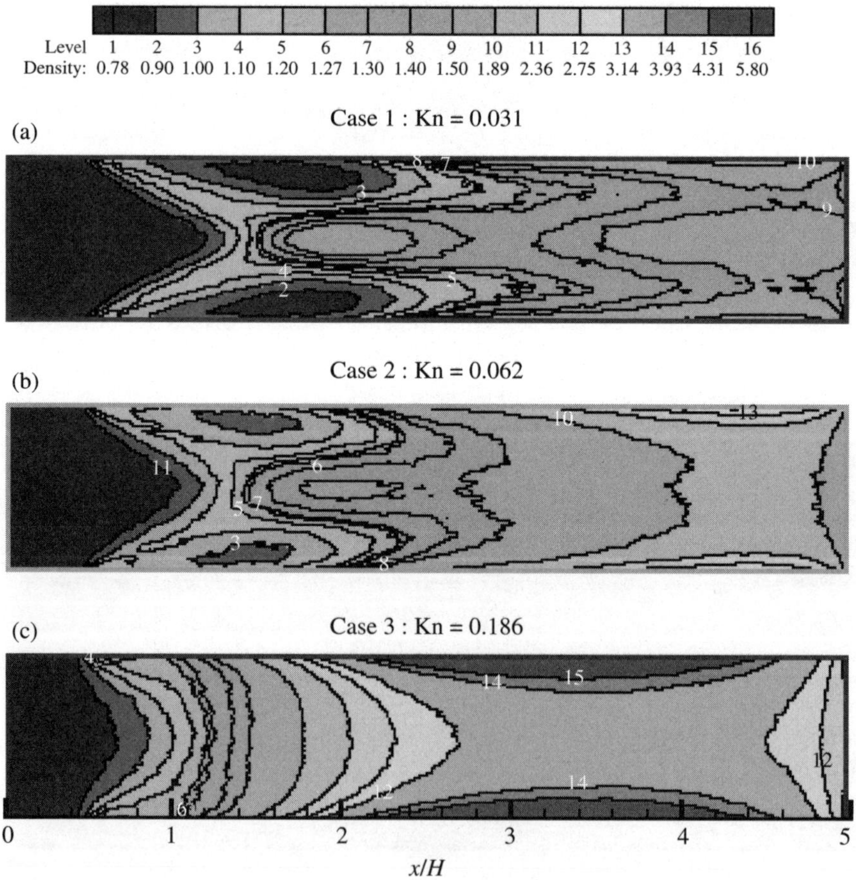

Figure 10.2.3 Density contours. (*Liou and Fang 2001.*)

For Kn = 0.186, the shocks are detached further from the leading edge and more diffused. The root of the bow shocks is nearly normal to the channel. The shock layer is much thicker than those for the lower Kn cases. The compression shocks spread almost over a half of the microchannel. As a result, there are no identifiable islands of low density near the wall, as was found in cases with lower Kn.

Figures 10.2.4*a*, *b* show the axial temperature T and mean velocity u profiles at $x/H = 0.5$, 1.5, 2.5, 3.5, and 4.5, respectively. The variation of the temperature profiles, Fig. 4*a*, reveals the development of the thermal boundary layers. The case with Kn = 0.031 shows two local peaks in all the profiles and these peaks tend to merge as the flow develops downstream. A fully developed state has not been reached at the downstream end of the channel. On the other hand for Kn = 0.062, the peaks merge at about $x/H = 4.5$ and the flow is becoming fully developed. For Kn = 0.186, the temperature profiles flatten out sooner, and the peak value at the centerline decreases significantly downstream. Axial velocity profiles are shown in Fig. 10.2.4*b*. There is a peak at the centerline for all the cases. The peak values vary little in the downstream direction in the Kn = 0.031 case. However, the peak values for Kn = 0.186 change quite significantly as the flow develops downstream. For Kn = 0.186, both the temperature and the kinetic energy reduce as the flow develops downstream.

The temperature jump at the wall is shown in Fig. 10.2.5. T_g denotes the gas temperature at the wall and T_w the constant wall temperature. Figure 10.2.5 shows that the temperature jump decreases as the flow develops downstream. The highest temperature jump occurs near the microchannel entrance. As Kn increases, the location of the peak temperature jump moves slightly downstream. Near the outlet of the simulated microchannels, the temperature jumps for the cases with Kn = 0.031 and 0.062 appear reaching an asymptotic value.

Figure 10.2.6 shows the temperature distribution along the centerline of the microchannels. The multiple plateaus of the centerline temperature for the cases with Kn = 0.031 and 0.062 are related to the intersects of the incident and the reflected shocks. For Kn = 0.186, the centerline temperature peaks near $x/H = 2.0$ and decreases monotonically thereafter. The value of centerline temperature peak increases with the increasing value of Kn. The location of the peak also moves upstream with the increasing Kn, conforming to the observation made earlier that the island of increasingly high temperature moves upstream as the Knudsen number increases.

Figure 10.2.7 shows the streamwise variation of the heat flux q at the wall. For Kn = 0.031 and 0.062, the heat flux decreases as the flow develops downstream. For Kn = 0.186, the streamwise variation is distinctively different from those of the lower Kn. After an initial

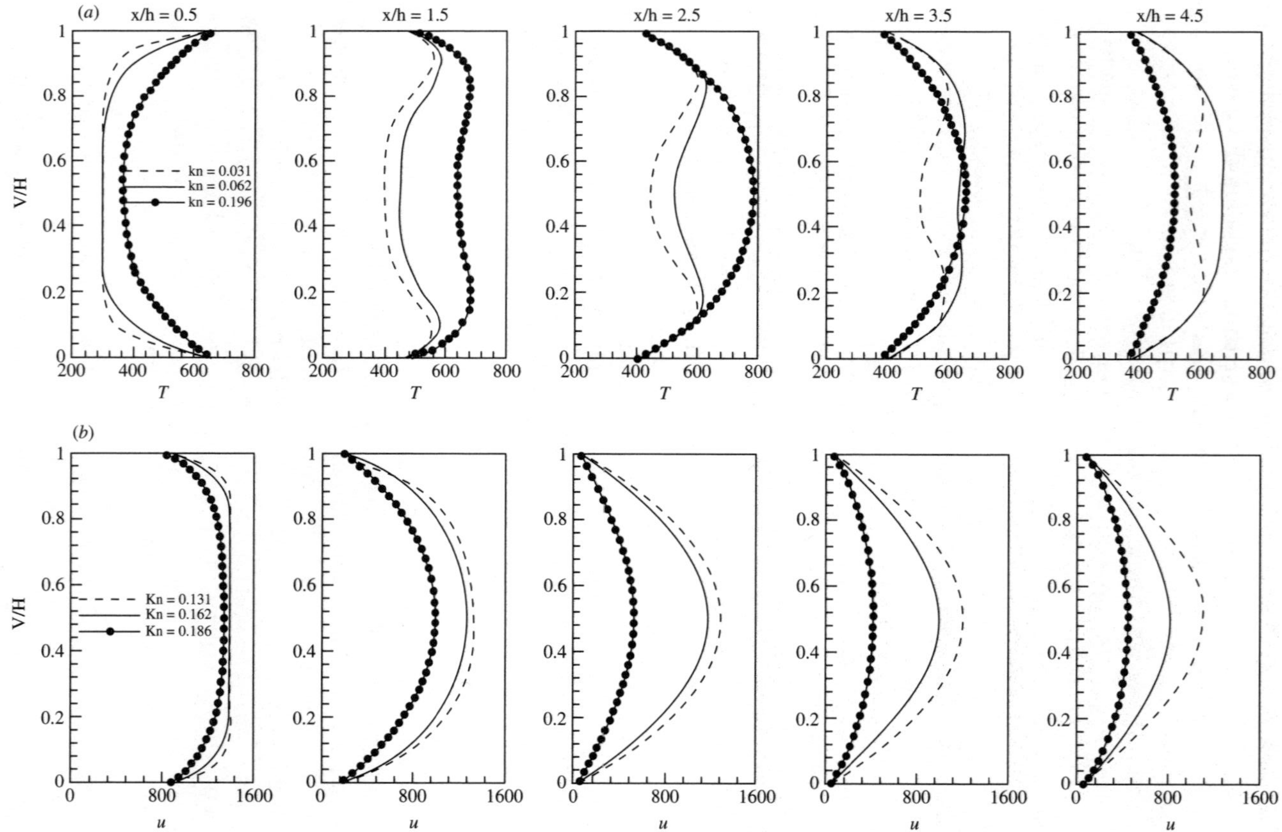

Figure 10.2.4 Property profiles: (*a*) Temperature; (*b*) Velocity. (*Liou and Fang 2001.*)

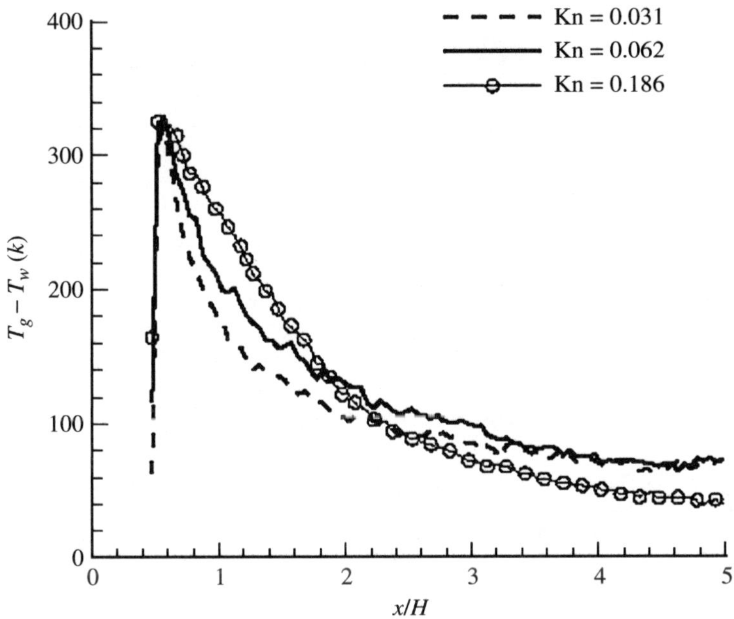

Figure 10.2.5 Temperature jump distributions. (*Liou and Fang 2001.*)

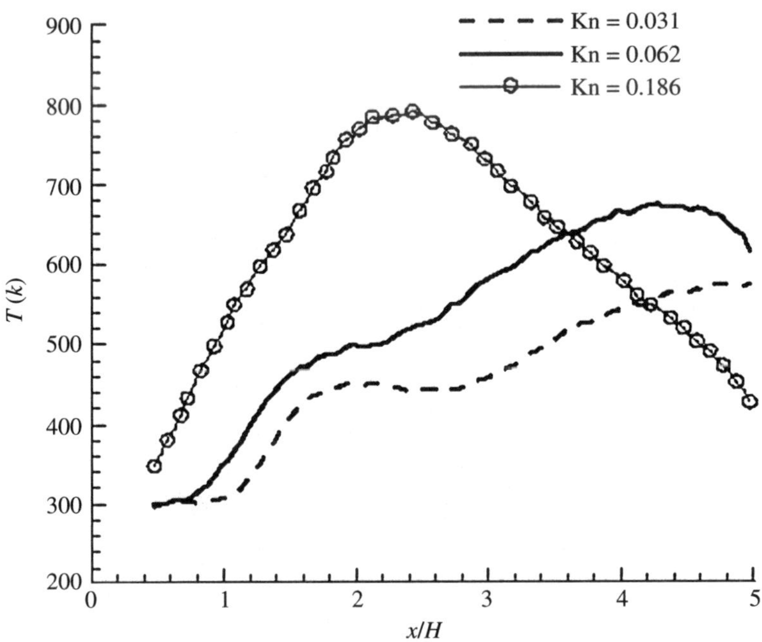

Figure 10.2.6 Centerline temperature distributions. (*Liou and Fang 2001.*)

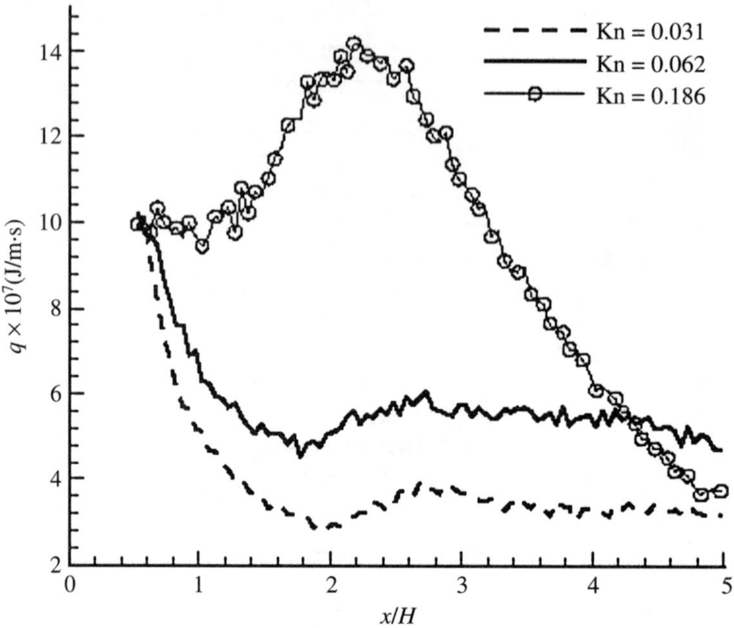

Figure 10.2.7 Heat flux distributions. (*Liou and Fang 2001.*)

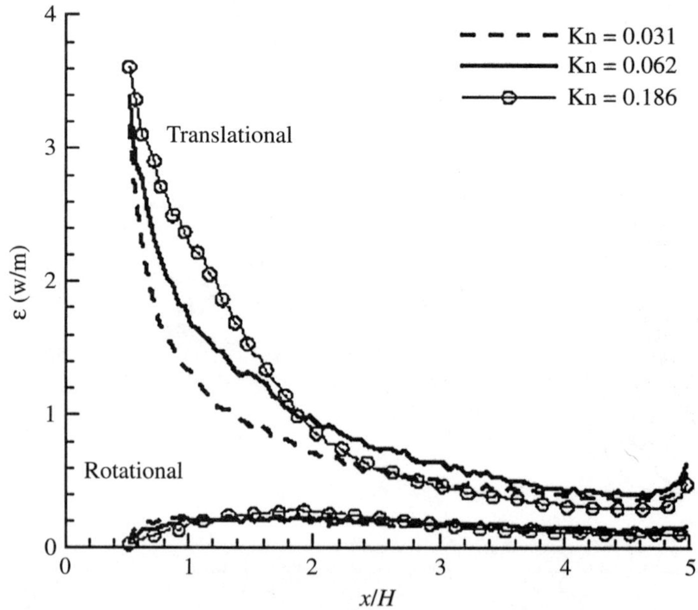

Figure 10.2.8 Distributions of averaged molecular energy deficit. (*Liou and Fang 2001.*)

region of small variation, the heat flux increases sharply in the midsection of the microchannel. The heat flux peaks at about 14×10^7 W/m. The wall heat transfer then decreases monotonically downstream. The maximum value of the wall heat transfer for Kn = 0.186 is about 3.5 and 2.3 times those for Kn = 0.031 and 0.062, respectively, representing a significant increase of heat transfer at high Kn.

Figure 10.2.8 shows the axial distributions of the averaged translational and rotational energy deficits, per molecule, ε normalized by 10^{-20}. The translational energy deficit increases slightly with the increasing Kn upstream of $x/H = 2.0$. On the other hand, the rotational energy deficit does not change in any significant way with Kn. The number rate of wall-impinging molecules, nondimensionalized by the number density of the free stream, 1.75×10^{25} m^{-3} is shown in Fig. 10.2.9. For Kn = 0.186, the distribution is quite different from the other two cases with lower Knudsen numbers. For both Kn = 0.031 and 0.062, there is a plateau near $x/H = 4.0$. The peak value of the number rate of molecules impacting the wall for Kn = 0.186 is roughly 1000, and those for Kn = 0.031 and 0.062 are 400 and 580, respectively.

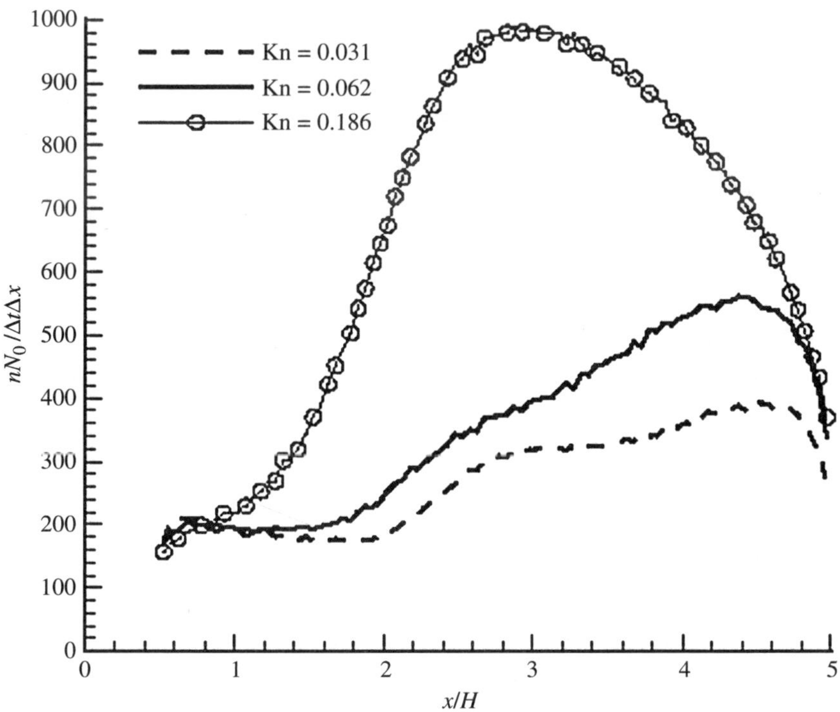

Figure 10.2.9 Number rate of molecules that has impinged on the wall. (*Liou and Fang 2001.*)

The increase of the number rate of molecules that impact the wall is apparently a dominating factor contributing to the observed increase of wall heat transfer. Recall that there are islands of low density near the walls for the lower Kn cases and these islands of low density were not observed for Kn $= 0.186$. It is believed that the high number rate of wall-impacting molecules for the high Kn case (0.186) is related to the general high density in the near wall regions that do not exist in the lower Kn cases.

The heat transfer characteristics of the high-speed microchannel flows can vary significantly with Knudsen numbers. For the two-dimensional supersonic microchannel flows shown above, there is a significant increase of wall heat transfer with Knudsen number. The results show that the enhanced wall heat transfer is mainly caused by the increased number rate of molecules that impact the wall. As the value of the Knudsen number for a typical MEMS device can be quite high, this unique behavior of the wall heat transfer observed in the present simulations may have a significant bearing on the development and the design of MEMS.

References

Bird, G.A., *Molecular Gas Dynamics and the Direct Simulation of Gas Flows*, Oxford University Press, New York, 1994.

Brouillette, M., Shock waves at microscales, *Shock Waves*, Vol. 13, pp. 3–12, 2003.

Dlott, D.D., Nanoshocks in molecular materials, *Acc. Chem. Res.*, Vol. 33, pp. 37–45, 2000.

Guckel, H., Christenson, T.R., Skrobis, K.J., Jung, T.S., Klein, J., Hartojo, K.V., and Widjaja, I., A first functional current exited planar rotational magnetic micromotor, *Proceedings of the Micro Electro Mechanical Systems* (Fort Lauderdale, Florida), IEEE, New Jersey, pp. 7–11, 1993.

Janson, S.W., Helvajian, H., and Breuer, K., MEMS, microengineering and aerospace systems, AIAA Paper 99-3802, 1999.

Liou, W.W. and Fang, Y.C., Heat transfer in microchannel devise using DSMC, *J. Microelectromech. syst.*, Vol. 10, No. 2, pp. 274–279, 2001.

Maxwell, J.C., On stresses in rarefied gases arising from inequalities of temperature, *Phi. Trans. R. Soc. Part 1*, Vol. 170, pp. 231–256, 1879.

Oh, C.K., Oran, E.S., and Sinkovits, R.S., Computations of high-speed, high Knudsen number microchannel flows, *J. Thermo. Phys. Heat Transf.*, Vol. 11, p. 497, 1997.

Ohashi, K., Kleinc, H., and Takayama, K., Characteristics of blast waves generated by milligram charges, in: F.K. Lu, ed. *Proceedings of the 22nd International Symposium on Shock Waves*. Fort Worth, Texas, July 2001.

Raju, R. and Roy, S., Hydrodynamic prediction of high speed microflows, AIAA paper 2003-4010, 2003.

Sobek, D., Senturia, S.D., and Gray, M.L., Micro fabricated fused silica flow chambers for flow cyclometry, *Solid-State Sensor and Actuator Workshop '94*, Hilton Head, South Carolina, 1994.

Yasuhara, M., Nakamura, Y., and Takanaka, J., Monte Carlo simulation of flow into channel with sharp leading edge, in: E.P. Monte, D.P. Weaver, and D.H. Campbell, eds. *Rarefied Gas Dynamics: Theoretical and computational Techniques*, Vol. 8: *Progress in Astronautics and Aeronautics*, AIAA, Washington, DC, pp. 582–596, 1989.

Chapter

11

Perturbation in Microflows

11.1 Introduction

For macroscale flows, in situations where there is little or no free stream forcing, small disturbances initially grow in a linear fashion. As the disturbances become amplified, nonlinear effects become important. New modes generated by the flow nonlinearity fill the disturbance spectrum and the flow becomes locally turbulent. This is usually called natural transition. When the free stream disturbances, for example, generated by free stream turbulence, are high, the linear stage of growth may be bypassed in the transition process. This is usually referred to as bypass transition. The fluid dynamics of either the natural or the bypass transition is highly nonlinear where energy of certain dominant modes of fluctuations cascades to their higher harmonic modes and eventually transfers to fluctuations of other scales. Many recent experiments in fluidic microchannels have reported significant difference in the heat and momentum transfer coefficients compared with those at the macroscales. For example, while the friction factor varies inversely with the Reynolds number when the Reynolds number is small, the proportionality constant does not agree with the conventional correlation. The friction factors were also found to diverge from this inverse proportionality, which at macroscale indicates a change of flow characteristics from that of a laminar flow to a turbulent flow, at smaller Reynolds numbers than those commonly observed in the corresponding large channels. This early change of flow characteristics has been attributed to the effects of, for instance, gas rarefaction and other surface mechanisms, the large change of the flow Reynolds number along the microchannels, and the likely experimental uncertainties in microscale measurements. Advanced measurement techniques such as molecular tagging can produce more detailed quantitative data for microflows, which may eventually lead to a better understanding of the apparent microflow transition. In the following, the characteristics of disturbances in a forced microCouette flow and in microRayleigh-Bénard flows are described. The discussions are aided by using results of DSMC and Navier-Stokes numerical simulations.

As the DSMC method is based on a molecular description of gases, there have been concerns whether DSMC is capable of resolving vortical fluid motions due to the lack of an exact conservation of angular momentum in collisions. Results (Nanbu et al. 1988) show that as long as the cell size is sufficiently small, the total angular momentum is nearly conserved. Bird (1987) has also shown that the nonconservation of the angular momentum has no significant effects on the DSMC solutions with the mean separation distance between collision partners being less than the local mean free path. During the past decade, the DSMC method has been employed to study the low-density limits of a number of flow instabilities. The centrifugal instabilities in Taylor-Couette flows have received the most attention (Riechelmann and Nanbu 1993; Stefanov and Cercignani 1993). The formation of Taylor vortices were clearly demonstrated for a range of Knudsen number in these two-dimensional (axisymmetric) flow computations. The convective instability associated with the Bérnard cells has also been studied.

The characteristics of microflow disturbances will be discussed by examining two microflows, i.e., a microRayleigh-Benard and a microCouette flow. In the first, the disturbances are associated with the buoyancy force. In the second, an artificial external force is applied to energize the flow and its disturbances.

11.2 Forced MicroCouette Flows

The Couette flow geometry considered is shown in Fig. 11.2.1. The flow develops between two diffuse walls in the y-direction. The top wall moves at a speed of U and the lower wall is brought to a stop instantaneously at the beginning of the calculations. The channel height is h. Periodic boundary conditions are applied in the x- and z-directions. In contrast to the microCouette flow cases studied in Chap. 9, an artificial gravitational forcing (AF)

$$\mathrm{AF} = c(\hat{u} - U)^2 \tag{11.2.1}$$

has been applied to the flow along the y-direction. $\hat{u}$ represents the fluid velocity component in the x-direction. c denotes a constant coefficient that is used to adjust the level of the external forcing on the flow.

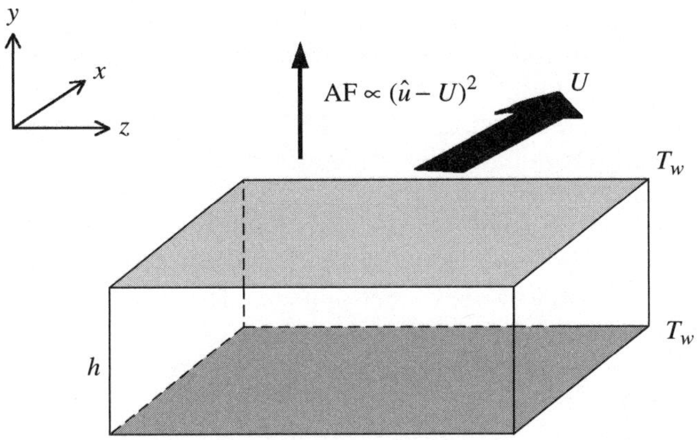

Figure 11.2.1 Forced microCouette flow geometry.

A similar imposition of the artificial acceleration has previously been developed (Malek Mansour 1990) in an analytical study of the onset of hydrodynamic instabilities and independently by Bird (2002) to study forced chaotic flows. Note that the sample averaged behavior of the microCouette flow without forcing has been discussed in Chap. 9.

The effect of the artificial gravitational force in the flow model is that when a fluid particle is perturbed and moves in the y-direction, an effective buoyant force favors further displacement of the particle in the y-direction. The scenario is similar to the instability mechanism associated with the centrifugal force in the Taylor-Couette and Gortler problems, where Taylor-Gortler vortices have been observed in experiments and computed using two-dimensional DSMC. As a result, it is reasonable to expect that the current simulations will also exhibit similar vortical structures or flow patterns. For a particle that moves from the bottom to the top wall in a straightline flight path, the work done by the artificial forcing can be written as

$$w = \int_0^h c(\hat{u} - U)^2 dy = cU^2 h \int_0^1 \left(\frac{\hat{u}}{U} - 1\right)^2 d\left(\frac{y}{h}\right) \tag{11.2.2}$$

Therefore, the work done by the forcing is positive definite. Its amount, however, depends on the actual vertical distribution of the velocity $\hat{u}$. For discussions, one can assume a linear velocity profile, where $\hat{u} = (y/h)U$, then

$$w = \frac{1}{3} cU^2 h$$

Therefore, the amount of energy put to the particle by the externally applied forcing, based on the simple analysis, is proportional to the value of the forcing coefficient, the square of the plate speed, and the channel height. The use of an artificial forcing, instead of a physical one, thus allows for an efficient adjustment of the strength of the disturbances by varying the values of these parameters.

DSMC allows for a direct physical simulation of gas flows. The minimum level of physical modeling is needed at the level of molecular models that have been used in kinetic theory of gases. However, a large number of particles are required in a DSMC simulation to reduce the inherent statistical scatter. The computer resource needed to perform three-dimensional flow simulations can be significant, which appears to be a major limiting factor for performing many necessary simulations to capture three-dimensional physical flow disturbances. By using an artificial forcing in the form of Eq. (11.2.1) instead of a physical one, it is possible to apply a wide range of forcing level to the flow without changing the flow geometry. Since the normalized root mean square of the disturbances is inversely proportional to the square root of the sample size (Landau and Lifshitz 1958), the present flow model allows for a more efficient use of the computer resources in capturing unsteady, three-dimensional disturbances by artificially energizing the flow. This type of first principle simulations of time-dependent three-dimensional flow disturbances has not been reported previously before the work of Bird (2002) and the authors (Liou et al. 2003*a*,*b*,*c*; Liou and Fang 2004).

Two types of sampling have been used in the study to obtain macroscopic, averaged, and disturbance components of the DSMC simulated flows. The "unsteady" property, say, for the vertical velocity component $\hat{v}$ is obtained by

$$\hat{v}(x, y, z, t) = \frac{1}{t_s} \int_{t-\frac{1}{2}t_s}^{t+\frac{1}{2}t_s} \sum_{i=1}^{M_c} c_{v,i}(t')dt' \tag{11.2.3}$$

where $\mathbf{c}_v$ denotes the vertical component of the particle velocity, M_c the number of particles in the cell at time t'. t_s represents the short time period for the unsteady sampling that is longer than the molecular collision time for obtaining a meaningful statistical sampling. It was set at about one order-of-magnitude larger than the mean collision time of the gas. The long-time-averaged or mean component of the flow is obtained by

$$v = \frac{1}{t - t_0} \int_{t_0}^{t} \sum_{i=1}^{M_c} c_{v,i}(t')dt' \tag{11.2.4}$$

where t_0 denotes the beginning time of the sampling. The sampling period for the long time average should be longer than t_s and, subsequently, the mean collision time. The difference between the unsteady sample average and the long time average is defined as the disturbance v'. Or

$$v' = \hat{v} - v \tag{11.2.5}$$

11.2.1 Flows in two-dimensional planes

Two-dimensional (2D) simulations of slices in the x-y and the y-z planes are described. The simulated flows exhibit highly organized discernable vortical structures. Fourier analyses of the vertical velocity signals show the existence of a hierarchy of fluctuation modes that consist of a fundamental and its harmonics.

Longitude planes. The 2D domain slices through the current model flow along a constant z-plane. The top plate moves in the positive x-direction. The size of the computational domain in the x and the y directions is 180×60, nondimensionalized by the initial mean free path λ. The Kundsen number (Kn) defined by the channel height h is then 1/60. The cell size is 90×30. Nearly 2 million computational molecules were used. The value of c in the artificial gravitational force formula is 0.1. The forcing is equivalent to a centrifugal forcing corresponding to a radius of curvature of 10 mean free path of the initial gas. Figure 11.2.2 shows comparisons of the vertical velocity perturbation contours and the pathlines between the results obtained by Bird's code (Bird 2001) and the present parallel DSMC results after the simulated flows appear to have reached a periodic flow pattern.

The results of the parallel runs using two different types of domain decomposition (1DD and 2DD) are shown: 1DD denotes one-directional decomposition and 2DD two-directional. Small statistical scatter can be observed in all three results. The pathlines show that the overall flow patterns are basically the same, both capturing a small region of vortical flows at the heel of the region of large, positive vertical velocity perturbation. Similar contour patterns are also observed in the 2DD results. Note that the boundaries are periodic and it is difficult to time the outputs so that comparable flow patterns appear at the same location for all three plots in Fig. 11.2.2.

The results shown in Fig. 11.2.2 indicate that the width of the domain 180 can be too small to contain more than one cyclical flow pattern. Figure 11.2.3 shows the vertical velocity perturbation contours and pathlines for a case of larger domain of 1200×100 and $c = 0.04$. Nearly 8 million computational molecules were used. There are three identifiable periodic flow patterns. The flow patterns are quite regular,

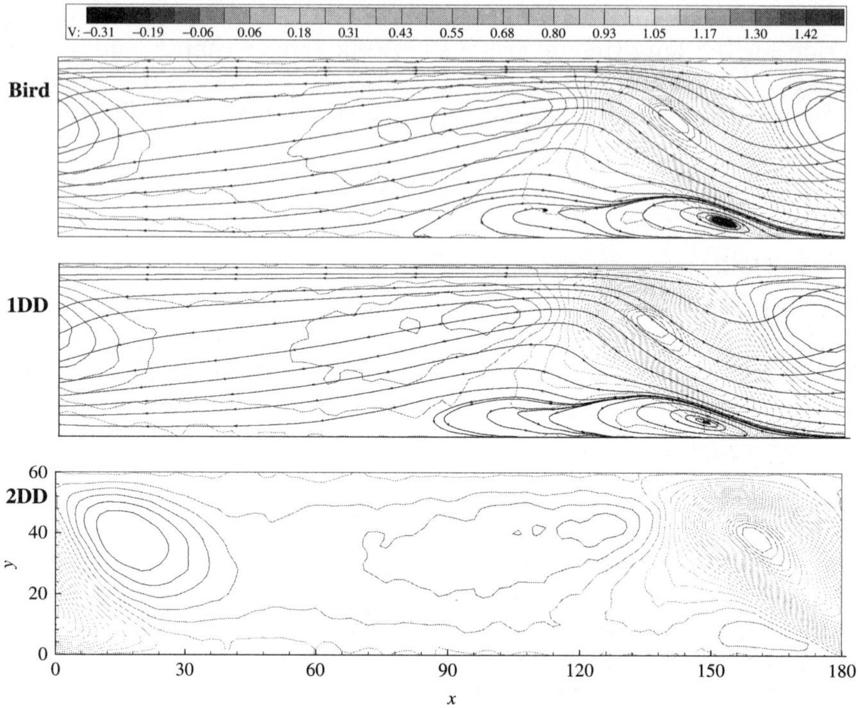

Figure 11.2.2 Comparisons between parallel and serial DSMC results.

with a wavelength Λ of about 400, and are rather smooth. In time, the cyclic flow pattern moves in the x-direction. Figure 11.2.4 shows the evolution of the unsteady specific kinetic energy in the cell. Or

$$\text{AKE} = \frac{1}{N}\sum_{\text{cell}}(\hat{u}^2 + \hat{v}^2) \tag{11.2.6}$$

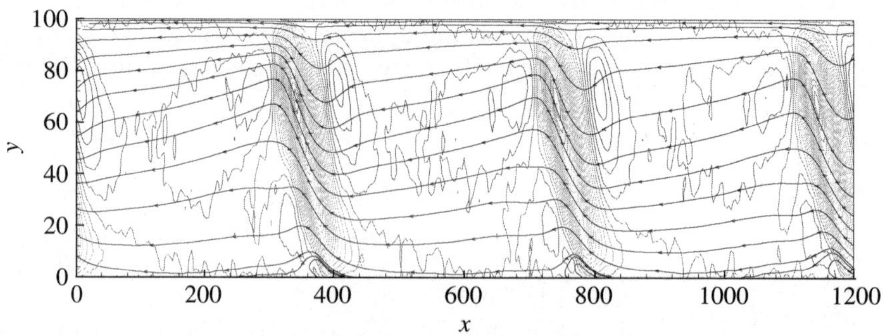

Figure 11.2.3 Vertical velocity perturbation contours and pathlines.

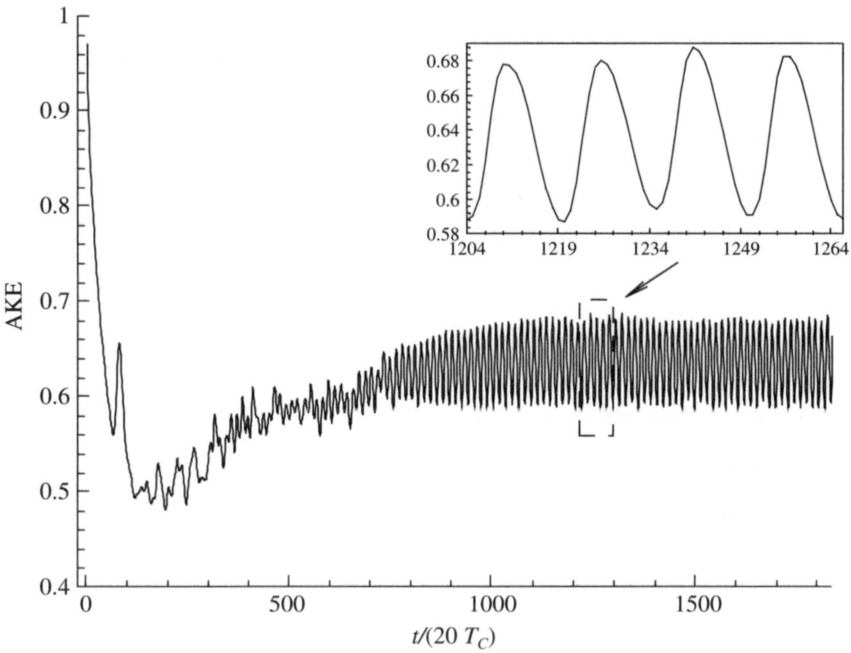

Figure 11.2.4 Evolution of cell-averaged specific kinetic energy.

where the summation is applied to all the cells and N denotes the total number of cells. The kinetic energy settles after an initial transient to finite amplitude oscillations. A part of the curve in the region of the finite amplitude oscillations has been enlarged and is also included in Fig. 11.2.4. It shows that the kinetic energy varies quite regularly and the flow has essentially reached a quasiperiodic state after 1000 print cycles, or 20,000 initial mean molecular collision time, T_C. It can be seen from Fig. 11.2.3 that the flow associated with the quasiperiodic state features flow structures with wavelength of about 400. Before the quasiperiodic state is reached, the simulated flow continuously evolves in time as it develops from the initially chaotic flow. The region near the bottom wall is strongly rarefied and first shows significant patterns of small-scale variations. As time advances, some small-scale patterns amalgamate and grow and others decay after their initial generation. Figure 11.2.5 shows the pathlines at $t = 2480T_C$, when the flow is in transition from the initial chaotic to the quasiperiodic states. Vortices of different sizes can be identified; some span across the height of the channel. The flow eventually evolves into a regular disturbance pattern as was shown in Fig. 11.2.3.

The distribution of the vertical velocity perturbation along the x-direction for $y = 66$, when the flow has become quasiperiodic, is shown

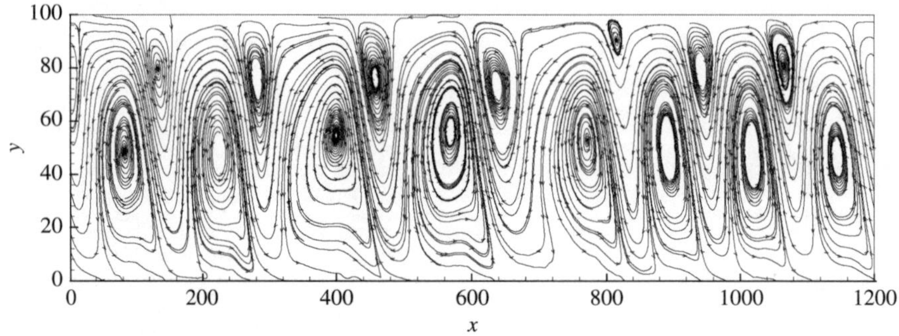

Figure 11.2.5 Pathlines for $t/(20T_C) = 124$.

in Fig. 11.2.6. The three concentrated regions of high, positive vertical velocity perturbations are associated with the three cyclical flow patterns in Fig. 11.2.3. The wavelength is about 400. The signals associated with the dominant flow pattern with wavelength 400 are quite smooth with little statistical scatter. In between, there appear to be waves of smaller wavelengths. The Fourier wave number spectrum of the vertical velocity perturbation signal given in Fig. 11.2.6 (along $y = 66$) is shown in Fig. 11.2.7. The Fourier spectrum of the initial, chaotic state has also been included for comparison. At the high wave number end, the spectrum is filled with low-amplitude noise, as can be expected from a statistical simulation such as DSMC. At the low wave number end,

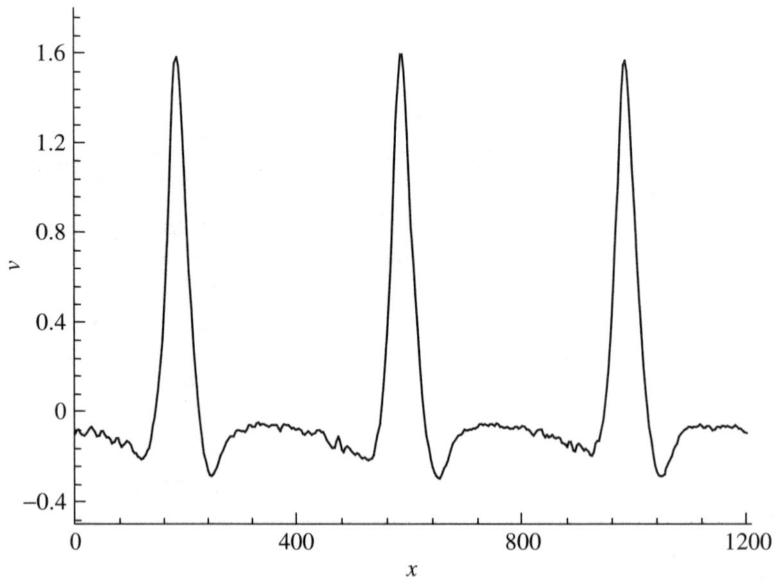

Figure 11.2.6 Distributions of vertical velocity perturbation along $y = 66$.

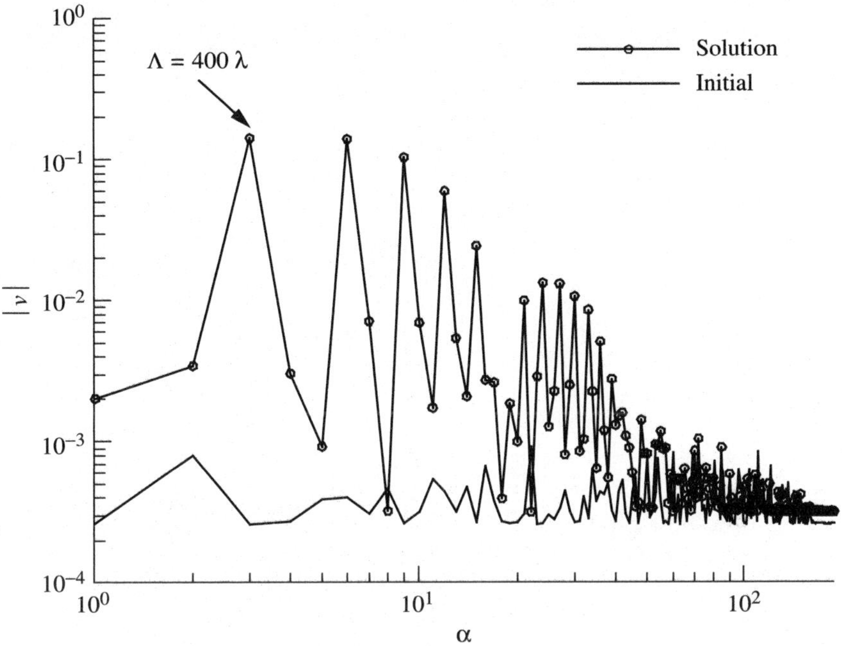

Figure 11.2.7 Fourier wave number spectrum of vertical velocity perturbation along $y = 66$.

multiple discrete peaks with amplitudes one to two orders of magnitude higher than those at the high wave number end appear. The first peak, with a wave number of 3, has a wavelength of 400, corresponding to the flow pattern described earlier in Figs. 11.2.3 and 11.2.6. The next peak has a wavelength of 400/2. If one refers to the wave with a wavelength of 400 as a fundamental mode, the wave with a wavelength of 400/2 is then its first harmonic. In all, about nine or more higher harmonics can be identified in Fig. 11.2.7 before their amplitudes drop off. Since neither the applied forcing nor the initial flow contains any preferred time and length scales, the generation of these harmonic waves is not associated with any externally imposed excitation, which is often applied in Navier-Stokes-based flow simulations. The wave hierarchy appears in the current DSMC simulations as a result of the nonlinearity of the simulated flow.

Figure 11.2.8 shows the time-dependent evolution of the Fourier amplitudes of four wave numbers. They are associated with the first three harmonic waves and a high wave number mode ($n = 60$). The harmonic modes arise from the initially chaotic flow and grow in amplitude as the flow develops. They then become quasiperiodic at large time. At high wave number ($n = 60$), the mode was initially amplified but become

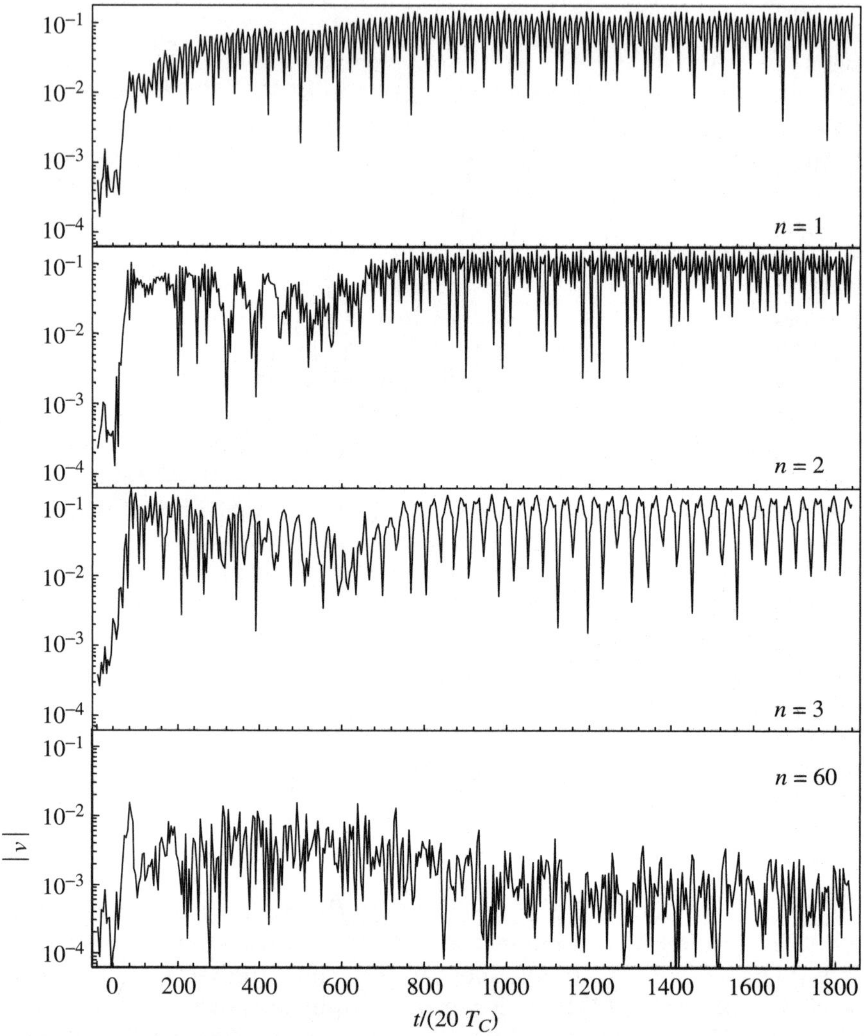

Figure 11.2.8 Time-dependent evolution of the Fourier amplitudes.

subdued with time. As the applied forcing is artificial, the growth of the harmonic waves and the resulting flow patterns may not be physically realistic. The results nevertheless show that DSMC can resolve a hierarchy of waves of a fundamental mode with wavelength $\Lambda = 400$ and its harmonics ($\Lambda = 400/n$).

Spanwise planes. In the spanwise planes, 2D simulations for a plane in the x-direction will be examined. As was noted earlier, the assumed

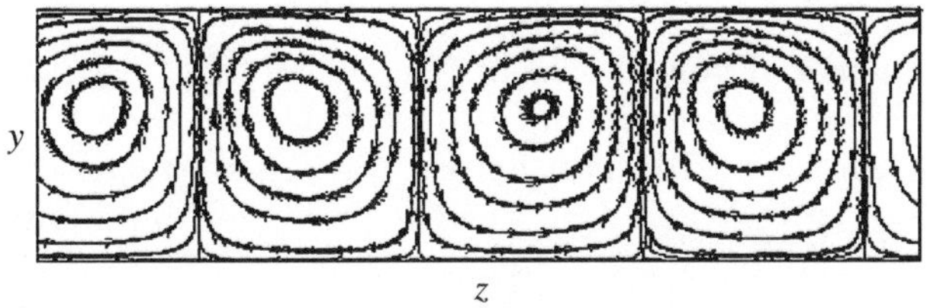

Figure 11.2.9 Steady flow pathlines. Kn = 1/60. (*Bird 2002.*)

forcing is similar to that of centrifugal force and Taylor-Gortler type of vortical motion are expected to be found in certain steady-state solutions of the current time-dependent simulations. The flows are analogous to the two-dimensional Taylor-Couette problem. Figure 11.2.9 shows the time-averaged pathline patterns for a case where vortices similar to that of the Taylor-Gortler vortices are observed. The computational domain size as shown is 60 × 180 in the y- and z-directions, respectively. The flow appears to be steady. The pathlines are based on an average from tU/h = 24 to 62. The flow pattern indicates a regular vortical flow almost identical to that of the Taylor-Couette flows. It shows that the current time-dependent DSMC is capable of capturing the Taylor-Gortler type of vortical flow motions.

The Knudsen number is 1/60 for the flow in Fig. 11.2.9. The flow patterns change as the Kundsen number decreases. The flow becomes varying with time and unsteady sampling must be used. Figure 11.2.10 shows the pathline patterns for two cases with Kn = 1/240 and 1/500,

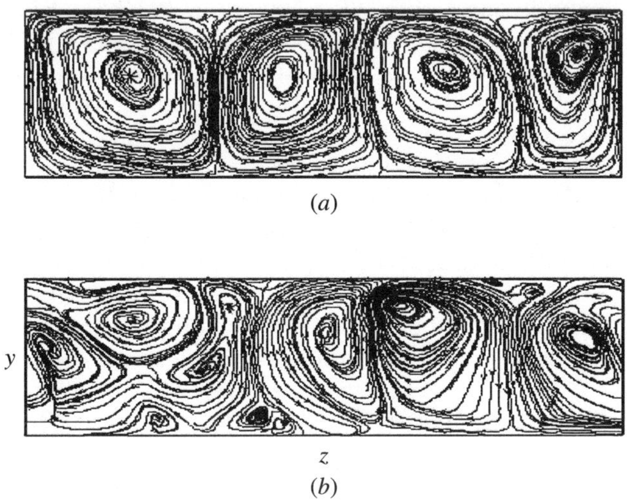

Figure 11.2.10 Flow pathlines: (*a*) Kn = 1/240, tU/h = 20; (*b*) Kn = 1/500, tU/h = 13. (*Bird 2002.*)

respectively. The corresponding domain sizes are 240×720 and 500×1500. Compared to the vortical flow shown in Fig. 11.2.9 (Kn $= 1/60$), the numbers of identifiable vortices in Fig. 11.2.10*a* for Kn $= 1/240$ are the same and the vortices still span across the entire height of the channel. The vortices in Fig. 11.2.10*a*, however, are deformed and irregular. With a further decrease of the Knudsen number, Fig. 11.2.10*b* (Kn $= 1/500$) describes a flow pattern that is irregular and contains vortices of different sizes that form and decay as time develops.

The 2D simulations performed on the two different cross-sectional cuts of the three-dimensional model flow have provided valuable insights into the applicability of DSMC to time-dependent flow simulations. They also serve as a guide to three-dimensional simulations, which are significantly more demanding on computational resources.

11.2.2 Three-dimensional flows

The three-dimensional (3D) simulations examined in this section are rich in physics and are demanding on computer resources. This is particularly true for the parallel DSMC simulations. Particle numbers in the order of 50 million have typically used for the larger domain simulations. Depending on the number of processors used, weeks of CPU hours are normally required to establish stationary states. We will begin by describing the flow parameters and the general features of these flows. This is followed by a more detailed description of the flow physics of the 3D flows, including the mean and the disturbances. The effects of the other flow parameters on the simulated flows are also discussed.

Two computational domain sizes, $800 \times 60 \times 320$ and $800 \times 320 \times 640$, nondimensionalized by the initial mean free path λ, in the x-, y-, and z-directions, respectively will be used. The value of U is set equal to the most probable molecular speed at the initial gas temperature of 1000 K. For hard-sphere gases, the inverse Knudsen number, based on the channel height, is then 60 and 320, respectively. The Reynolds numbers are 108 and 576, based on the channel height and U. The number of the computational molecules used is about 16 million for the small domain cases ($h = 60$) and about 48 million for the large domain case ($h = 320$). The average separation distances are roughly 1.4 times the mean free path. The following set of figures showing the general features of the sample average flow and the disturbances are for $h = 60$ and $c = 0.04$.

Figure 11.2.11 shows the time history of the instantaneous and the long-time-averaged or mean velocity components for a cell that is located at the center of the computational domain. The data were obtained by using two different time-steps. The time-steps are 20 percent ($\Delta t^* = 0.2$) and 10 percent ($\Delta t^* = 0.1$) of the mean molecular

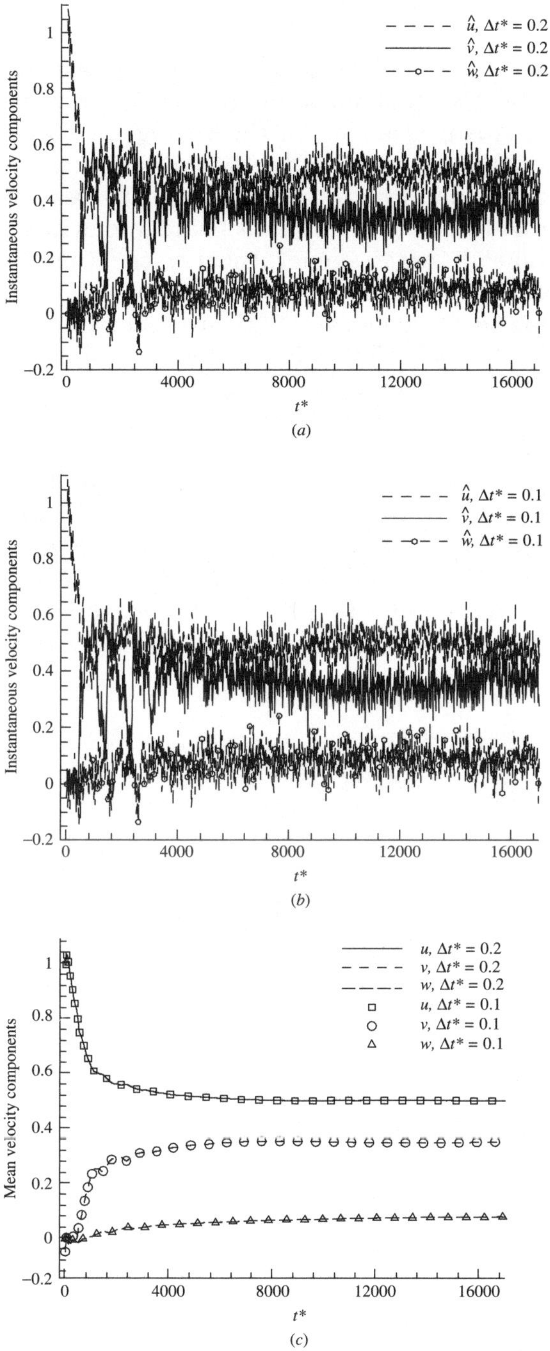

Figure 11.2.11 Evolution of velocity components. $h = 60$, $c = 0.04$: (*a*) Instantaneous using $\Delta t^* = 0.2$; (*b*) Instantaneous using $\Delta t^* = 0.1$; (*c*) Long-time-averaged values. (*Liou and Fang 2004*.)

collision time of the initial gas (T_c). The instantaneous velocity signals are practically identical, indicating that the simulations are independent of the time-steps used. This is also true for the averaged velocity components (Fig. 11.2.11*c*) obtained by using the two different time-step sizes, in that their evolutions in time are identical. The results show that the current parallel DSMC simulation tools are numerically accurate. The results also indicate that the general observation that, as long as the time steps are small compared with the mean molecular collision time, DSMC simulations are independent of their exact values, is valid for not only the stationary state of a flow but also its development in time. The following results were obtained by using $\Delta t^* = 0.2$.

Figure 11.2.12 shows the time history of the cell-averaged specific kinetic energy on an x-plane located at $x = 350$. After an initial transient, the flow appears to have become stationary and the statistical oscillations are small, as is shown in the insert in Fig. 11.2.12. Figure 11.2.13 shows the pathlines of the long-time-averaged flow on three planes at different x locations. Note that the top wall at $y = 60$ moves along the x-direction. The pathlines indicate that a fluid flow exhibiting pairs of counter rotating vortical motion may exist in the 3D simulation,

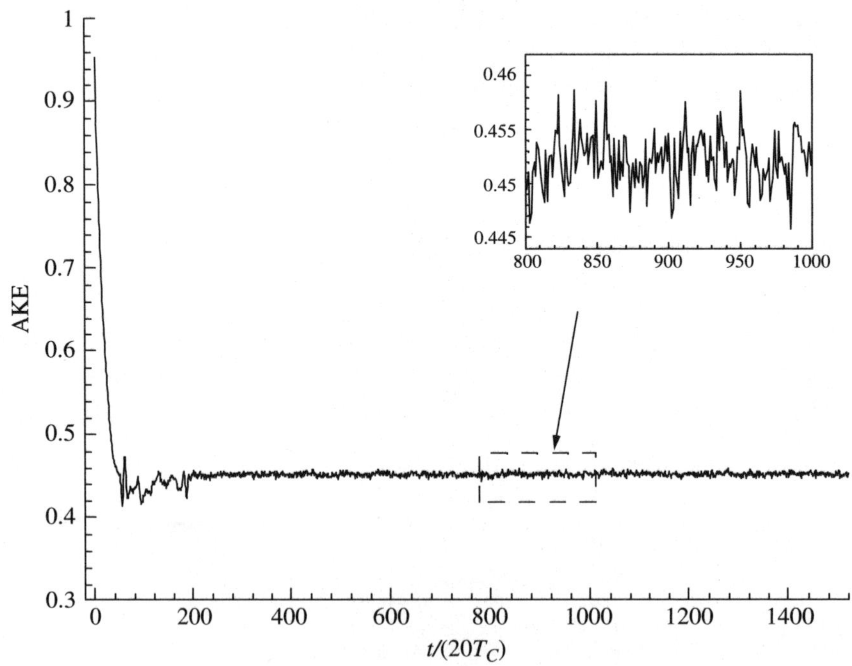

Figure 11.2.12 Evolution of cell-averaged specific kinetic energy.

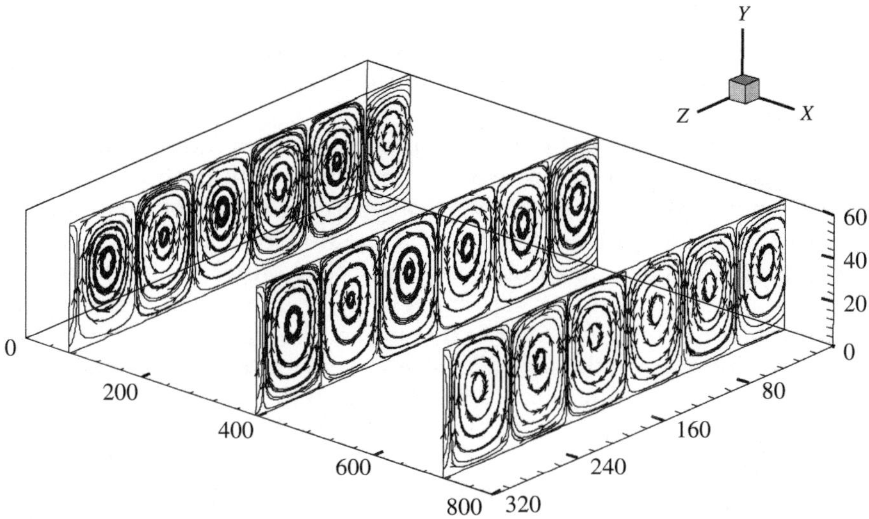

Figure 11.2.13 Time-averaged vertical velocity contours and in-plane pathlines.

a characteristic of the Taylor vortices. Note that the results of the 2D simulation of the same Knudsen number also shows similar patterns of pairs of vortices. The value c is 0.1 in that case.

Figure 11.2.14 shows the contours of the time-averaged vertical velocity in the three planes perpendicular to the coordinate axes,

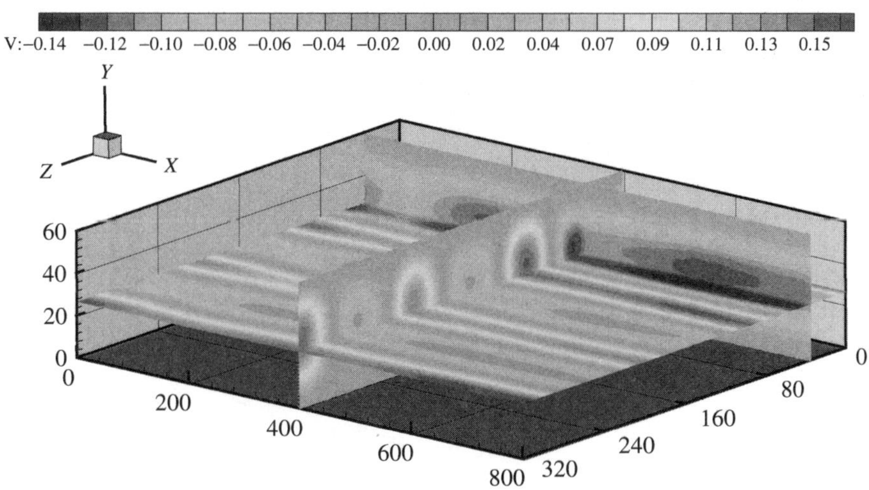

Figure 11.2.14 Sampled vertical velocity contours on three orthogonal planes.

respectively. The contour pattern in the y-plane shows streamwise streaks, indicating that the vortical motions exhibited in Fig. 11.2.13 are indeed imprints of pairs of counter rotating streamwise structures, or the Taylor vortices.

Figure 11.2.15 shows the in-plane pathlines in the three planes perpendicular to the coordinate axes, respectively. The pathline pattern in the z-plane shows that the vortical motions have not advanced a complete rotation, probably because of the limited domain size in the x-direction. The pathlines in the y-plane either bundle up underneath the regions of high positive vertical velocity or spread out underneath the region of high negative vertical velocity. The pattern suggests that fluid is swept in from the side and ejected upward in the up-wash part of the Taylor vortex motion. Similarly, the fluid is pushed downward and spread out sideways near the wall in the down-wash part of the Taylor vortex motion.

Figure 11.2.16 shows the Fourier spectrum of the steady specific kinetic energy where K represents the magnitude of the wave number vector. At the low wave number end, the spectrum is dominated by a number of discrete peaks, each apparently corresponding to the coherent vortical flow pattern shown in the previous figures. Beyond $K = 10$ at the higher wave number end, the spectrum shows, in general, a monotone decrease. The magnitudes are orders-of-magnitude smaller than the spectral peaks in the low wave number range, which are associated with identifiable flow signals.

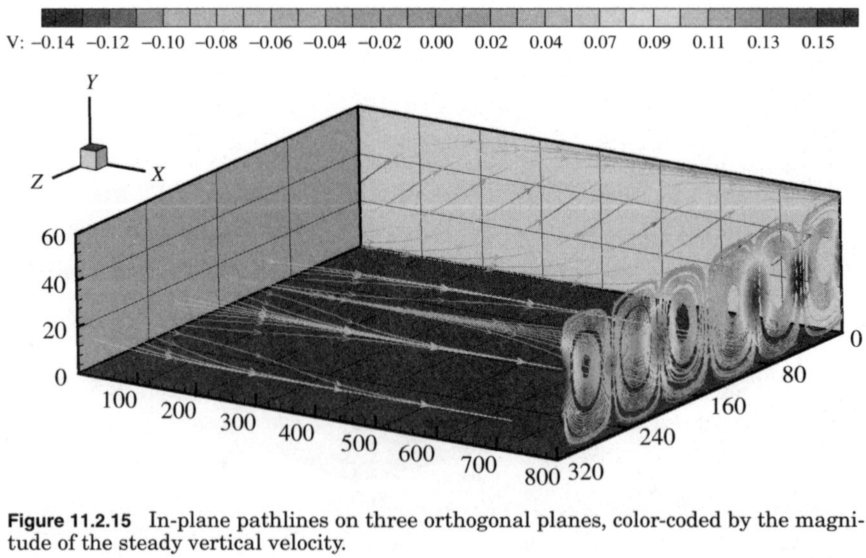

Figure 11.2.15 In-plane pathlines on three orthogonal planes, color-coded by the magnitude of the steady vertical velocity.

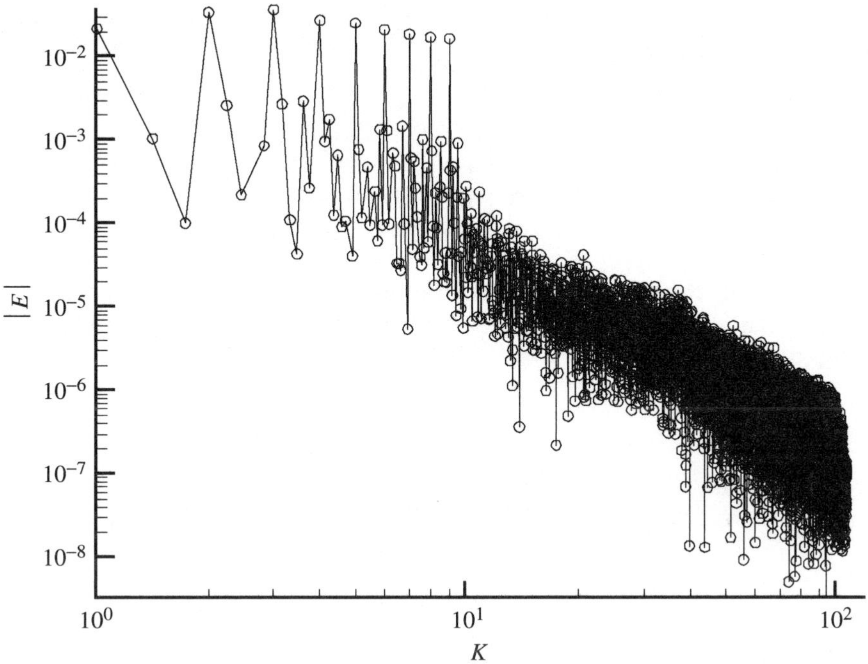

Figure 11.2.16 Specific kinetic energy spectrum.

Effects of the forcing coefficient. We will use the case with $h = 60$ to describe the effect of the changes of the forcing coefficient c on the forced microCouette flow.

$c = 0.0$. The DSMC results will be compared with those obtained by using a continuum slip flow model (Liou et al. 2003*c*). The model uses the Navier-Stokes (NS) equations and slip boundary conditions.

In the NS solver, a fifth-order compact upwind difference scheme (Tolstykh and Lipavskii 1998) is used for the spatial convective terms and a third order low-storage Runge-Kutta scheme (Williamson 1980) for the time advancement of the NS equations. High-order compact schemes have attracted much attention due to their spectral-like accuracy (Lele 1992; Hixon et al. 1998) and the demonstrated ability to deal with flows in a complex domain and to resolve shocks. Tolstykh and Lipavskii (1998) developed a different high-order compact upwind difference scheme by correcting the conventional first-order upwind difference scheme and applying a high-order *Pade* series to the correction. Their studies show that the scheme can capture wiggle-free steady-state solutions with steep gradients without having to use additional numerical dissipation. For the viscous terms, fourth-order finite difference schemes were used for the discretization for the interior nodes and the boundary nodes. The applied forcing is added to the NS equations with a forcing term,

$$\mathbf{F} = \frac{1}{J}\begin{pmatrix} 0 \\ 0 \\ c\rho(u-U)^2 \\ 0 \\ c\rho(u-U)^2 v \end{pmatrix}$$

The Maxwell-Smoluchowski slip boundary conditions were used to determine the slip velocity components (u and w) and the temperature jump at the surfaces.

$$u_s - u_w = \frac{2-\sigma_v}{\sigma_v}\mathrm{Kn}\left(\frac{\partial u}{\partial n}\right)_{\mathrm{wall}} \tag{11.2.7}$$

$$T_s - T_w = \frac{2-\sigma_T}{\sigma_T}\frac{2\gamma}{\gamma+1}\frac{\mathrm{Kn}}{\mathrm{Pr}}\left(\frac{\partial T}{\partial n}\right)_{\mathrm{wall}} \tag{11.2.8}$$

The pressure gradient normal to the surfaces are determined by applying the momentum equation locally, which gives, for instance, on the lower surface,

$$\frac{\partial p}{\partial y} = c\rho(u-U)^2 \tag{11.2.9}$$

Both the NS and the DSMC calculations ran in a time-accurate manner. The NS calculation converges to a solution with the flow properties being homogeneous in the x- and z-directions. The DSMC simulation develops a statistically stationary state that also shows homogeneity in the x- and z-directions. Figure 11.2.17 gives the vertical distributions of the streamwise velocity component, pressure, and density. The DSMC solutions shown are the long-time-averaged macroscopic flow properties. The NS solutions are the converged results. The velocity profiles agree quite well and both show a slight departure from linear distributions. The inflectional, as opposed to a linear, velocity profile is consistent with many microchannel flow computation results that show the velocity deviating from a linear distribution as the Knudsen number increases. Both the NS and DSMC model predict a constant pressure of 1.05 in the vertical direction. Figure 11.2.17*c* shows the computed density distributions in the y-direction. There is a fairly good agreement between the two solutions.

Figure 11.2.18 shows the spanwise (z) distributions of the velocity slip and the temperature jump on the plate surfaces. On the upper plate the NS model predicts well the velocity slip. The NS model slightly underpredicts the slip velocity on the lower plate. For the NS solution,

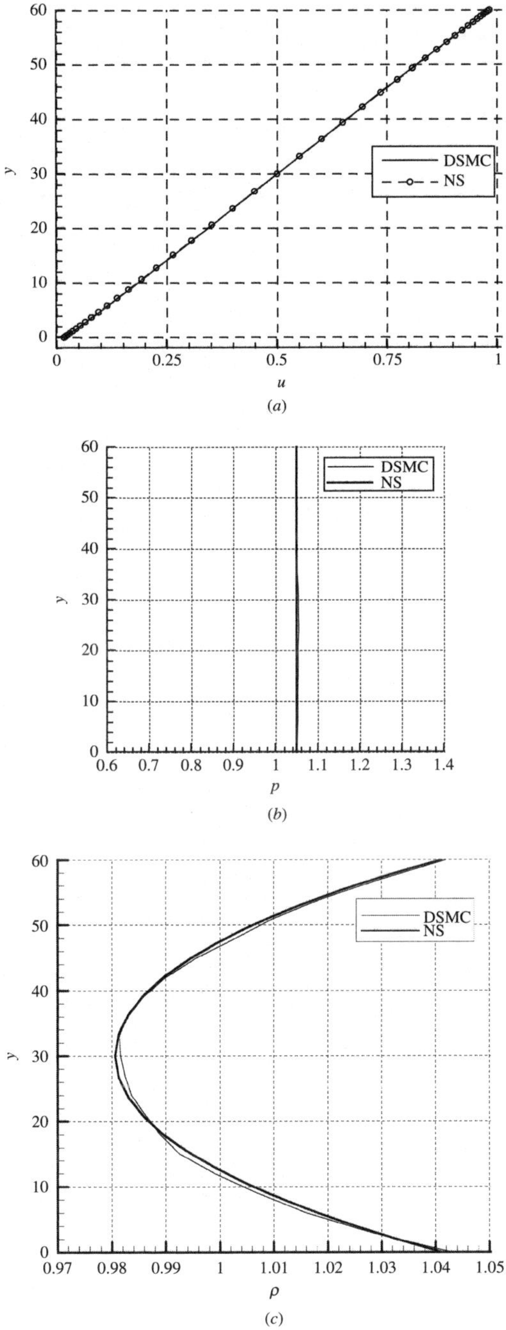

Figure 11.2.17 Comparisons of the calculated flow with $c = 0$: (a) Streamwise velocity component; (b) Pressure; (c) Density.

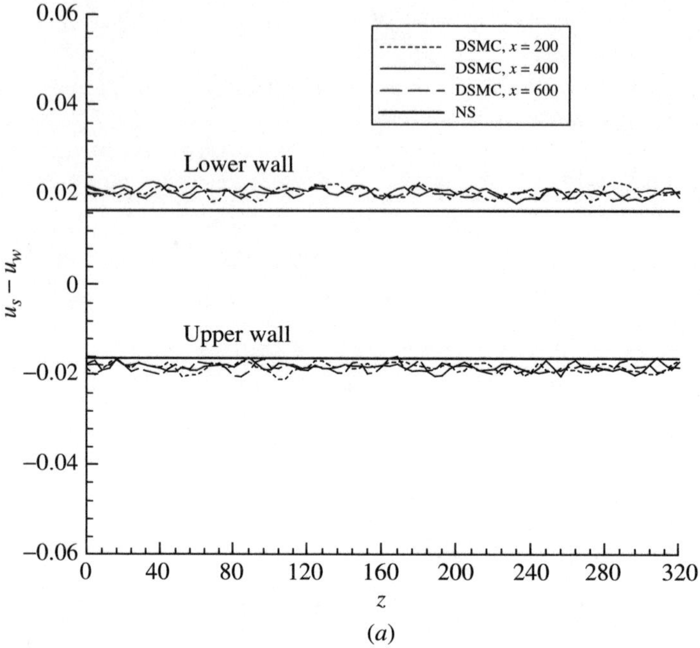

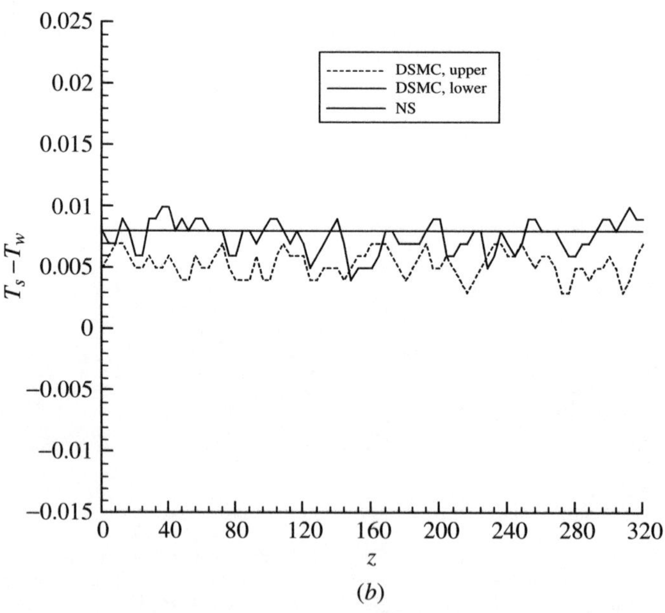

Figure 11.2.18 Comparison of slip boundary conditions: (*a*) Velocity; (*b*) Temperature, $x = 400$.

the temperature jump is the same on both surfaces, which agree well with the DSMC results.

Overall, in the absence of the external forcing, the gas microflow computed by using the NS model agrees well with that by using the DSMC method. The good agreement between the two model results in this essentially two-dimensional flow with low Knudsen number is generally observed in other similar comparisons.

$c = 0.001$. In the following, the external forcing is activated. For $c = 0.001$, the computed flow reaches a stationary state. There are less than 3 percent changes of density, pressure, and temperature compared with those of the no-forcing case. The averaged streamwise velocity profiles are nearly linear and homogeneous. Figure 11.2.19 shows the vertical distributions of the averaged spanwise velocity at three spanwise locations. The figure shows that there are no secondary motions in the computed flow and the flow is essentially two-dimensional.

$c = 0.005$. For $c = 0.005$, the computed flow also reaches a stationary state. In contrast to the $c = 0.001$ case, the stationary flow is highly three-dimensional. Figure 11.2.20 shows the in-plane pathlines on three planes at different x locations. Note that the top wall at $y = 60$ moves along the x-direction. The pathlines indicate a fluid flow with

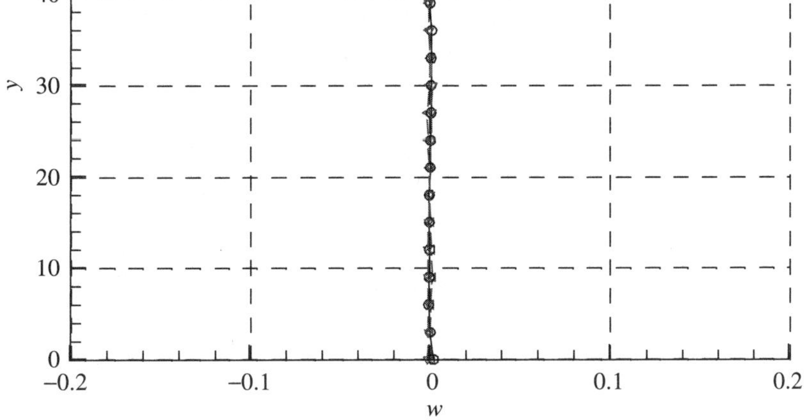

Figure 11.2.19 Averaged spanwise velocity distributions. $c = 0.001$.

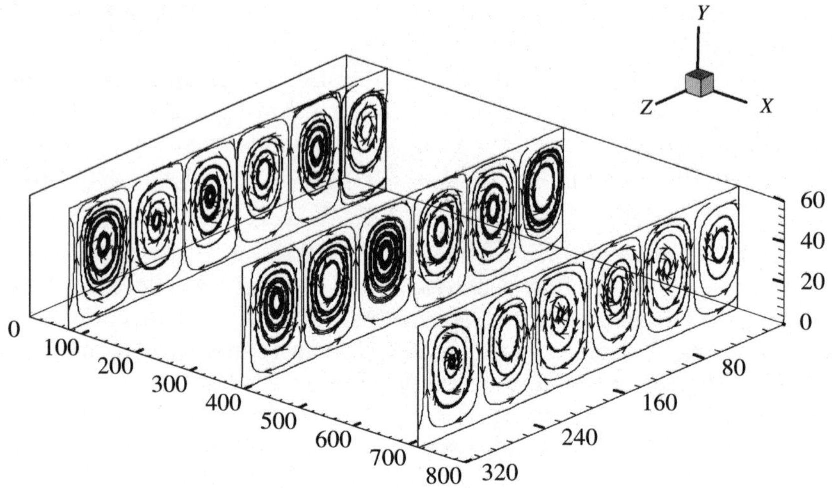

Figure 11.2.20 In-plane pathlines of the averaged flow. $c = 0.005$.

three pairs of counter rotating vortical motion. This is characteristic of the Taylor vortices and the vortical flow patterns appear the same as that of the corresponding 2D chaotic flow simulation. The unsteady quantities exhibit practically the same vortical flow patterns.

Figure 11.2.21 shows the spanwise distributions of the averaged energy at three streamwise locations along the midplanes ($y = 30$) and

$$e = \frac{1}{2}(u^2 + v^2 + w^2) \tag{11.2.10}$$

The distributions are rather smooth. The cyclic distributions reflect the vortex pairs observed in Fig. 11.2.20. The amplitudes of the energy vary slightly from 0.15 to 0.14. Figure 11.2.21 also shows that the simulated flow is nearly homogeneous in the x-direction.

$c = 0.01$. For $c = 0.01$, the in-plane pathlines pattern are similar to the $c = 0.005$ case, with three pairs of vortical flow structures. The amplitudes of the energy vary slightly from 0.18 to 0.55, slightly higher than that for $c = 0.005$. Figure 11.2.22 shows the vertical distributions of the streamwise velocity at three spanwise locations. Compared with the no-forcing case, the streamwise velocity profiles deviate significantly from a linear distribution due to the strong rarefaction in the lower part of the flow, which is consistent with the DSMC result of similar flows (Fang and Liou 2002).

Figure 11.2.23 shows the spanwise distributions of v at $y = 30$ and $x = 400$ for two results of the DSMC simulations. The physical and the numerical parameters of the two simulations are identical. The only

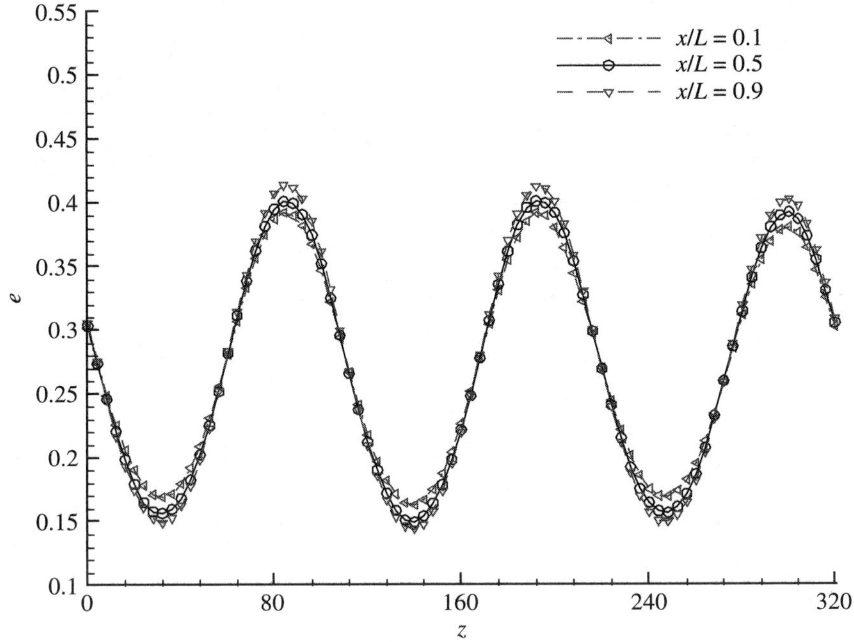

Figure 11.2.21 Spanwise energy distributions. $c = 0.005$.

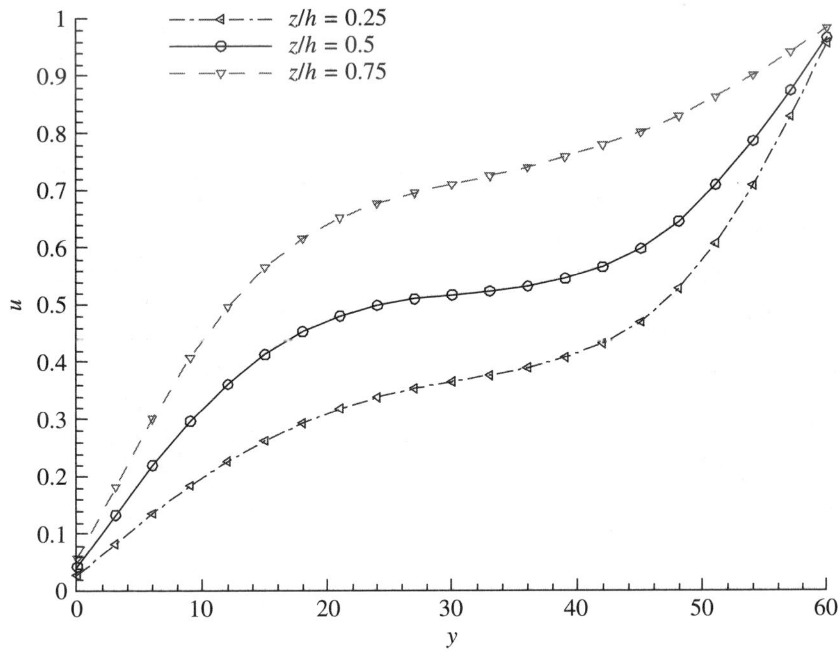

Figure 11.2.22 Averaged streamwise velocities. $c = 0.01$.

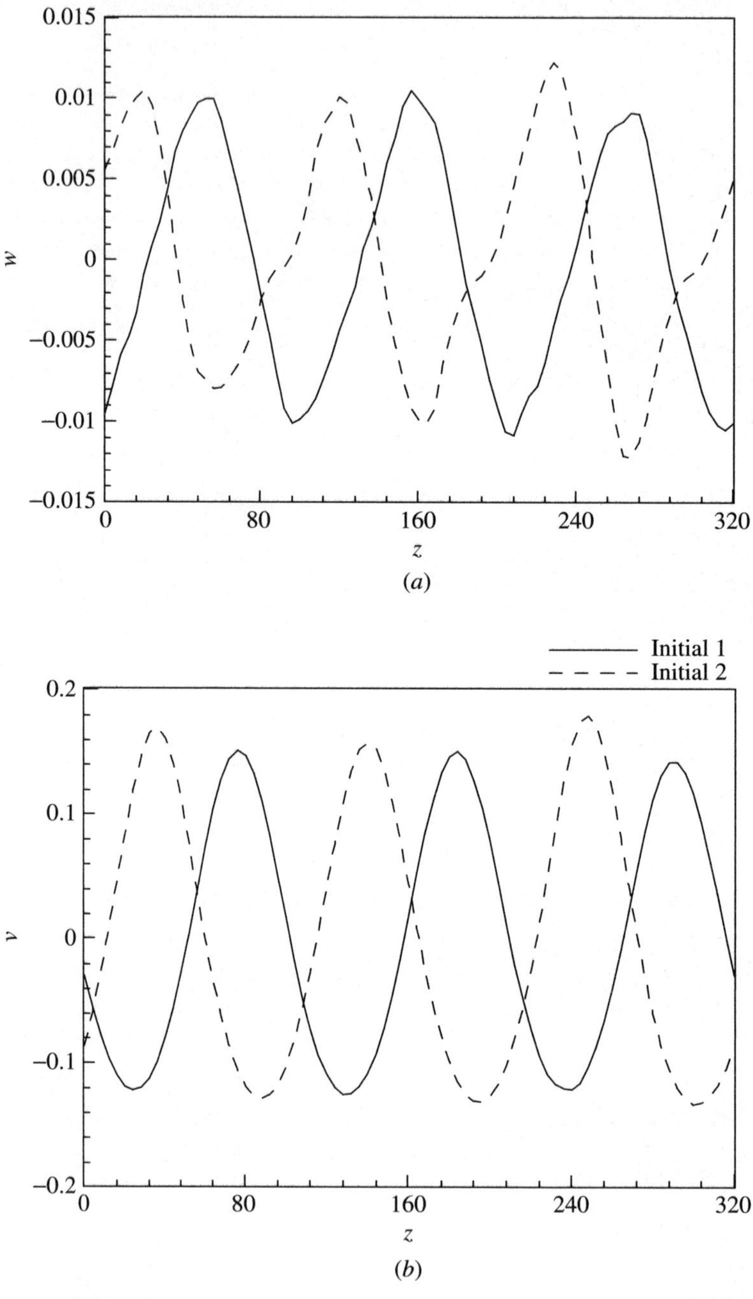

Figure 11.2.23 Spanwise distributions of velocity components using DSMC. $c = 0.01$: (*a*) v; (*b*) w.

difference is that two different methods have been applied to the initial seeding of the random variables in the different processors, one (initial 1) with the same seed for all the processors that are used in the simulations and varying in the second (initial 2). Except for an apparent phase difference, the flow patterns predicted, as can be seen in Fig. 11.2.23, are essentially the same. The nonuniqueness of the DSMC solutions is, perhaps, not too surprising in their numerical aspects, since the numerically enforced flow boundary conditions are periodic. The DSMC results, nevertheless, suggest that for the current flow model the stationary macroscopic flow properties depend on the initial conditions at the microscopic level. The results shown are obtained using the uniform seeding method.

Figure 11.2.24*a* shows the frequency spectra of the instantaneous (unsteady), the long-time-averaged, and the disturbance kinetic energies for $c = 0.01$. Figure 11.2.24*b* shows the frequency spectra of the cross correlation of the vertical and the spanwise velocity components. The long-time-averaged component of the instantaneous velocity field, which shows highly spatially coherent patterns (Fig. 11.2.22), is temporally dominated by the zero frequency ($=m/(20T_c)$), or stationary mode. At other frequencies, the spectra suggest mainly statistical noise. There does not appear to be very strong correlations between the vertical and the spanwise velocity components for the any of the three signals.

Comparisons can also be made with the NS calculations. The forcing is applied in the y-direction as described earlier and has been implemented explicitly in the NS solutions. The NS calculation converges to steady-state solution. The DSMC solution also eventually becomes stationary. Both solutions, as are shown shortly, exhibit organized flow pattern in the spanwise direction. To facilitate the comparison between the DSMC and the NS models, the DSMC results in the following figures were translated in the z-direction by a constant value so that there is an alignment with the vortical structure in the NS solution. The anchoring vortex center is located about $z = 52$. Figure 11.2.25 shows the in-plane streamlines on three planes at different x locations. Note that the top wall at $y = 60$ moves along the x-direction. The streamlines in Fig. 11.2.25*a* indicate a fluid flow with three pairs of counter rotating vortical motion. This is characteristic of the Taylor vortices and the vortical flow patterns appear the same as that of the corresponding 2D flow simulation. The instantaneous quantities exhibit practically the same vortical flow patterns with some statistical scatter. Figure 11.2.25*b* shows the in-plane streamlines at the same axial locations based on the NS solutions. The NS solutions also show three vortex pairs. The streamline patterns are virtually identical to those obtained by using DSMC.

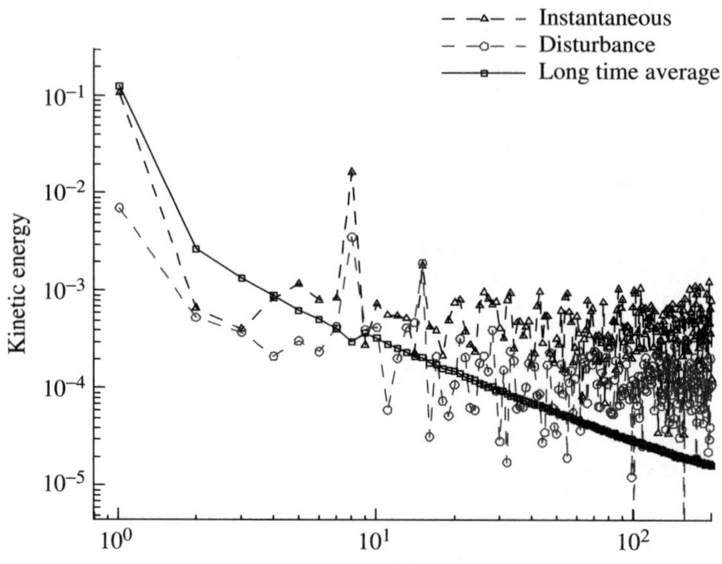

(*a*)

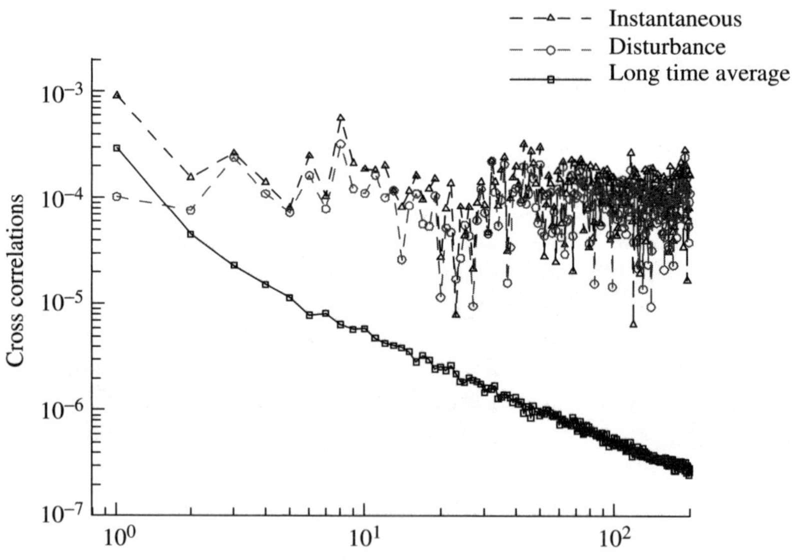

(*b*)

Figure 11.2.24 Fourier frequency spectra for $c = 0.01$, $h = 60$, and $U = 1$: (*a*) Energy; (*b*) Cross correlation.

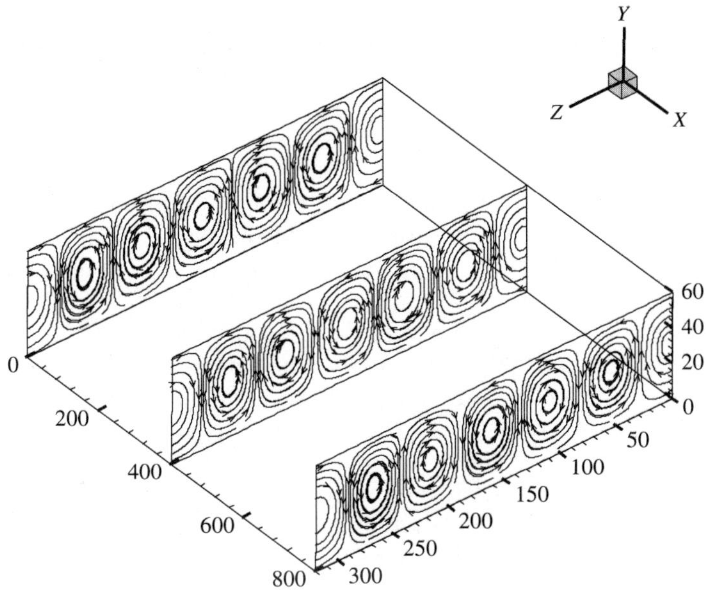

(a)

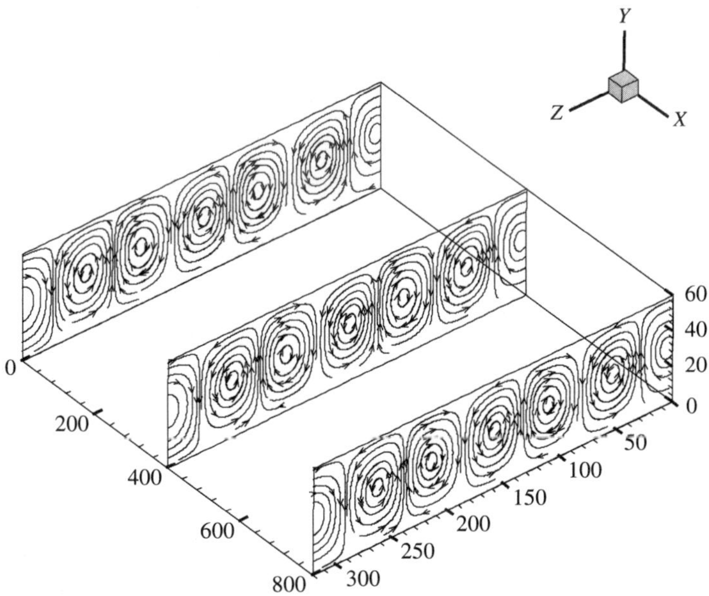

(b)

Figure 11.2.25 In-plane streamlines of the steady flow: (*a*) DSMC model; (*b*) NS model. (*Liou and Fang 2004.*)

The spanwise distributions of the calculated streamwise velocity component on the lower plate ($y = 0$), the centerline of the channel ($y = 30$), and the upper plate ($y = 60$) are compared in Fig. 11.2.26 for $x = 400$. There is a fairly good agreement in both the phase and amplitude of the periodic patterns of the slip velocities on the plates between the two model solutions.

In the following figures the characteristic of a computed vortex will be extracted from each of the two solutions and compared. There are three vortex pairs. Consequently, each vortex spans vertically between the plates and occupies about 320/6 in the z-direction. Figure 11.2.27 shows the spanwise distributions of the calculated spanwise velocity component at $y = 0$, 30, and 60, respectively. Similar to what has been shown for the u slip-velocity component, there are small differences in the amplitude and phase between the computed periodic patterns. Overall, the spanwise slip velocity component calculated by the NS model agrees reasonably well with the molecular DSMC model.

Figure 11.2.28 shows the vertical distributions of u at three spanwise locations, $z = 36$, 52, and 68, which correspond closely to the quarter locations of the anchoring vortices extracted from each solution. The

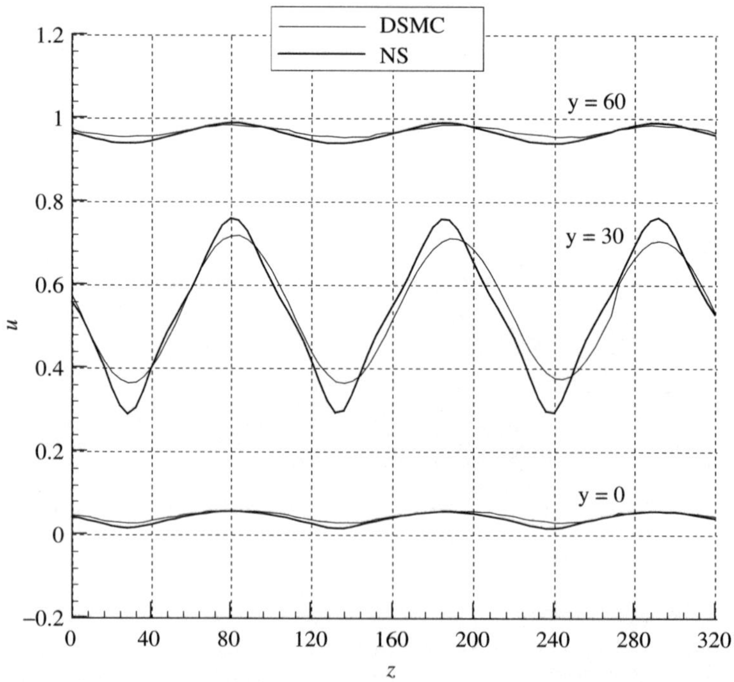

Figure 11.2.26 Distributions of the streamwise velocity component. (*Liou and Fang 2004.*)

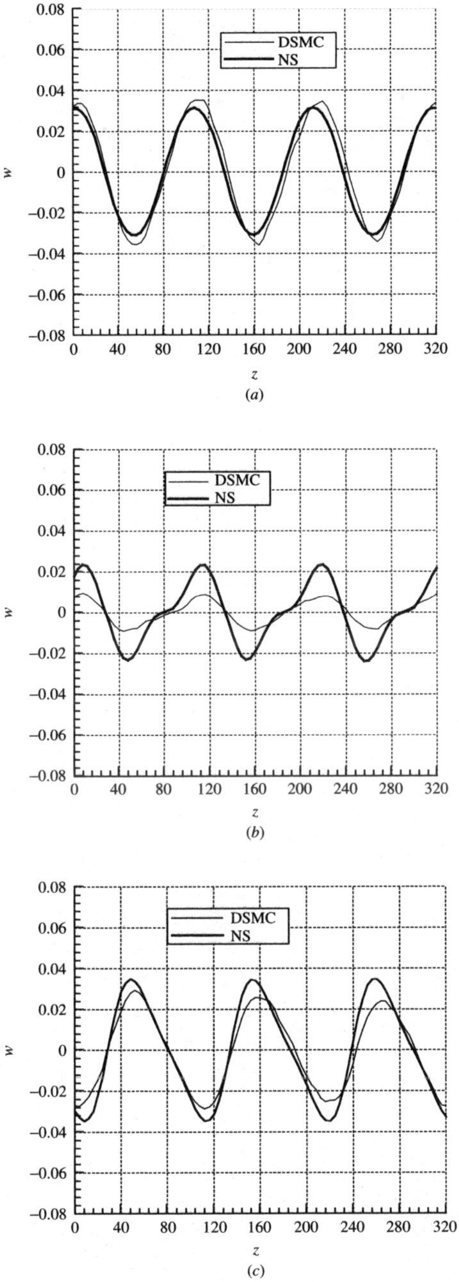

Figure 11.2.27 Spanwise distribution of w: (a) $y = 0$; (b) $y = 30$; (c) $y = 60$.

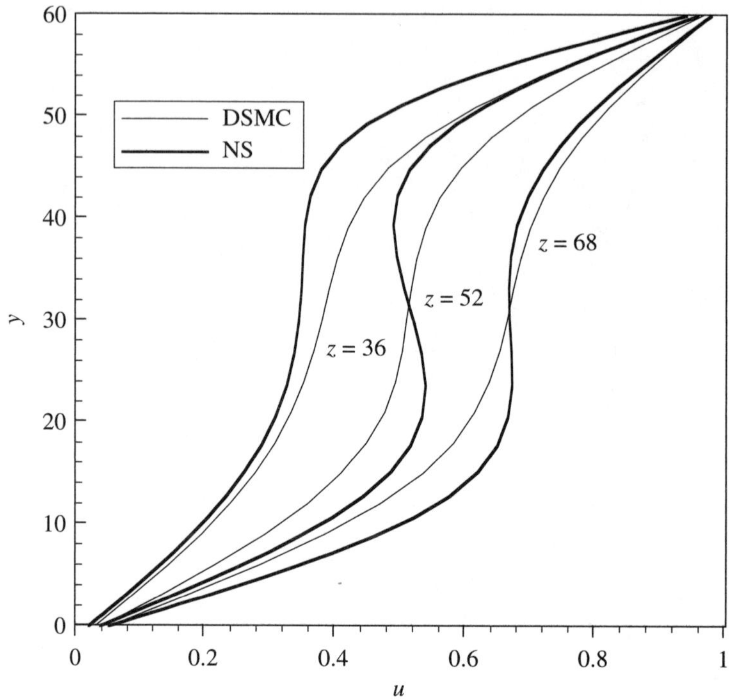

Figure 11.2.28 Streamwise velocity components in the anchoring vortices.

overall trend of variation with the locations in the vortices appears to be similar. Both the NS and DSMC capture highly inflectional distributions. Compared with the no-forcing results in Fig. 11.2.17*a* for the same Knudsen number, u has more significantly deviated from the linear profile. This large deviation apparently is a result of the secondary microflow motion driven by the applied forcing. There are marked differences, up to 10 percent of U, between the NS and DSMC distributions at all three locations, which are not observed in the no-forcing case result.

Figure 11.2.29 shows the vertical distributions of v at the same quarter locations in the anchoring vortices as those in Fig. 11.2.28. In general, the trend of variation is similar between the two model predictions. The NS solutions show larger vertical speeds on both sides of the vortex in comparison with the DSMC solutions.

Figure 11.2.30 shows a composite plot of v and w. The figure shows the distributions of v along the horizontal center line, denoted by $v(z)$, and those of w along the vertical centerline $w(y)$ of the anchoring vortices. The anchoring vortices exhibit clockwise rotation motion. For both v and w, the NS solutions have larger values than the DSMC solutions. It appears to indicate that the vortex computed by the NS model is

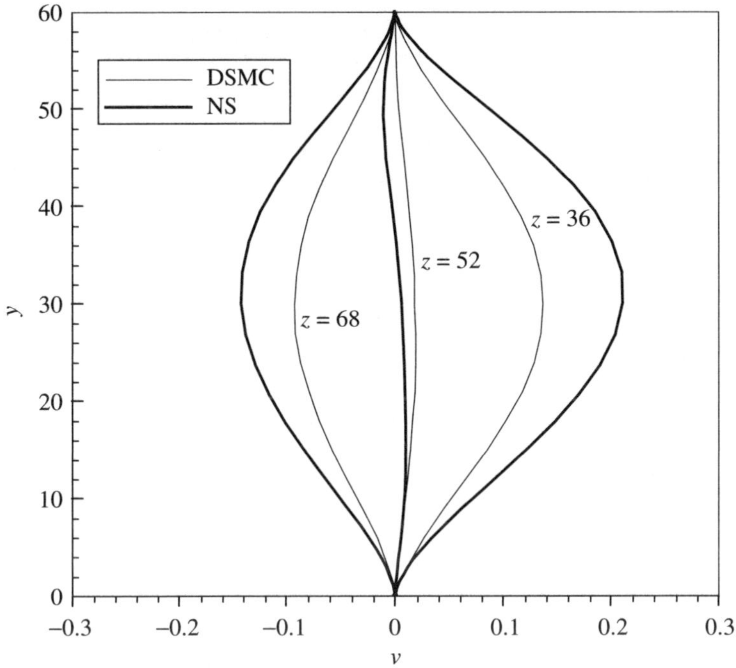

Figure 11.2.29 Vertical velocity distributions in the anchoring vortices.

rotating at a higher rate than that computed by the DSMC model. A rough estimate of the amount of rotation can be obtained by evaluating the streamwise vorticity associated with the vortices using the vortex centers and the maximum v's and w's in the figure. The simple calculation shows that the estimated vorticity for the vortices computed by using the NS model is about 60 percent higher than that computed using the DSMC model. The vertical density, pressure, and temperature distributions at the quarter locations of the anchoring vortices are shown in Fig. 11.2.31. In general, the agreement is satisfactory between the two solutions. There is a certain degree of rarefaction in the lower half of the channel where the applied forcing is strong. As a result, the density and pressure are higher in the upper half of the channel. The temperature distributions show an increased thermal velocity compared with the no-forcing case, particularly in the upper half of the channel. The NS solutions also show similar trends.

The results indicate that microflow behavior simulated by using the molecular DSMC model is significantly different from that using the continuum Navier-Stokes model, when the forcing is applied. There could be a number of factors that contributed to the results. Effects of the slip boundary conditions and the numerical resolutions of the NS

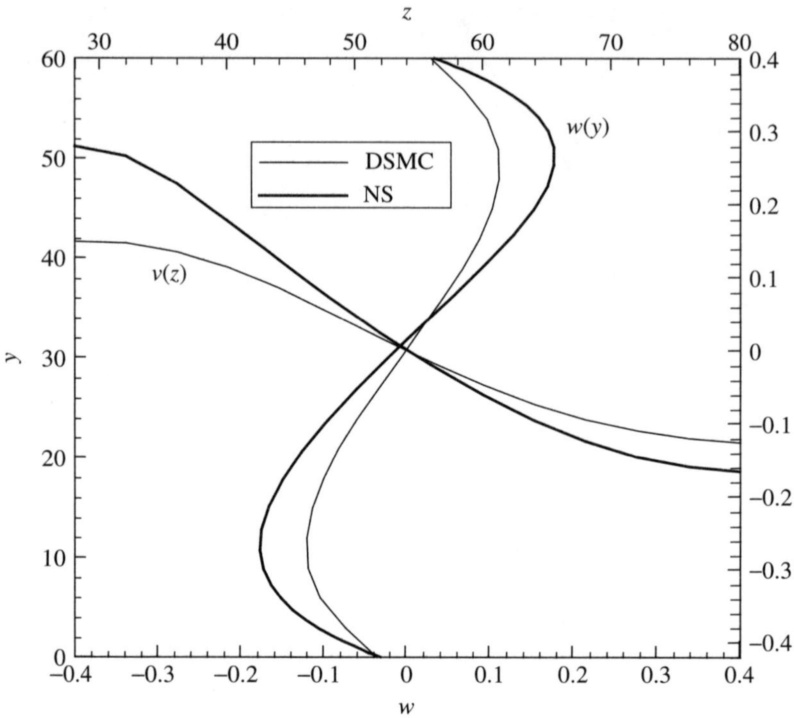

Figure 11.2.30 Composite comparisons of v and w in the anchoring vortices.

results may be examined. There are strong 3D flows in both solutions, which might also have caused the observed differences.

$c = 0.04$. With $c = 0.04$, the flow develops into a stationary state and the in-plane pathline patterns of the long-time-averaged flow show three pairs of vortical motion, same as the $c = 0.005$ case and has been shown in Fig. 11.2.13. As was mentioned earlier, the vortical structures appear as a result of the externally applied forcing in the y-direction that mimics the centrifugal forcing in Taylor-Couette problems. Figure 11.2.32 shows a comparison of the mean spanwise velocity (w) distributions in the spanwise direction at three x-stations and $y = 30$. The spanwise velocity distributions also exhibit wavy patterns that are periodic in the z-direction and fairly homogeneous in the x-direction.

Figure 11.2.33 shows the Fourier spectra of the averaged spanwise velocity components given in Fig. 11.2.32. The distinct peaks indicate that the averaged flow is dominated by fluid motion of discrete wave numbers in the spanwise direction. The first peak has a wave number of 3 with a wavelength Λ of 107 ($\sim$320/3), which corresponds to the dominant flow pattern observed in Fig. 11.2.32. The statistical noises

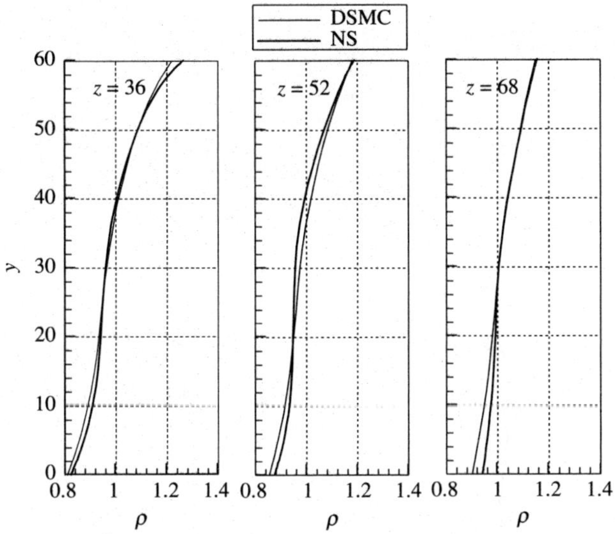

(a)

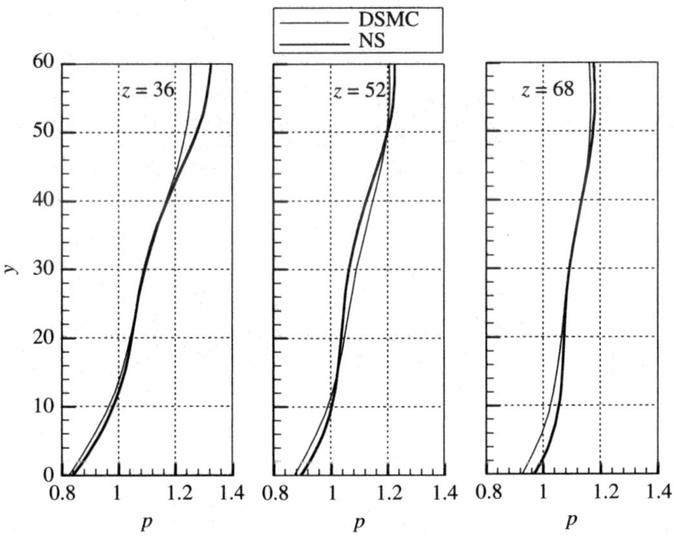

(b)

Figure 11.2.31 Vortex properties: (*a*) Density; (*b*) Pressure; (*c*) Temperature.

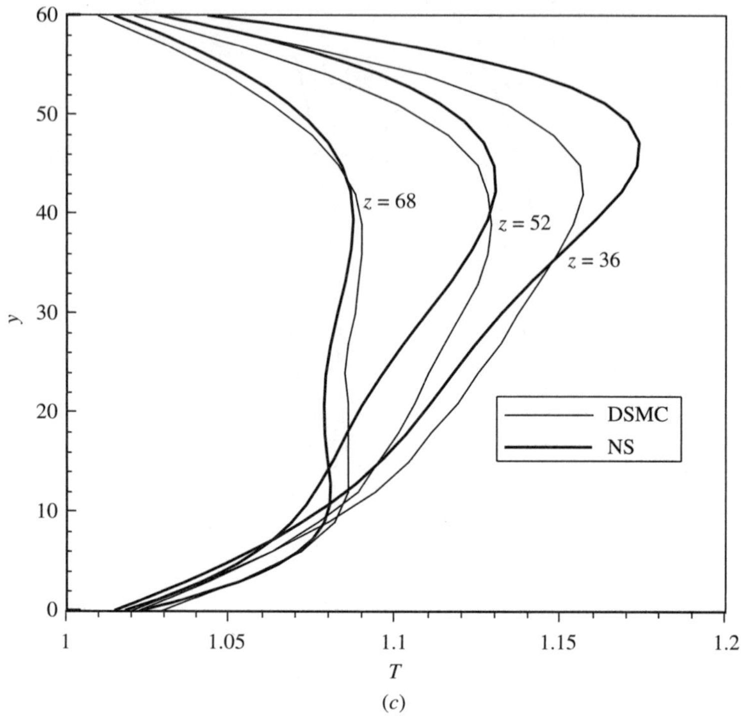

Figure 11.2.31 *(Continued)*

are orders of magnitude lower than the coherent flow signals. If $n = 1$ denotes the first peak, the second peak corresponds to $n = 2$ with a wave number of 6 and a wavelength of 320/6. The 3D DSMC simulation is able to capture harmonic modes with no preferential forcing of these modes in the initial field. In the absence of the external forcing, on the other hand, no distinguishable secondary, vortical flow patterns can be found in a 3D simulation with the same initial field. The flow pattern observed arises as a result of both the applied forcing and the nonlinearity in the simulated flow.

Figure 11.2.34 shows the evolutions of the Fourier amplitudes with time for the first three modes and for $n = 12$. Figure 11.2.34 provides a view of how the averaged secondary vortical patterns develop as the flow evolves with time. The amplitudes of all four modes pick up quickly at the beginning of the simulations and then decrease until $t/(20T_c) = 100$. Their magnitudes fluctuate significantly and there is not a definite dominant mode. The flow initially contains motions of a wide range of length scale. The fundamental mode ($n = 1$) amplitude begins to increase after $t/(20T_c) = 120$ and becomes the dominant mode at $t/(20T_c)$ about 200. The first harmonic mode ($n = 2$) also grows but

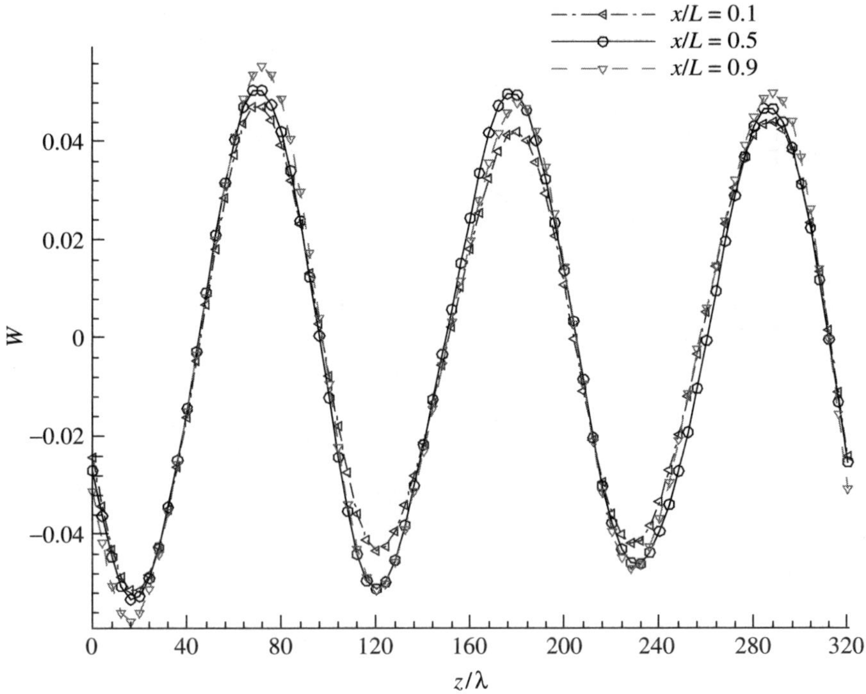

Figure 11.2.32 w distributions. $h = 60$, $c = 0.04$. (*Liou and Fang 2004.*)

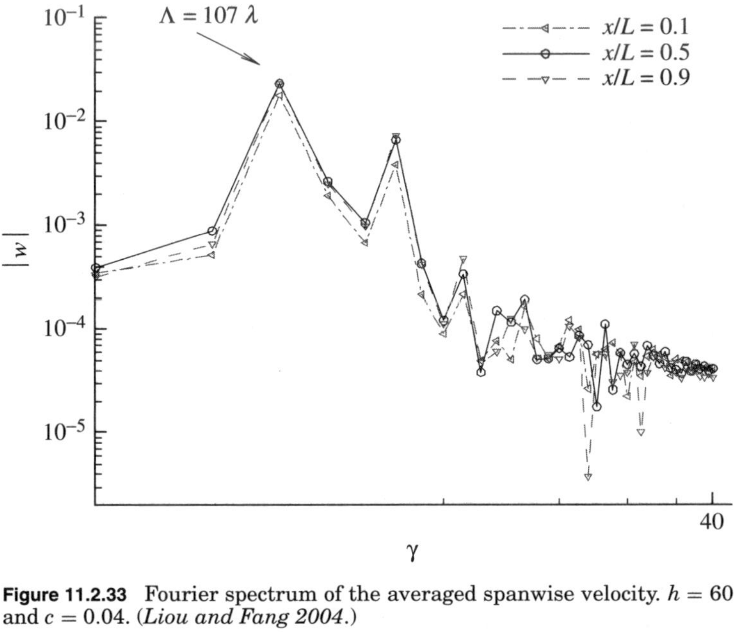

Figure 11.2.33 Fourier spectrum of the averaged spanwise velocity. $h = 60$ and $c = 0.04$. (*Liou and Fang 2004.*)

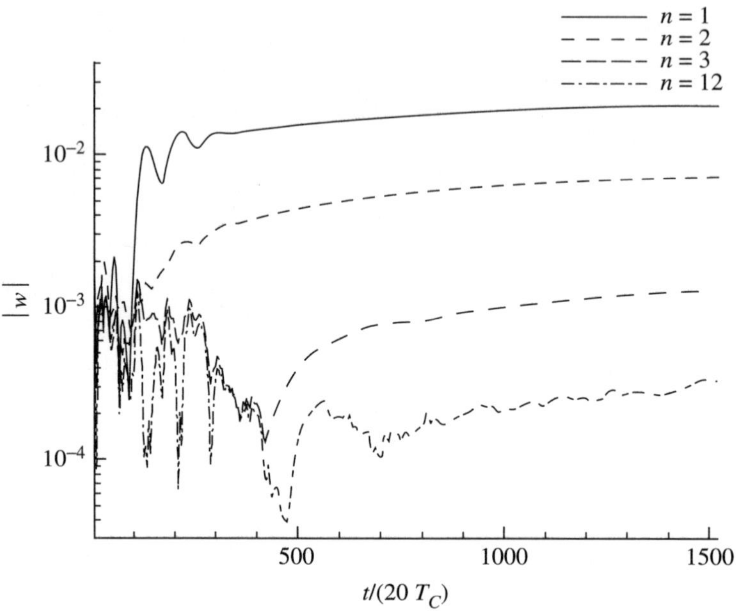

Figure 11.2.34 Evolution of Fourier amplitudes. $h = 60$, $c = 0.04$. (*Liou and Fang 2004.*)

asymptotes to a lower amplitude than the fundamental mode. During this period, the amplitudes of the other three modes continue to drop. The amplitude decaying process for the high wave number modes continues until $t/(20T_c)$ reaches about 400. At which point, the $n = 3$ mode begins to pick up its amplitudes. At a later time of 450, the amplitude of the $n = 12$ mode begins to increase. It is apparent that the low wave number fundamental mode is more efficient in extracting energy from the applied forcing and becomes the dominant mode in the averaged flow. The Fourier amplitude of the highest harmonic is small and is more likely to be associated with statistical noises than physical flow signals.

Figure 11.2.35*a* shows the frequency spectra of the instantaneous (unsteady), the long-time-averaged, and the disturbance energy at the center of the simulation domain. Figure 11.2.35*b* shows the frequency spectra of the cross correlation of the vertical and the spanwise velocity components. The long-time-averaged component of the instantaneous velocity field, which shows highly spatially coherent patterns (Figs. 11.2.32–35), is temporally dominated by low frequency ($=f/(20T_c)$) modes. On the other hand, the disturbance and the instantaneous components appear to be mainly statistical and do not have any preferential modes. It is interesting to note that DSMC is able to

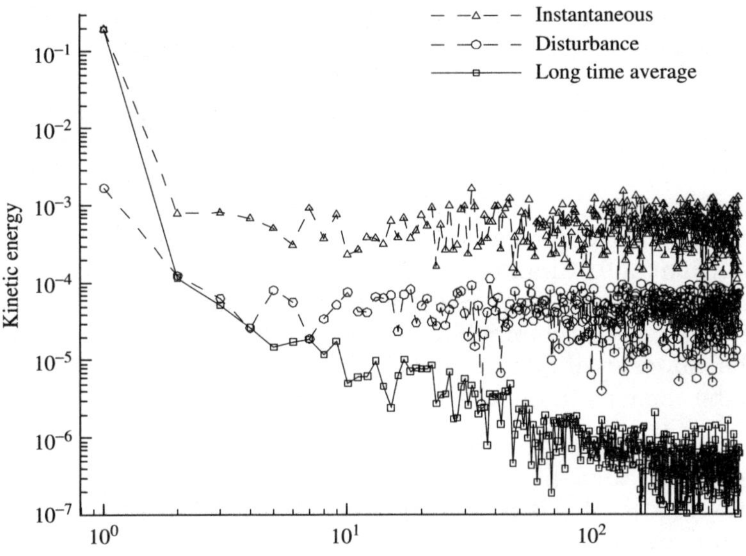

(*a*)

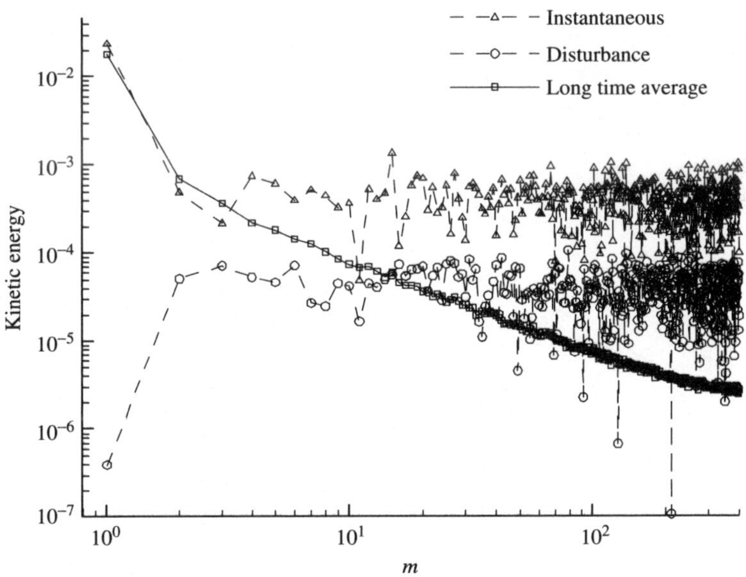

(*b*)

Figure 11.2.35 Frequency spectra. $h = 60$, $c = 0.04$: (*a*) Energy; (*b*) Cross correlation. (*Liou and Fang 2004.*)

capture the very small amplitude, but fairly organized, pattern of the long-time-averaged flow in the apparent statistical signals. The long-time-averaged velocity components also correlate well at the low frequency. There is no preferred mode of correlations for the instantaneous and the disturbance components.

Effects of domain sizes. To illustrate the effects of the domain sizes, we will examine the results for the cases with $h = 320$ for two values of the forcing coefficient of 0.001 and 0.01. They should be compared with the $h = 60$ cases with the same values of the forcing coefficient presented in the last section. For this channel height, the Reynolds number is about 550 and the Knudsen number is 1/320. The flow operating conditions used were the same as those for the previous $h = 60$ cases.

$c = 0.001$. For the large domain, 3D vortical structures appear in the averaged component of the stationary flow for $c = 0.001$, which have not been observed at the same forcing level for $h = 60$. Figure 11.2.36 shows the in-plane pathlines for the averaged flow. There are two pairs of vortical structures.

Figure 11.2.37 shows the spanwise distributions of the averaged energy at three streamwise locations along the midplanes ($y = 60$). The energy is distributed roughly between 0.37 and 0.67. Note that for $c = 0.001$ there is no secondary motion in the $h = 60$ case. In fact, on the average, the energy for the large domain case is nearly two times that for the flow with $c = 0.005$ and $h = 60$ (Fig. 11.2.21).

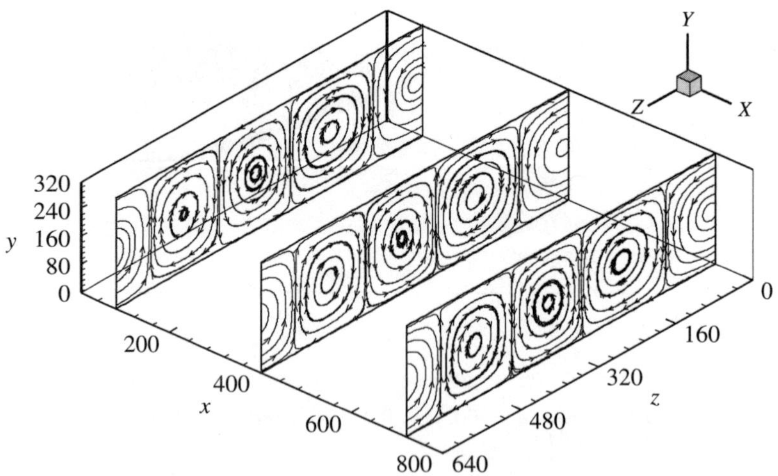

Figure 11.2.36 In-plane pathlines. $c = 0.001$. (*Liou and Fang 2004.*)

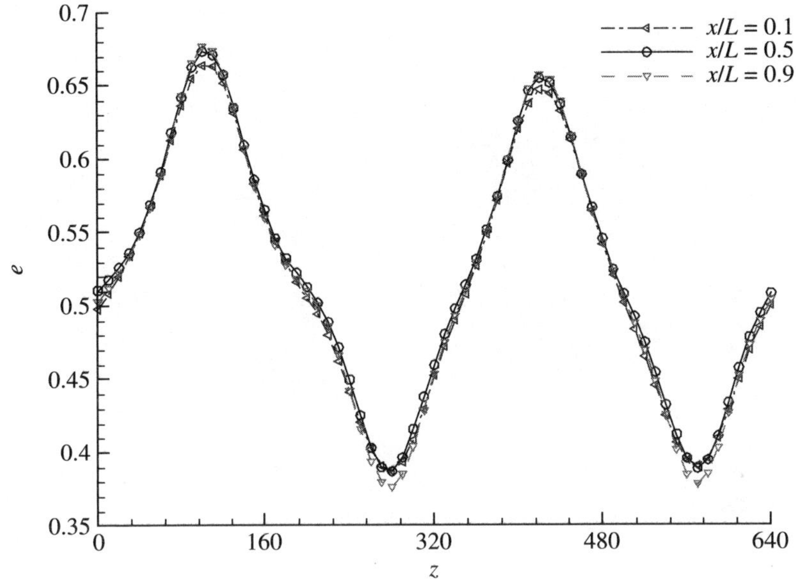

Figure 11.2.37 Spanwise energy distributions. $c = 0.001$.

$c = 0.01$. Figure 11.2.38 shows four snap shots of the in-plane pathlines for the averaged flow for $c = 0.01$. The long-time-averaging of the flow begins at the initiation of the simulations. At the nondimensional time of $tU/h = 7.5$, the average flow shows the formation of vortices of different sizes from a chaotic initial flow. The time scale h/U equals to $(640/\sqrt{\pi})T_c$. The secondary flows first appear near the bottom plate due to the applied forcing. The flow is also far from being homogeneous in the streamwise direction. The vortical flow develops further at a later time. New regions of small-scale vortical motion can also be identified. Some grow as they merge with the neighboring ones. The vortices appear to deform in different ways. Since these are in-plane pathlines, the appearance of a convergence to a point in a pathline indicates a strong fluid motion in the third or x direction. At $tU/h = 15.5$, a large vortex forms near the periodic boundaries in the z-direction. The centerline region is occupied by two connected vortices. Some small vortices seen in the earlier time have disappeared. At $tU/h = 33.2$, the flow develops into two large vortices. Except for the center region of the large vortical motions, small-scale motions have dissipated. The averaged flow varies slowly onward. At $tU/h = 63.4$, the flow shows only two large counter rotating vortical structures.

The long time averages of the simulated flows show distinctively pairs of counter-rotating structures that are aligned with the direction

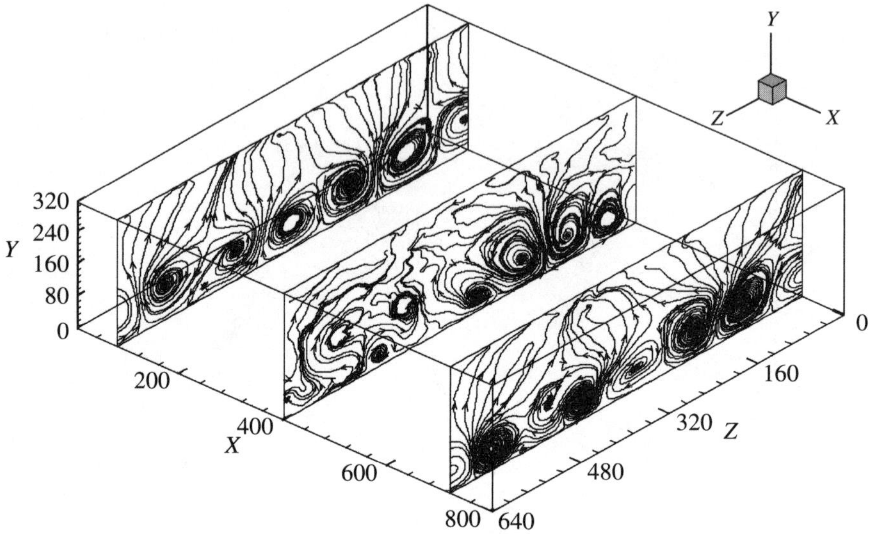

(a)

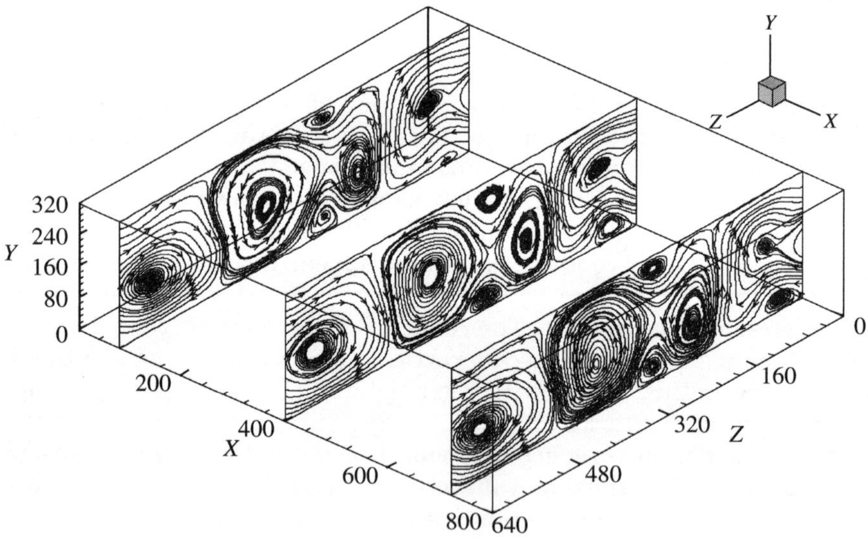

(b)

Figure 11.2.38 In-plane streamlines of long-time-average flow. $c = 0.01$: (*a*) $tU/h = 7.5$; (*b*)15.5; (*c*) 33.2; (*d*) 63.4.

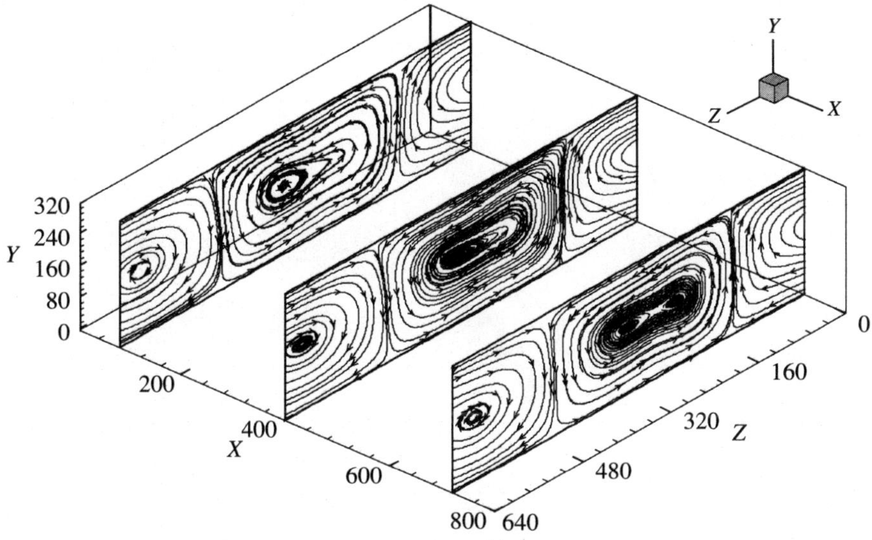

(c)

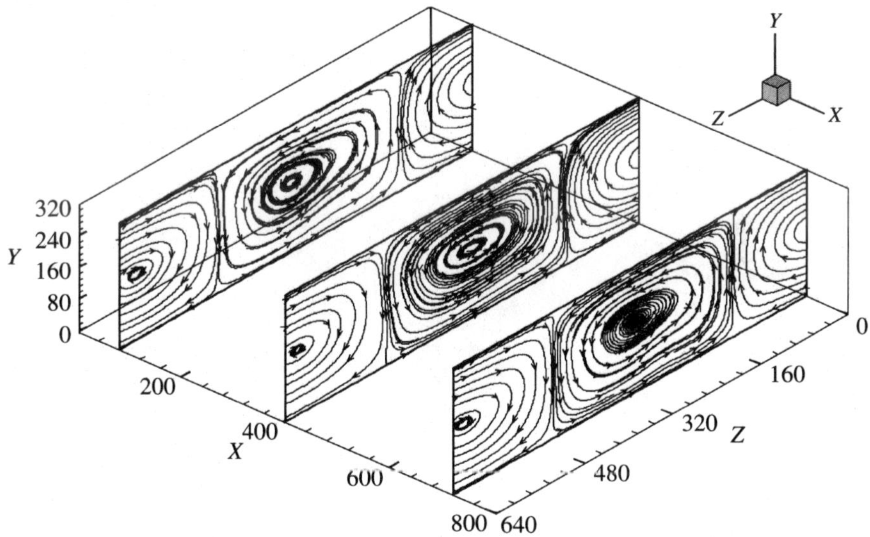

(d)

Figure 11.2.38 (*Continued*)

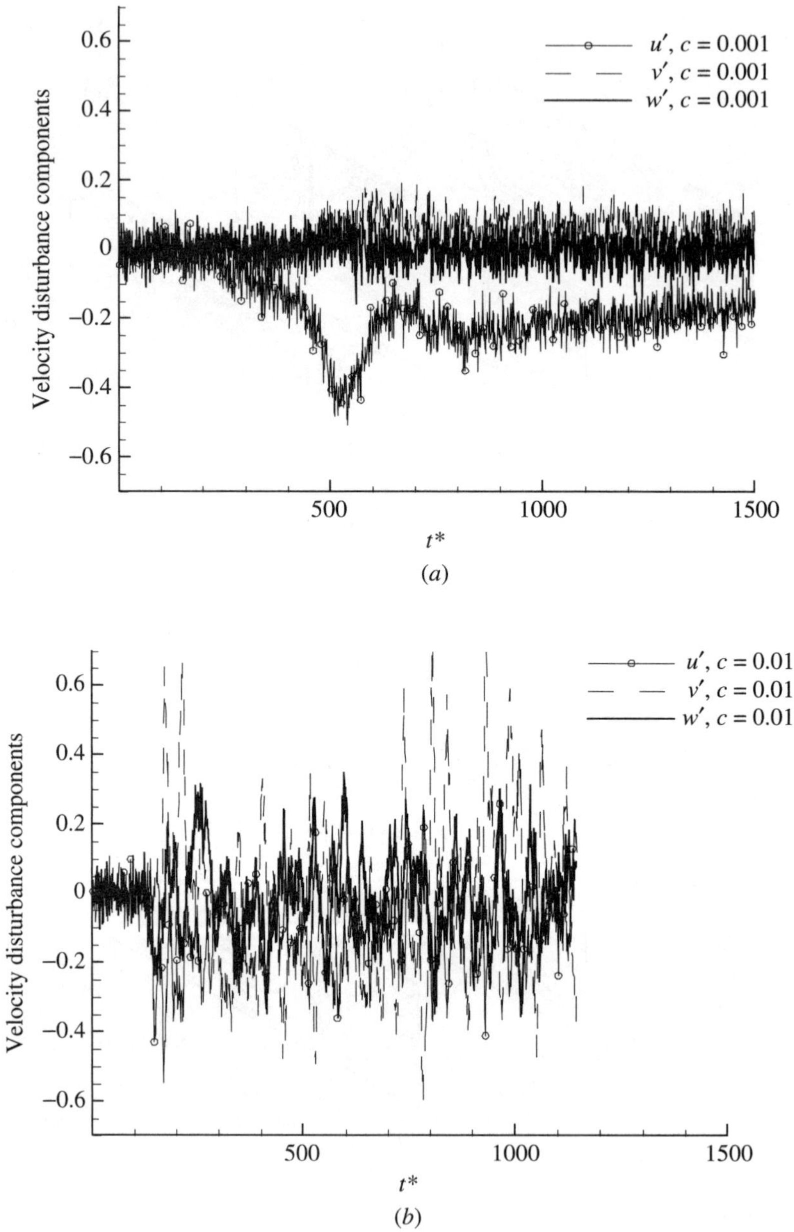

Figure 11.2.39 Evolution of the velocity disturbance components. $h = 320$: (a) $c = 0.001$; (b) $c = 0.01$. (*Liou and Fang 2004.*)

of movement of the top wall. The characteristics of the long-time-averaged flows are similar to what has been shown in Figs. 11.2.32 to 11.2.34 for the $h = 60$ cases. It is likely that the periodic boundary conditions have dictated the flow pattern of the averaged flow for $c = 0.01$ and $h = 320$, where there is only one pair of counter-rotating vortices found as the flow becomes stationary.

For the disturbances, Fig. 11.2.39 shows the time history of the velocity component signals for the cell that is located at the center of the simulation domain. Note that a disturbance has been defined in the present study in Eq. (11.2.5) as the difference between the unsteady sample average and the long time average. For $c = 0.001$, all three disturbance velocity components are oscillatory, which shows the statistical nature of DSMC. It is also apparent that the signals develop mean values, about which the oscillations occur, particularly for the streamwise disturbance velocity component u'. Similar oscillatory behavior of the disturbances can also be observed in Fig. 11.2.39*b* ($c = 0.01$) immediately after the commencement of the simulation. For $c = 0.01$ the disturbance velocity behavior changes at a later time. There are frequent occurrences of large amplitude excursions that did not exist for $c = 0.001$. The apparent differences in the disturbance velocity between the two cases can also be viewed by examining their spectral content.

Figure 11.2.40 shows the kinetic energy spectra for the disturbance time-traces in Fig. 11.2.39. For comparison, the energy spectra for the instantaneous and the long-time-averaged flows are also included. Compare to the $h = 60$ cases where the spectra distributions are basically random, for $h = 320$ and $c = 0.001$ the disturbance spectra in the low frequency region (e.g. $m < 10$) are distinguishably higher than those in the high frequency range that is dominated by the sampling error of DSMC. Figure 11.2.40*b* shows the energy spectra for the instantaneous, the long time average, and the disturbances for $c = 0.01$. The long-time-averaged component contains predominantly the zero-frequency mode and is stationary. Similar to the previous case with $c = 0.001$, the disturbance spectra in the low frequency range are significantly higher than those in the higher range. Moreover, the disturbance energy spectra are at least one order of magnitude higher than the statistical scatter at the high frequency end of the spectra and this is apparent in a wider region of low to medium frequency (e.g. $m < 60$).

Figure 11.2.41*a* shows the cross correlation frequency spectrum for $c = 0.001$. Similar to the case with $h = 60$ and $c = 0.04$ (Fig. 11.2.35*b*), the long-time-averaged flow correlates well at low frequency range and the disturbances are mainly statistical noises. With a larger forcing coefficient $c = 0.01$, Fig. 11.2.41*b* shows that the cross correlation spectrum of the vertical and the spanwise disturbance velocity components is dominated by low frequency modes.

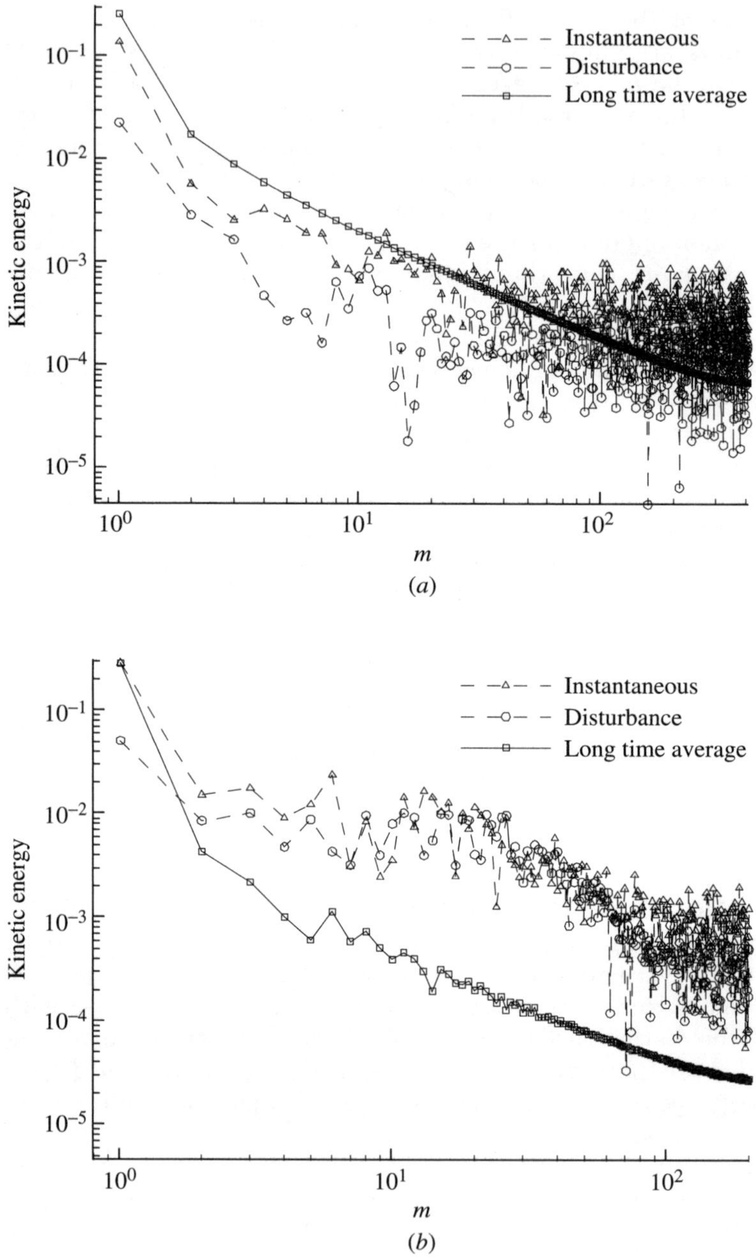

Figure 11.2.40 Kinetic energy spectra. $h = 320$: (a) $c = 0.001$; (b) $c = 0.01$. (*Liou and Fang 2004.*)

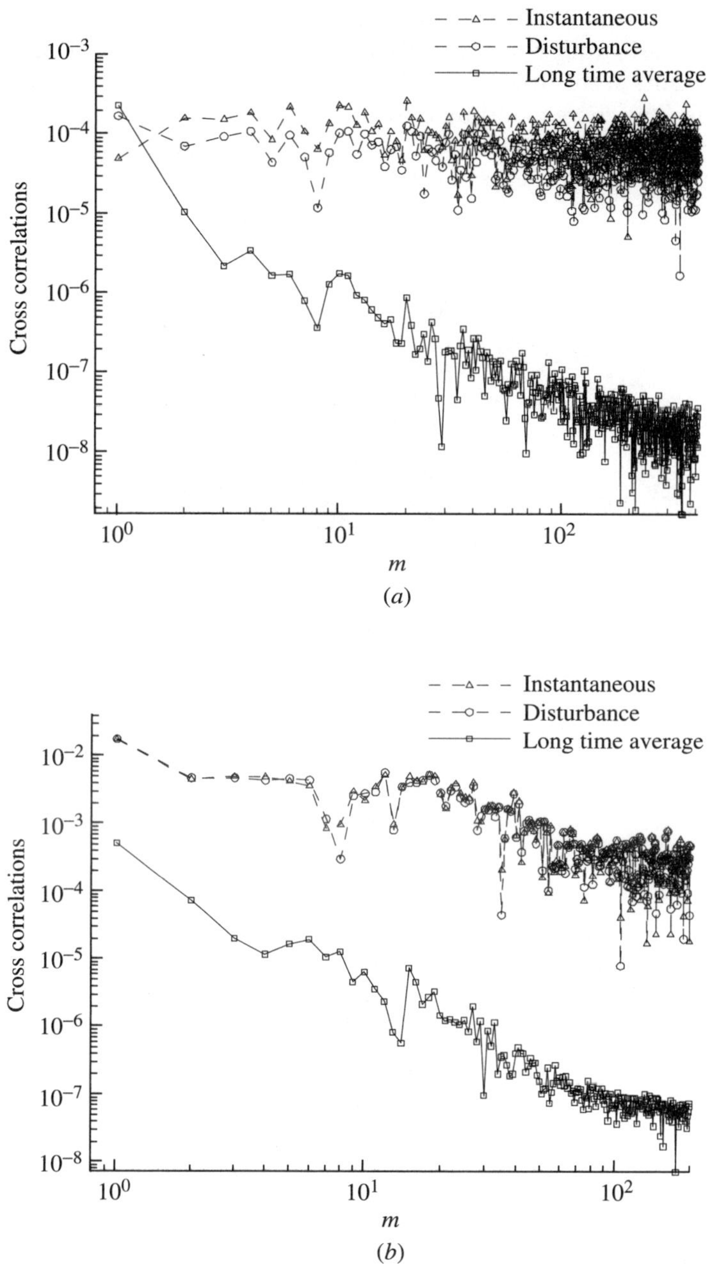

Figure 11.2.41 Cross correlation spectra. $h = 320$: (*a*) $c = 0.001$; (*b*) $c = 0.01$. (*Liou and Fang 2004.*)

Figures 11.2.40*b* and 11.2.41*b* show that, with $c = 0.01$, the disturbances are energetic and well-correlated. Figures 11.2.42 and 11.2.43 show the snap shots of the in-plane pathlines for the disturbances at two instants of time. The corresponding unit in t^* are roughly 600 and 1140, respectively. Figure 11.2.42 is for those on the *yz*-planes and Fig. 11.2.43 for the *xy*-planes. Together, Figs. 11.2.42 and 11.2.43 provide a view of

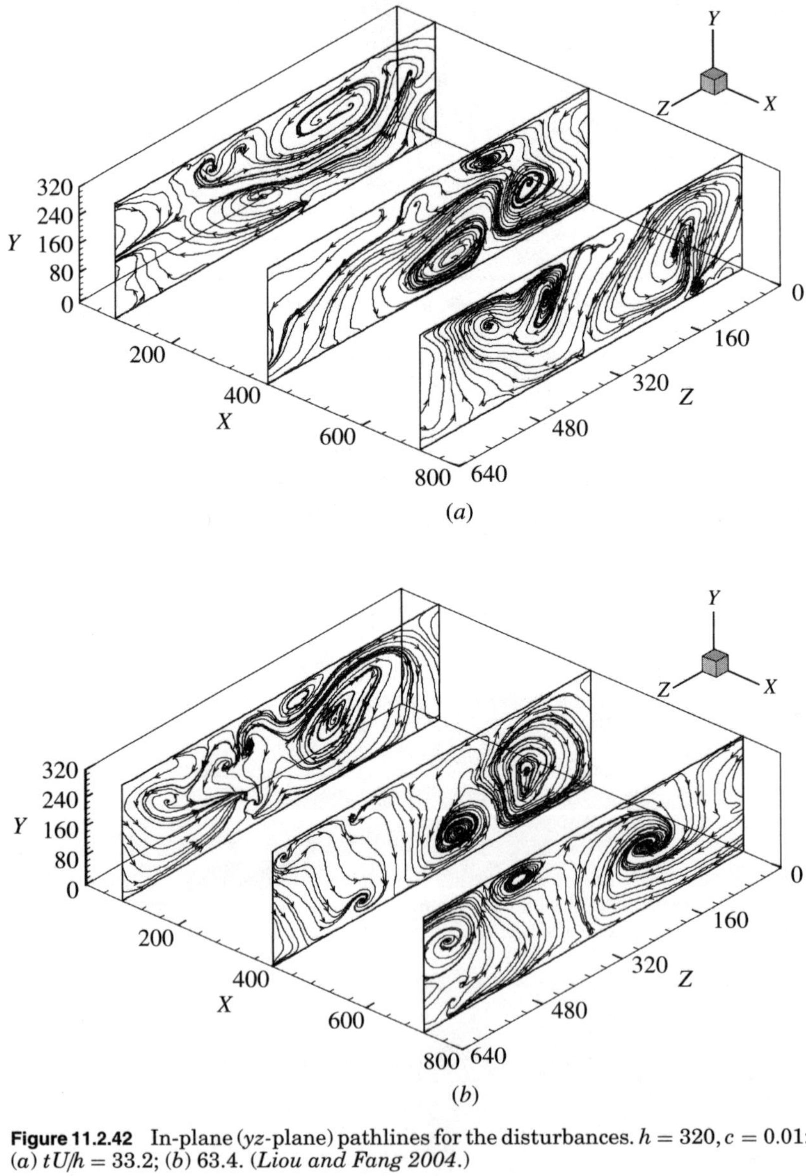

Figure 11.2.42 In-plane (*yz*-plane) pathlines for the disturbances. $h = 320, c = 0.01$: (*a*) $tU/h = 33.2$; (*b*) 63.4. (*Liou and Fang 2004.*)

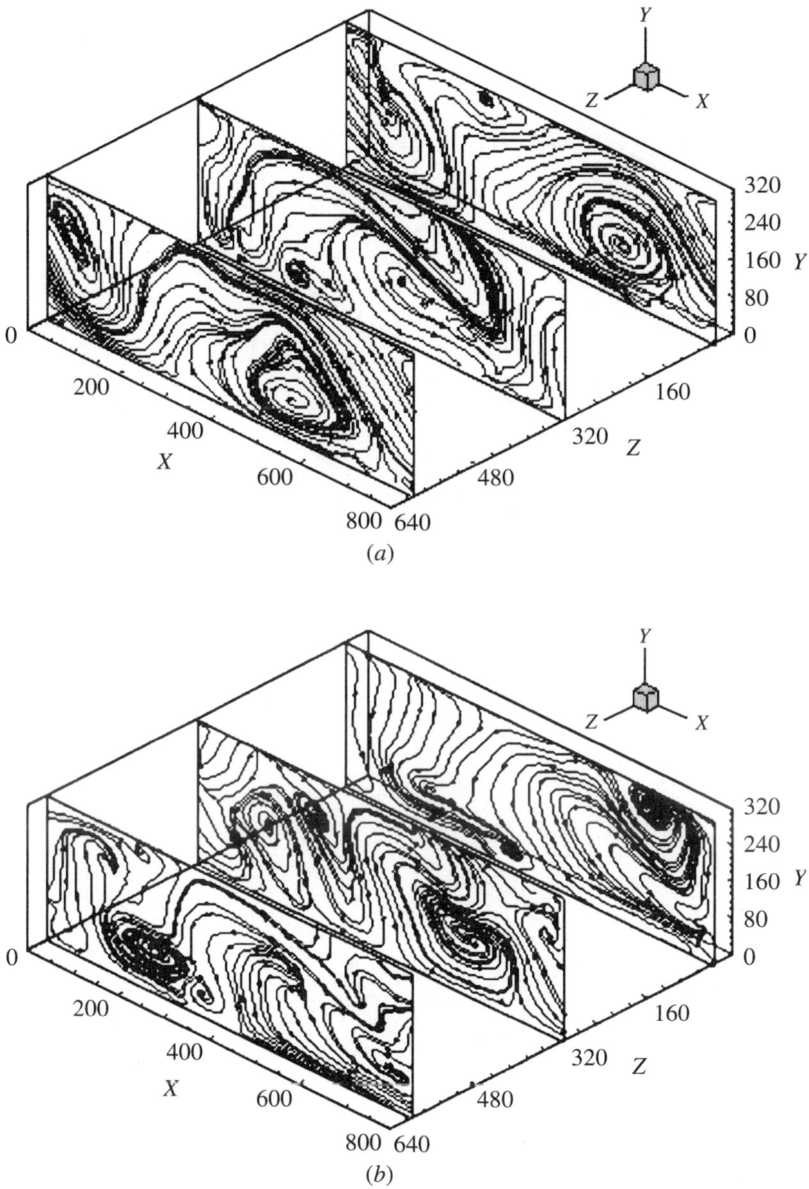

Figure 11.2.43 In-plane (*xy*-plane) pathlines for the disturbances. $h = 320, c = 0.01$: (*a*) $tU/h = 33.2$; (*b*) 63.4. (*Liou and Fang 2004.*)

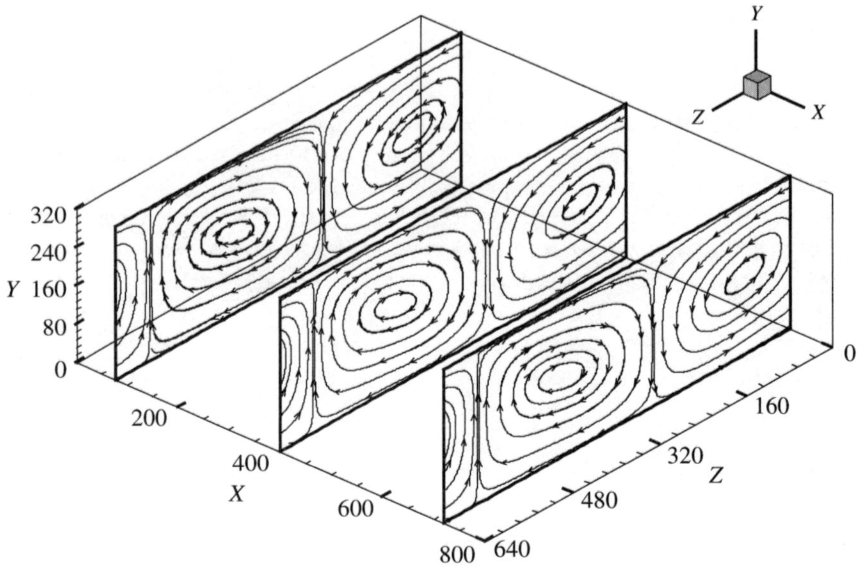

(a)

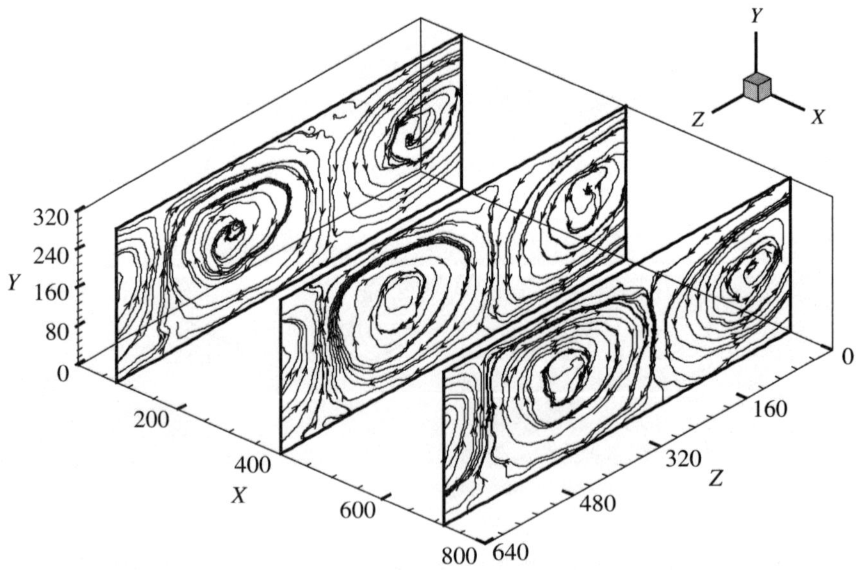

(b)

Figure 11.2.44 In-plane pathlines of flowfield for $c = 0.001$, $h = 320$, and $U = 2$: (a) Long-time-average flow; (b) Disturbances.

the spatial structure of the disturbances. The disturbances are far from being random as one may infer from the appearance of the traces of the individual disturbance velocity components in Fig. 11.2.39. In fact, they are well-correlated and well-organized. The disturbance pathlines indicate that the disturbances are three-dimensional and contain events with scales as large as the channel height as well as smaller ones. These highly correlated disturbances of large and small scales continue to evolve with time. The three-dimensional vortical motions go through growth, amalgamation, and decay after their initial formation. The time and the spatial scales of most of the small-scale events are in the order of the molecular scales and the flow cannot be computed by using the NS equations.

Effects of plate speeds. Two cases are simulated with $U = 2$ for $c = 0.001$ and $c = 0.01$. The operating conditions are the same as those for the previous $h = 320$ cases. The Reynolds numbers are about 1156. Figure 11.2.44 shows the in-plane pathlines of the long-time-averaged flow and the disturbances at three x-locations for $c = 0.001$ at $tU/h = 83.08$. Figure 11.2.44*a* shows that mean flow contains one large vortex pair compared with two in the corresponding $U = 1$, $c = 0.001$ case. Figure 11.2.44*b* shows that, despite some statistical noise, there is also one pair of identifiable vortical structure in the disturbances.

Figure 11.2.45 shows the energy and the cross correlation spectra for the instantaneous, the long time average, and the disturbances. For both the mean and the disturbance, the energy spectra contain regions of low to medium frequency with Fourier amplitudes that are at least one order of magnitude higher than the statistical scatter apparent at the high-frequency end of the spectra. The zero-frequency mode, in particular, is nearly four order-of-magnitude higher than those in the high-frequency end. The kinetic energy spectra are nearly linearly increasing toward the low-frequency end. The correlation spectra peak at the zero-frequency and appear random for other modes.

Figure 11.2.46 shows evolutions of the disturbance velocity components for $c = 0.001$ and 0.01, respectively. For $c = 0.001$, the disturbance velocity components oscillate about a mean value as the flow develops and appears to be converging toward the end of the simulations. For $c = 0.01$, occurrences are significantly larger in amplitudes compared with the $c = 0.01$ case.

Figure 11.2.47 shows the in-plan pathlines for the mean and the disturbance flow fields for $c = 0.01$. One pair of vortical structures appears in the mean flow as shown in Fig. 11.2.47*a*. The disturbance (Fig. 11.2.47*b*) contains three-dimensional structures of different sizes. Compared with the $U = 1$ case, the disturbance field in the current $U = 2$ case contains more small scale events. Note that the secondary

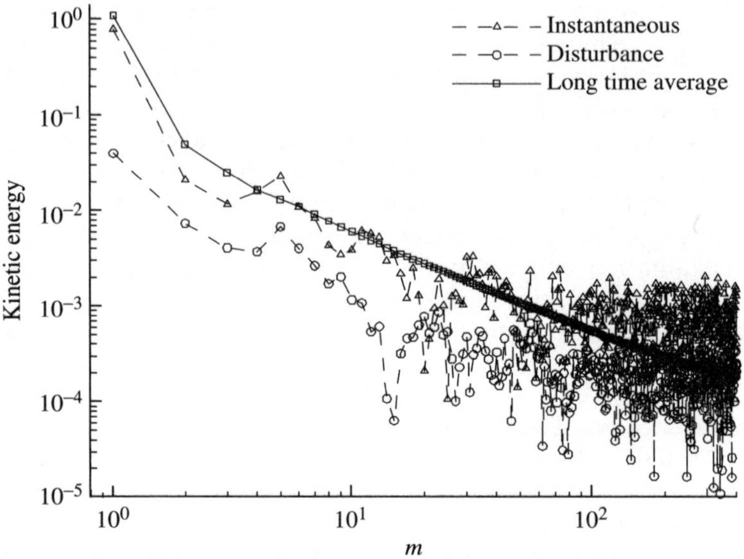

(*a*)

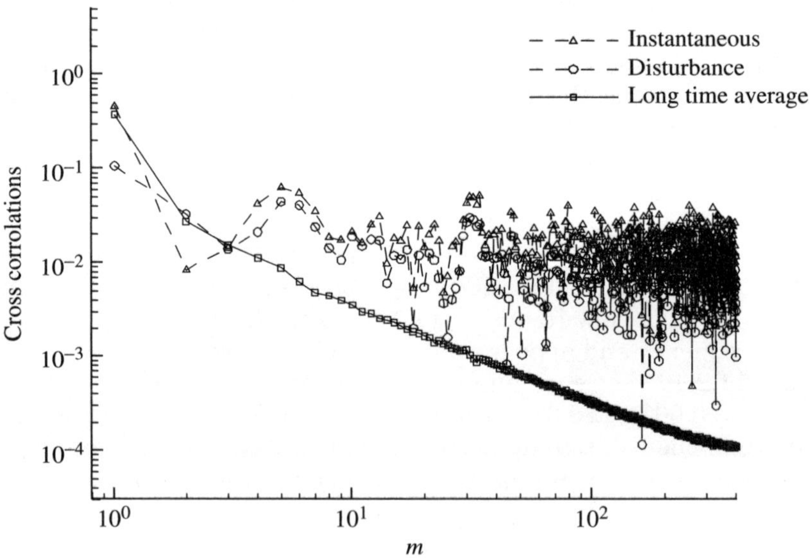

(*b*)

Figure 11.2.45 Spectra for $c = 0.001$, $h = 320$, and $U = 2$: (*a*) Energy; (*b*) Cross correlation.

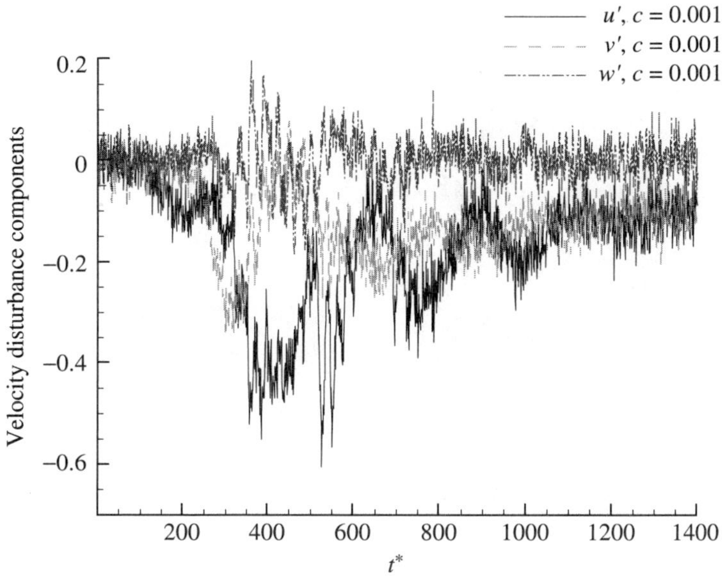

(a)

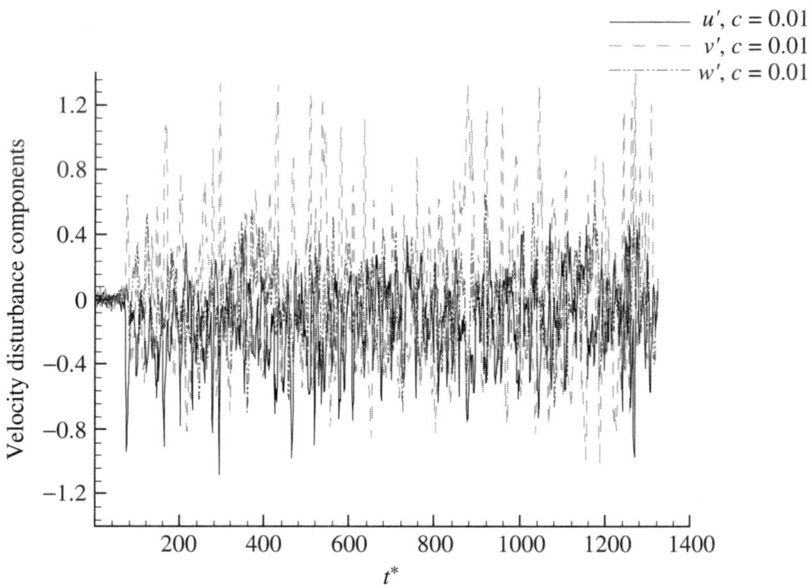

(b)

Figure 11.2.46 Evolutions of disturbance velocity components for $h = 320$ and $U = 2$: (*a*) $c = 0.01$; (*b*) $c = 0.001$.

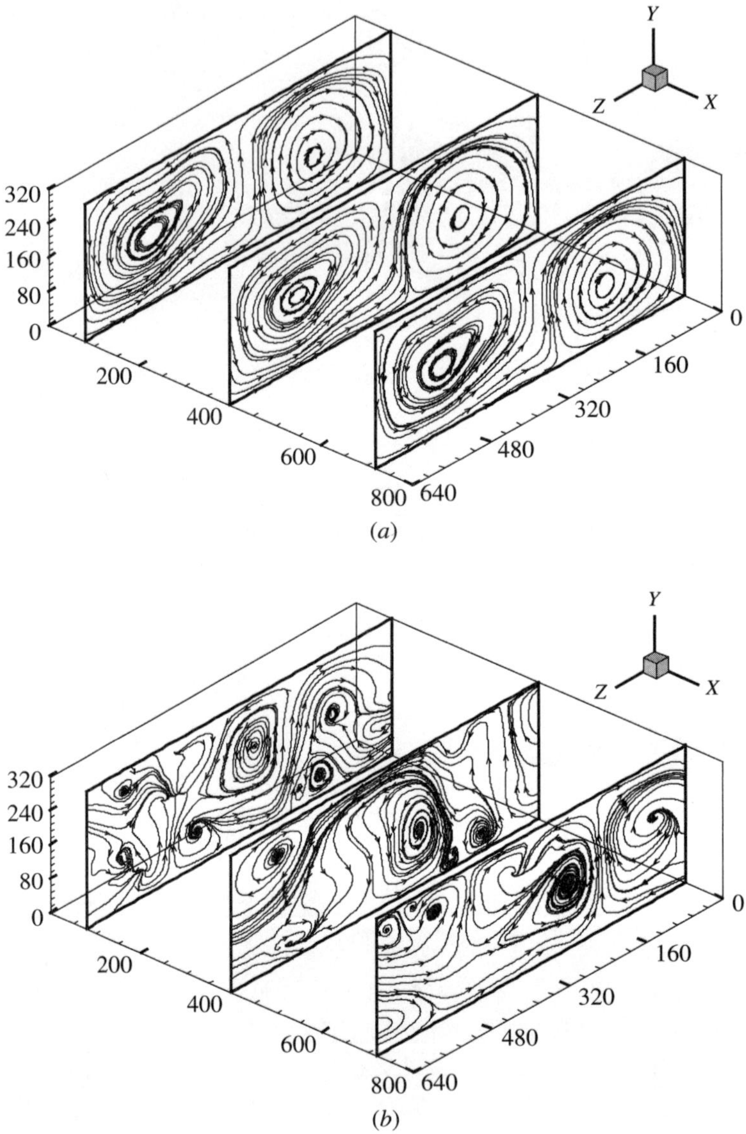

Figure 11.2.47 In-plane pathlines of the flowfield for $c = 0.01$, $h = 320$, and $U = 2$: (*a*) Long time average; (*b*) Disturbance.

flow in the mean and the three-dimensional disturbances do not exist in the initial conditions and are not preferentially perturbed by any means.

Figure 11.2.48 shows the kinetic energy and the correlation spectra for $c = 0.01$. For the mean flow, the spectrum for the zero-frequency

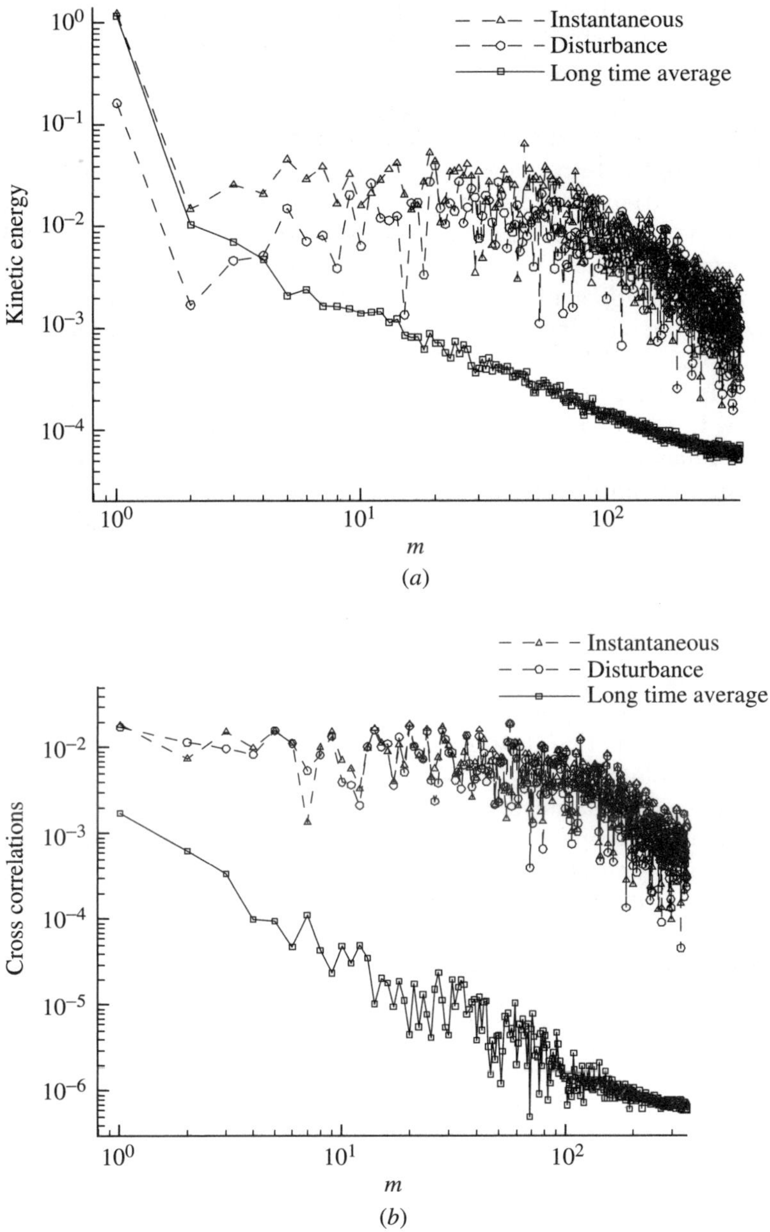

Figure 11.2.48 Fourier spectra for $c = 0.01$, $h = 320$, and $U = 2$: (a) Energy; (b) Cross correlation.

mode is significantly higher than all the other modes. The disturbance and the instantaneous kinetic energy spectra are not random. The spectra are fuller in the midfrequency than those for the low- and the high-frequency modes, where there is an apparent drop-off. Similar appearance can also be seen in the correlation spectra.

To summarize, the first principle DSMC method has been used as a numerical simulation tool to study the structures and the dynamics of an artificially forced three-dimensional microCouette flow. The applied forcing is proportional to the square of a local velocity difference (local x-velocity less the upper plate velocity). The parallel DSMC code used has been validated with independent serial code results, by comparisons with the continuum NS model results, and by varying the sample sizes. The simulations contain no preferred scales in the initial states. The results show that the long-time-averaged, or the mean, flows contain a hierarchy of harmonics with the fundamental mode being the dominant one. The Fourier amplitudes of these modes develop in a sequential manner in time. The pathlines for the flows show that these modes are associated with spatially coherent, counter rotating vortex pairs. The results also show that the flow disturbances captured by DSMC can be well-correlated in space and time. The evolution of the pathline pattern indicates that the flow disturbances are three-dimensional.

As the forcing being applied is not physical, the disturbance flow pattern captured may not be realistic. The results, nevertheless, provide a verification that the DSMC method can capture, in a time-dependent manner, the development of the stationary mean component and the unsteady three-dimensional disturbance component of a flow, which has not been previously reported. Due to the computer constraints, there is still a large parameter space, for instance, in Knudsen and Reynolds numbers, that is yet to be examined.

In the following section, a microRayleigh-Bénard convection problem will be examined. The gas considered is being confined between two walls and is heated from below by a hot, lower wall. Compared with the forced microCouette flow studied in this section, the buoyant force naturally arises in the fluid in the Rayleigh-Bénard flow due to heat transfer and is therefore physical. The results will show that DSMC can also capture the physical disturbances in the microRayleigh-Bénard flows.

11.3 MicroRayleigh-Benard Flows

The Rayleigh-Bénard convection (RBC) is a well-known phenomenon in traditional continuum fluid dynamics, in which a viscous fluid layer is enclosed between two infinite parallel flat plates, with the lower plate at a higher temperature than the upper plate. If the temperature difference between the two plates exceeds a certain critical value, there exists a transition from thermal conduction to natural convection. The point is reached when the buoyancy forces released by the heating overcome the inertial viscous forces in the fluid. At higher temperature gradients, cellular flow pattern develops and results in the Rayleigh-Bénard cells, with cellular vortical motions in the plane normal to the plates. The first theoretical solution was provided by Rayleigh (1916), upon which many modern theories of RBC are built. What followed were many experimental studies (Ahlers and Behringer 1978; Behringer and Ahlers 1982; Wu et al. 1995) and numerical simulations using the incompressible NS equations (Kessler 1987; Ozal and Hara 1995) and the Boussineq equations (Godreche and Manneville 1998). The Rayleigh-Bénard problem was also studied using the lattice Boltzmann method at the molecular level (Shan 1997; He et al. 1998; Watanabe 2004). Many recent studies focused on probing the onset of turbulent convection in the system. A recent extensive review can be found in Bodenschatz et al. (2000).

In the following sections, the results of 2D and 3D micro Rayleigh-Bénard convection (MRBC) problems are presented. The length is normalized by λ, the mean free path of the initial gas. The 2D and 3D MRBC have the same aspect ratios. The velocities are normalized by using the most probable speed of molecules at T_H, density by the initial state value, and temperature by T_H. DSMC is applied in a time-dependent manner. The time t is the real physical time normalized by $\lambda/\sqrt{2R}$, where R is the gas constant of monatomic molecules. The short time period t_s is set 200 times the mean collision time T_c of the initial gas and the time step is $0.2T_c$. In all the simulated cases, the wall temperature ratio $r(=T_C/T_H)$ is 0.1 and the Knudsen number is 0.01. The Froude number is 3.5. The resulting Rayleigh number is 10,159. The simulations are computationally intensive and the parallel DSMC algorithm described in Chap. 6 has been used and the parallel computations were performed on the Beowulf cluster at the CFD laboratory of Western Michigan University.

Evolutions of the convective flow patterns are studied in detail, and their corresponding pressure and heat transfer on the wall are calculated in terms of the averaged pressure coefficient, the heat transfer coefficient, and the Nusselt number. Three stages of flow development can be identified. In the first two stages, the difference in the flow development between the cases is not significant. In the third, all the 2D cases and the 3D case with the smallest aspect ratio develop into steady flows. For the 3D cases with the two larger aspect ratios, however, the cellular vortex rolls are found to translate at constant speeds. Such an advective phenomenon of the vortex rolls in the three-dimensional, unsteady Rayleigh-Bénard convection problem, to the best of our knowledge, has not been previously reported in the open literature prior to Fang et al. (2004).

In continuum fluid dynamics, a measure generally used for the flow transition from thermal conduction to natural convection is the Rayleigh number, Ra,

$$\mathrm{Ra} = \frac{\alpha g \Delta T h^3}{\kappa \nu} \tag{11.3.1}$$

where α denotes the thermal expansion coefficient, g the gravitational acceleration, ΔT the temperature difference between the upper and lower plates, h the distance between the two plates, κ the thermal diffusivity, and ν the kinematic viscosity. The critical Rayleigh number Ra_c for the onset of convection is taken to be 1708.

On the other hand, there have been far less number of reports on the studies of the Rayleigh-Bénard convection in micro gas flows, in rarefied and transitional gas flows. In such flows, the rarefied gas and compressibility effects can be important and should be taken into account. Such studies in kinetic theory might be of great importance for the purpose of understanding the fundamental phenomena of flow instability and self-organization from first principles. Studies have been carried out on the RBC in rarefied gas flows by using DSMC, compressible NS equations with slip boundary conditions (Stefanov et al. 2002*a*,*b*), and the linearized Boltzmann (BGK) equation (Sugimoto et al. 1995; Sone et al. 1997). These calculations have been for 2D flows in a near-continuum regime, and most of the concerns were with the onset of the natural convection with rarefied gas effects. However, three-dimensional investigations of RBC for a rarefied gas can be valuable to the discovery of the transition from laminar to a turbulent regime (Stefanov et al. 2002*b*). The previous section has shown that unsteady DSMC is capable of resolving a hierarchy of waves, consisting of a fundamental and its harmonic modes and the time-dependent DSMC simulations also provide a description of the nonlinear dynamic evolutions of the wave hierarchy. Compared to the artificially forced microCouette flow described there, the buoyancy force that naturally arises in the current RBC problem.

2D and 3D MRBC problem and the transition to natural convection in microflows will be examined based on time-dependent DSMC simulation results. The aspect ratio of the computational domain ranges from 2:1 to 6:1. The MRBC problem can be described by the Boltzmann equation. Evolutions of the convective flow patterns are studied in detail, and the corresponding pressure and heat transfer on the wall are calculated in terms of the averaged pressure coefficient, the heat transfer coefficient, and the Nusselt number.

Basic equation for microRayleigh-Bénard convection problem. The MRBC can be studied from the microscopic point of view in kinetic theory.

A monatomic gas is considered enclosed between two infinite, horizontally parallel plates with two different temperatures, T_C on the upper wall and T_H on the bottom wall. The geometry is shown in Fig. 11.3.1. The gas is heated from the bottom plate ($T_H > T_C$), and there acts a gravitational force $\vec{F} = (0, -mg, 0)^T$, which accelerates the gas molecule downwards, with m the mass of the gas molecule. Periodic boundary conditions are applied in the x- and z-directions, and the two plates are modeled as diffuse walls. The physical dimension of the flow geometry is small, with length scales ranging from 100 to 600 molecular mean free paths.

The MRBC problem is described by the time-dependent Boltzmann equation,

$$(nf) + \mathbf{c} \cdot \frac{\partial}{\partial \mathbf{x}}(nf) - g\frac{\partial}{\partial c_y}(nf) = \Delta[Q(f)] \qquad (11.3.2)$$

where f is the velocity distribution function, n the number density, and t the time. $\mathbf{x} = (x, y, z)^T$ and $\mathbf{c} = (c_x, c_y, c_z)^T$ are the physical space vector and velocity space vector, respectively. The collision integral operator in Eq. (11.3.2) is

$$\Delta[Q(f)] = \int_{-\infty}^{\infty}\int_{0}^{4\pi} n^2(f^* f_1^* - f f_1)c_r \sigma d\,\Omega d\mathbf{c}_1 \qquad (11.3.3)$$

where $c_r = |\mathbf{c} - \mathbf{c}_1|$, σ denotes the differential cross section describing the interactions between particles, and Ω the solid angle.

Equation (11.3.2) is coupled with the following initial and boundary conditions. At the beginning of calculation ($t = 0$), we assume equilibrium state with temperature T_H and

$$f(0, \mathbf{x}, \mathbf{c}) = f_0(\mathbf{x}, \mathbf{c})\big|_{T=T_H} \qquad (11.3.4)$$

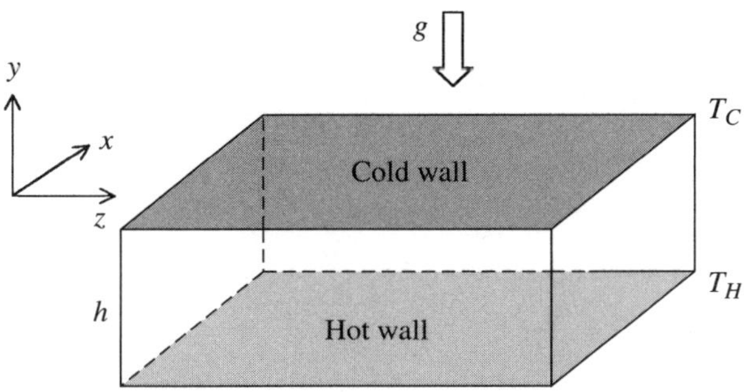

Figure 11.3.1 Flow geometry for microRayleigh-Bénard problem.

That is, at the initial state, the number density of molecules is uniformly distributed; the velocity components of mean flow are zero, and the flow temperature is set as T_H. At the two parallel plates we assume diffusive reflection with complete thermal accommodation according to the temperature T_C and T_H, i.e.,

$$f(t, \mathbf{x}, \mathbf{c}) = M_{T_H}(\mathbf{c}) \int_{c_y>0} f(x, 0, z, \mathbf{c}_1) c_y \, d\mathbf{c}_1 \quad (c_y \geq 0) \tag{11.3.5}$$

$$f(t, \mathbf{x}, \mathbf{c}) = M_{T_C}(\mathbf{c}) \int_{c_y<0} f(x, h, z, \mathbf{c}_1) c_y \, d\mathbf{c}_1 \quad (c_y \leq 0) \tag{11.3.6}$$

where M_T is half-space Maxwellian

$$M_T(\mathbf{c}) = \frac{1}{2\pi RT} \exp\left(-\frac{\mathbf{c}'^2}{2RT}\right) \tag{11.3.7}$$

Rayleigh number in microflows. The critical Rayleigh number Ra_c was obtained (Rayleigh 1916) by using a perturbation of the Boussineq equations, which are a linearized form of the Navier-Stokes equations with the Boussineq approximation. This approximation assumes the physical properties of the fluid are independent of temperature. Except for the density change that produces the buoyancy force, the fluid is otherwise assumed to be incompressible. These assumptions hold in the classical Rayleigh-Bénard problem, where the temperature ratio T_C/T_H is typically close to unity because of the large scale of its characteristic length.

In microflows, however, there exist rarefied gas and compressibility effects and the Boussineq approximation might not hold. The Rayleigh number is expected to relate to the Knudsen number Kn because of the rarefied gas effects that produce velocity slip and temperature jump.

For monatomic molecules with the hard sphere model, the Knudsen number Kn can be defined based on the mean free path $\lambda = (\sqrt{2}\pi n_0 d^2)^{-1}$ as

$$\mathrm{Kn} = \lambda / h \tag{11.3.8}$$

n_0 denotes the number density of molecules in the initial state and d the molecule diameter. If the thermal diffusivity κ is defined in the form of

$$\kappa = k/(\rho_0 c_p) \tag{11.3.9}$$

where k represents the heat conduction coefficient, ρ_0 the gas density at the initial state and c_p specific heat at constant pressure, the Rayleigh number can be derived from the first-order approximation of the Chapman-Enskog theory (Chapman and Cowling 1970) as

$$\mathrm{Ra} = \frac{512 h\alpha \Delta T g}{75\pi \mathrm{Kn}^2 (2RT)} \tag{11.3.10}$$

where T is the system temperature and R the gas constant. Using the bottom wall temperature T_H as the reference temperature and thermal expansion coefficient taken as

$$\alpha = 2/(T_H + T_C) \tag{11.3.11}$$

we can write Eq. (11.3.10) as

$$\mathrm{Ra} = \frac{1024}{75\pi}\frac{1-r}{1+r}\frac{1}{\mathrm{Fr}\cdot \mathrm{Kn}^2} \qquad r = \frac{T_C}{T_H} \tag{11.3.12}$$

where the Froude number is defined as

$$\mathrm{Fr} = \frac{V_H^2}{gh} \qquad V_H = \sqrt{2RT_H} \tag{11.3.13}$$

Based on the most probable speed of molecules on the bottom hot wall, the Froude number Fr relates the thermal speed with the gravitational acceleration. Equation (11.3.12) shows that the Rayleigh number Ra is a function of the characteristic parameters, Kn and Fr, of the micro Rayleigh-Bénard problem, and $\mathrm{Ra} \propto \mathrm{Kn}^{-2}$; i.e., the natural convection becomes less likely with increasing Kn.

Due to its complexity, especially the collision integration term, the Boltzmann equation (Eq. 11.3.2) is not solved directly. Instead, the DSMC method will be used. The DSMC method does not solve the Boltzmann equation, however, it does provide solutions that are consistent with the Boltzmann equation as long as it obeys the principles and boundary conditions of the Boltzmann equation for a specific problem. More detailed implementation of DSMC to obtain a stationary solution of the Boltzmann equation can be found in Stefan and Cercignani (1993).

11.3.1 Two-dimensional flows

Three cases have been calculated in 2D microchannels with the aspect ratios (AR) of 2:1, 4:1, and 6:1. As shown in Table 11.3.1, the number of rectangular field cells are 60×30, 120×30, and 180×30, respectively. The number of simulated molecules is 2 millions in all cases.

Heat transfer. The heat transfer coefficient c_h is defined as

$$c_h = q\bigg/\left(\frac{1}{2}\rho_0 V_H^3\right) \tag{11.3.14}$$

TABLE 11.3.1 Parameters in 2D Flow Simulations

	Case 1	Case 2	Case 3
AR	2:1	4:1	6:1
Grid size	60 × 30	120 × 30	180 × 30
Molecules	2 million	2 million	2 million
Processors	1	1	1
R	0.1	0.1	0.1
Fr	3.5	3.5	3.5
Kn	0.01	0.01	0.01
Ra	10,159	10,159	10,159

with the wall heat transfer rate q defined as

$$q = \frac{\sum_{i=1}^{N_s}\left(\frac{1}{2}mc_i^2\right)^{\text{in}} - \sum_{i=1}^{N_s}\left(\frac{1}{2}mc_i^2\right)^{\text{re}}}{t_s \cdot \Delta A} \tag{11.3.15}$$

where N_s denotes the total number of molecules impacting the wall element during the short sampling time period t_s, ΔA the area of the wall element, m the mass of the molecule. The superscripts "in" and "re" denote the values associated with the incident and the rebounding parts of the collision of a molecule with the wall, respectively. The history of heat transfer coefficient on the hot wall is shown in the log-linear plot of Fig. 11.3.2. For all the three 2D cases, there appears to be three main stages associated with the development of the wall heat transfer. In the first two stages, the wall heat transfer continuously increases nearly linearly with time in the log scale with two distinctly different slopes. The growth trend stops in the third stage. The heat transfer coefficients become oscillatory with an average value around -0.062. For AR $= 2$ and 4, the change occurs at about $t = 50$, and at about 80 for AR $= 6$. There is an apparent jolt in the heat transfer coefficient around $t = 225$ for AR $= 4$, as it develops toward a stationary value. The wall heat transfer gradually decreases first and two abrupt changes follow, each reversing the directions of change, before it gradually settles to a stationary state.

Figure 11.3.3 shows the evolutions of Nusselt number (Nu) on both the hot and the cold walls, where

$$\text{Nu} = q\Big/\left(k\frac{\Delta T}{h}\right) \tag{11.3.16}$$

It can be seen that the Nusselt numbers for both the hot and the cold walls approach 3.33 after the flows develop to stationary states. For the 4:1 case shown in Fig. 11.3.3*b*, the development of the stationary

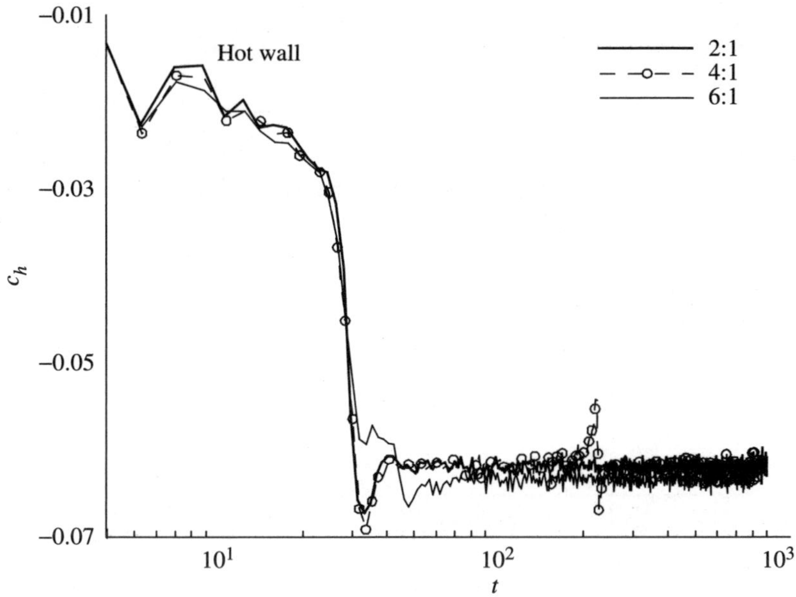

Figure 11.3.2 Evolution of hot-wall heat transfer coefficients for 2D cases, with aspect ratio of 2:1, 4:1, and 6:1.

state appears to be interrupted by the sudden and large-amplitude jolt described in Fig. 11.3.2. Nu is about 3.29 before the jolt and about 3.32 afterward.

The pressure coefficient c_p distributions on the cold wall are shown in Fig. 11.3.4, where

$$c_p = p\Big/\left(\frac{1}{2}\rho_0 V_H^2\right) \tag{11.3.17}$$

Pressure p is defined as

$$p = \frac{\sum_{i=1}^{N_s}(mc_{n,i})^{\text{in}} - \sum_{i=1}^{N_s}(mc_{n,i})^{\text{re}}}{t_s \cdot \Delta A} \tag{11.3.18}$$

where the subscript "n" denotes the molecule velocity component normal to the wall. It can be seen that the calculated pressure coefficients decrease to a constant about 0.31 for all the three 2D cases. Similar to the wall heat transfer shown in Figs. 11.3.2 and 11.3.3*b*, the pressure coefficient evolution also appears to indicate two stationary states after the initial transition for the case with AR = 4. The pressure is low initially and later becomes more comparable to those for the 2:1 and the 6:1 cases.

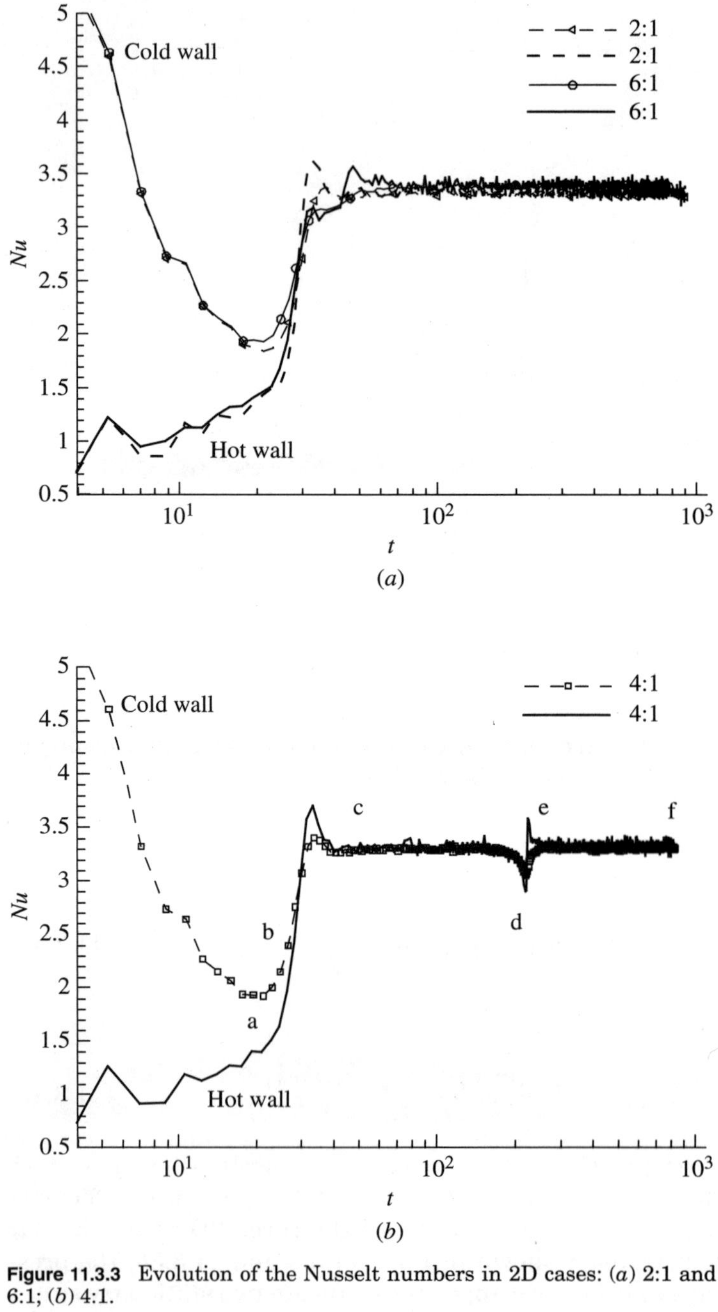

Figure 11.3.3 Evolution of the Nusselt numbers in 2D cases: (*a*) 2:1 and 6:1; (*b*) 4:1.

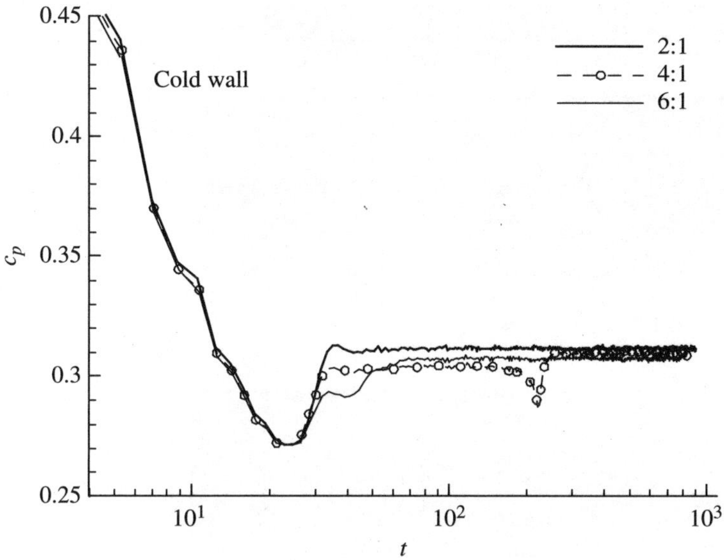

Figure 11.3.4 Evolution of pressure coefficient on the cold wall in 2D cases, with aspect ratio of 2:1, 4:1, and 6:1.

Transition to stable stationary flow pattern. Figure 11.3.5 shows the pathlines of the 2D steady-flow field at time $t = 567.19$, sampled after time $t = 88.62$. The contours of the flow vorticity are also shown in the background. The vorticity is defined as

$$\omega_z = \frac{1}{2}\left(\frac{\partial u}{\partial y} - \frac{\partial v}{\partial x}\right) \tag{11.3.19}$$

There are one, two, and four pairs of vortices, in the flow field with AR $=$ 2, 4, and 6, respectively. In Figs. 11.3.5*a* and 11.3.5*c*, the vortex pairs are uniformly distributed in space, with about the same maximum vorticity magnitude. In Fig. 11.3.5*b* for AR $= 4$, the second vortex pair from the left, which consists of the first and the fourth vortices, is slightly deformed. The vorticity magnitude is also smaller than that of the first vortex pair.

Figures 11.3.6 and 11.3.7 show the corresponding density and temperature fields. In the density fields, there are one, two, and four pairs of mushroom-like areas for the 2:1, 4:1, and 6:1 cases, respectively. The centers of the vortices, or convection rolls, can be found on either sides of the stem of the mushrooms. It can be seen in Fig. 11.3.7 that, other than the regions where there are strong, vertical downwash (upwash) of high (low) temperature gas, the temperature is about 0.5 in a large part of the flow.

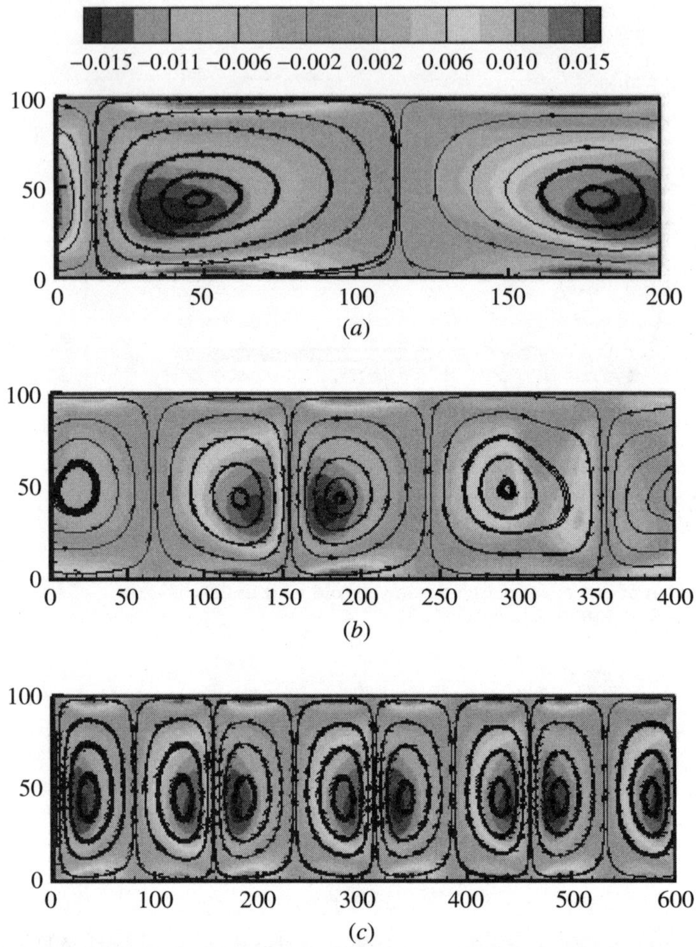

Figure 11.3.5 Pathlines and vorticity contours for the long-time averaged 2D flow field cases at time $t = 567.19$: (*a*) 2:1; (*b*) 4:1; (*c*) 6:1.

Snapshots of the instantaneous velocity vector field at five instants of time are presented in Figs. 11.3.8*a–f* for the case with AR = 4, marked by the vorticity levels. The wall heat transfer values that correspond to these times are marked by (*a*) to (*f*) in Fig. 11.3.3*b*. The horizontal arrow depicted in the top left corner of each figure represents the reference magnitude for the velocity vectors in that figure. The simulation time and the number of output cycles are shown in the upper right corner. The initial gas temperature is T_H with zero average velocity. The cold wall heat transfer decreases as the simulation commences and that for the lower hot wall increases. Vortical motion can be seen near the upper cold wall at time $t = 17.72$ (point "a" in Fig. 11.3.3*b*) as shown

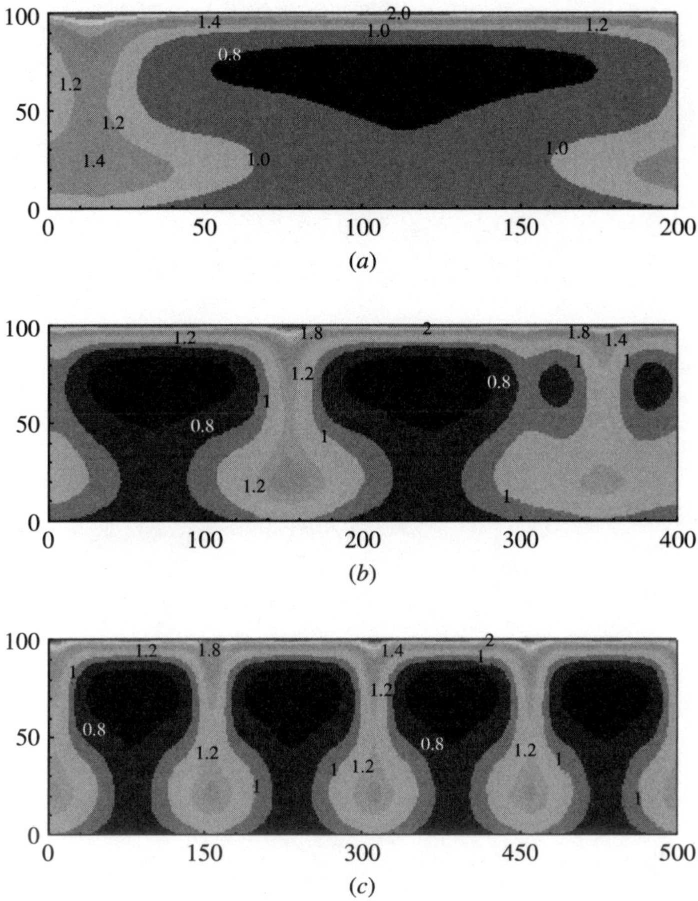

Figure 11.3.6 Density contours for the long-time averaged 2D flow cases at time $t = 567.19$: (*a*) 2:1; (*b*) 4:1; (*c*) 6:1.

in Fig. 11.3.8*a*. In the time range between 17.72 and 50 (point "c" in Fig. 11.3.3*b*), the vortical structures grow quickly in size and strength, and extend toward the lower hot wall. As more high temperature gas is brought from the lower hot wall, the heat transfer coefficient on the upper cold wall reverses its decreasing trend and begins to increase. For the lower hot wall, the heat transfer coefficient continues its increasing trend, and at a faster pace, during this period. After slight overshoots near $t = 40$, the heat transfer coefficients on both walls stabilize and converge to the same value of about 3.29. The flow develops into a seemingly stationary pattern with three pairs of vortices as shown in Fig. 11.3.8*c*. This flow pattern stays for a time period of about 160 before one of the three pairs of vortical flow structures near the

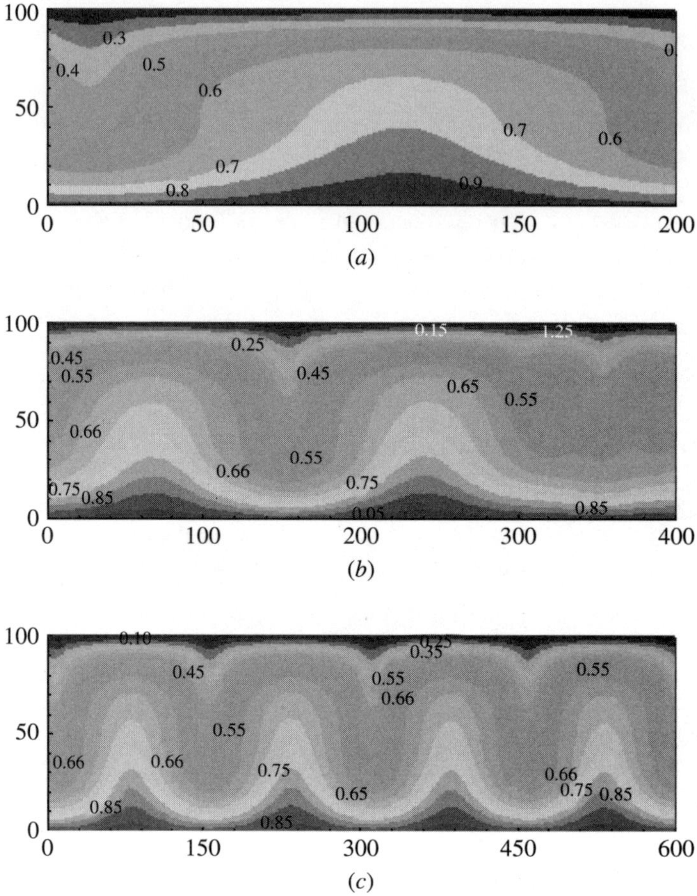

Figure 11.3.7 Temperature contours for the long-time averaged 2D flow field cases at time $t = 567.19$: (*a*) 2:1; (*b*) 4:1; (*c*) 6:1.

right flow boundary begins to shrink in size with reducing strength, as is shown in Fig. 11.3.8*d*. This vortex pair quickly weakens and disappears, and the flow is left with only two larger vortex pairs (shown in Fig. 11.3.8*e* at $t = 221.56$). During this transformation of flow patterns, the wall heat transfer dips first (point d in Fig. 11.3.3*b*) as the vortex pair is diminishing and overshoots when the remaining structures grow to fill the space left behind (point e in Fig. 11.3.3). The jolts of the wall heat transfer observed in Fig. 11.3.3*b* and of the wall pressure in Fig. 11.3.4, therefore, are caused by the change of the dominant structures of the flow. The newly formed flow pattern of two vortex pairs shown in Fig. 11.3.8*f* is not found to change any further and appears to be stationary.

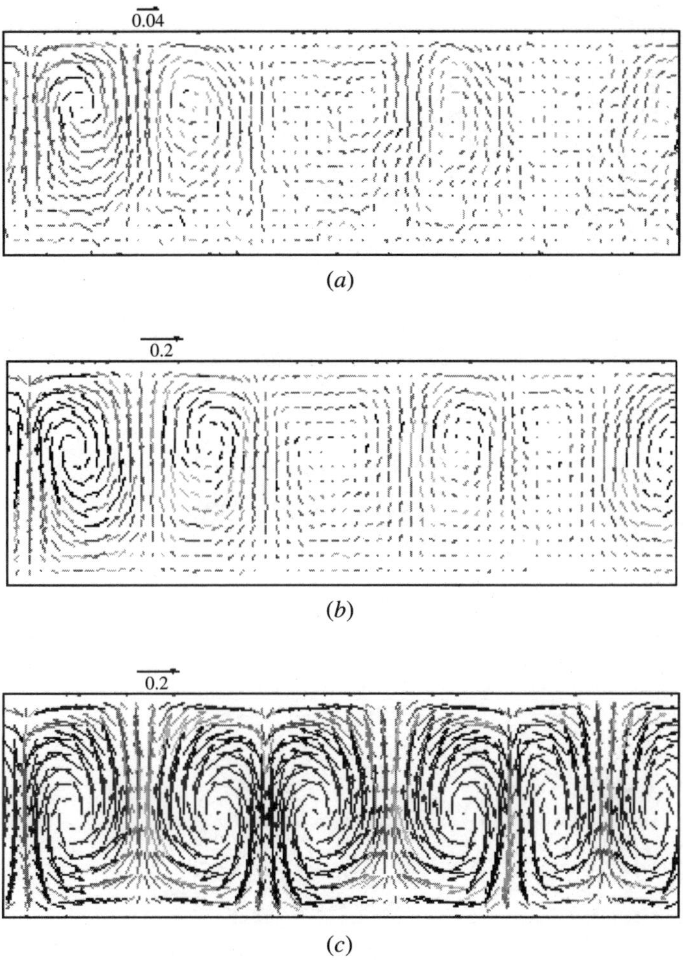

Figure 11.3.8 Snapshots of the instantaneous velocity vector field for the 2D simulation with the aspect ratio of 4:1, with color representing the flow vorticity magnitudes. Corresponding heat transfers are shown in Fig. 11.3.3*b* at time point *a–f* ($t = 17.72 - 834.83$).

Stationary flow patterns. Once a flow develops into a stable stationary flow pattern, the steady-flow properties sampled after the transition should not change with further sampling. Figure 11.3.9*a* shows the comparisons of long-time-averaged velocity profiles of u and v at $y = 20$ and at two different times of $t = 177.25$ and 912.81 for the 2:1 case. The sampling begins at $t = 88.62$ when the flow appears to have become fully developed. The associated flow pattern has been shown in Figs. 11.3.5*a*. The profiles obtained over the large time span overlap well, indicating that the flow has become stationary. Figure 11.3.9*b*

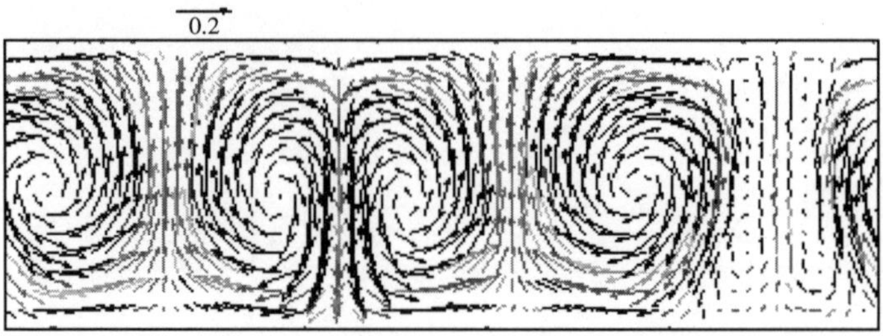

(*d*)

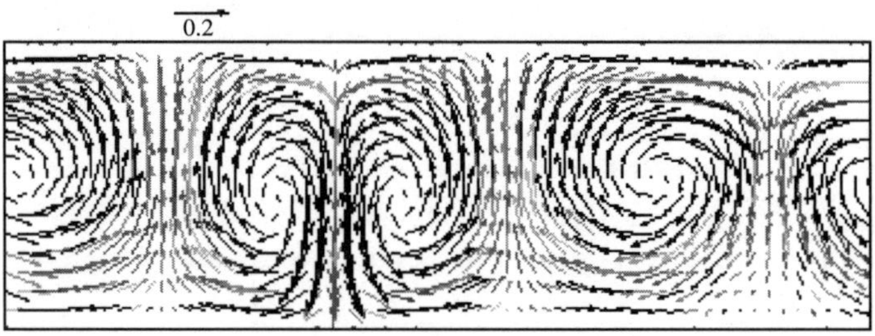

(*e*)

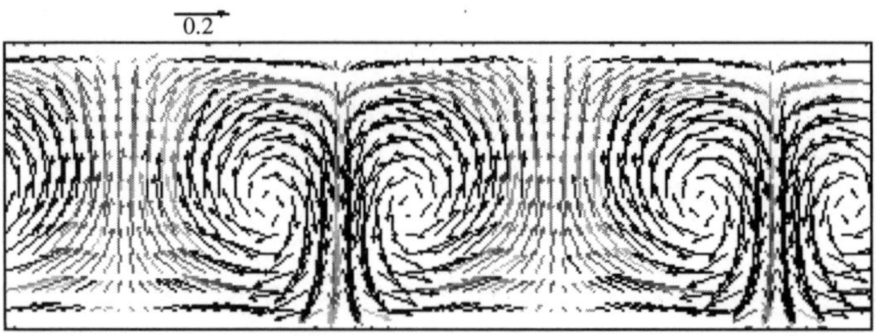

(*f*)

Figure 11.3.8 (*Continued*)

shows the similar comparison for the 6:1 case. The profiles agree well, which indicates that the flow shown in Fig. 11.3.5*c* for the 6:1 case has also become stationary.

For the 2D cases simulated, it can be seen that the aspect ratio does not have significant effects on the wall heat transfer rate and the averaged wall pressure. On the other hand, the final flow patterns and the process of their formation are significantly different among the cases with the different aspect ratios. In all of the three cases simulated, stationary states were found and these stable flows contain cells of vortical flow structures, similar in appearance to the convection roll in the Rayleigh-Bénard instability problem.

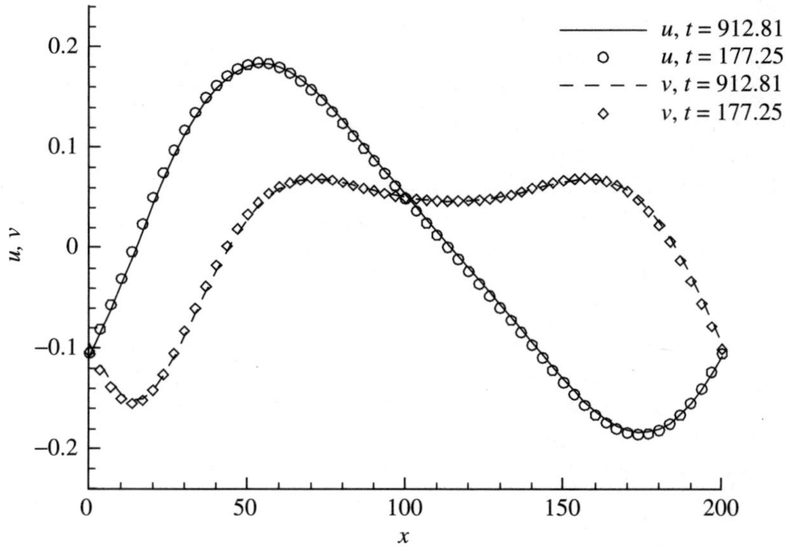

(a)

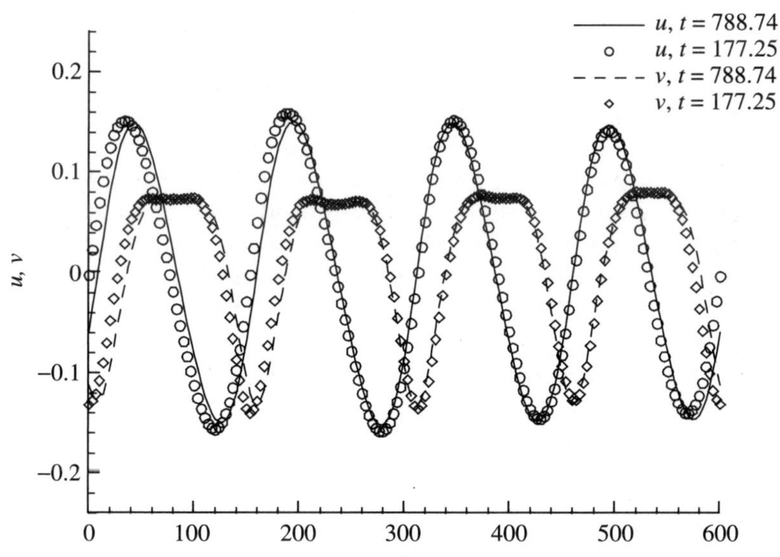

(b)

Figure 11.3.9 Comparisons of the steady flow velocity at $y = 20$: (a) 2:1 at $t = 177.25$ and 912.81; (b) 6:1 at $t = 177.25$ and 788.74.

11.3.2 Three-dimensional flows

Three cases will be examined in 3D microchannels with the size ratio of 2:2:1, 4:4:1, and 6:6:1, with the first two numbers representing the box dimensions in the two directions with periodic boundary conditions (x- and z-directions) and the third the direction normal to the walls (y-direction). The simulation parameters are given in Table 11.3.2. The physical flow conditions are the same as those used in the previous 2D cases, including the temperature ratio, the Froude, Knudsen, and the Rayleigh numbers. The numbers for the rectangular field cells are $20 \times 20 \times 20$, $40 \times 40 \times 20$, and $60 \times 60 \times 20$, respectively. The numbers of simulated molecules are 2 million, 8 million, and 18 million. Simulations were performed using 1, 4, and 9 processors for AR = 2, 4, and 6, respectively. In the following, the heat transfer and pressure on the walls will first be described. The structural features of the simulated flow will be examined.

Heat transfer and pressure. The evolutions of the heat transfer coefficients c_h on the hot wall are shown in Fig. 11.3.10 for all of the three 3D simulations. The three-stage developments of the wall heat transfer for the three 3D cases are very similar to those found in the 2D simulations. Initially, the wall heat transfer increases nearly linearly due to thermal conduction. At about $t = 28$, the second stage begins with a much larger rate of increase than that in the first stage. At the end of the second linear stage, the wall heat transfer coefficients c_h for the three flows start to branch off. For the 2:2:1 case, this third stage, which begins at about $t = 58$, gives an average heat transfer coefficient of -0.062 after an initial overshoot has subsided, which is about the same as those for the 2D cases when the flows have become stationary. The jolt found in the 2D, 4:1 case during the third stage did not appear in the present 3D, 4:4:1 case. Instead, the heat transfer decreases slightly before it reaches a steady average value of -0.059, about 5 percent lower than that for the 2D cases.

TABLE 11.3.2 Parameters in 3D Flow Simulations

	Case 4	Case 5	Case 6
AR	2:2:1	4:4:1	6:6:1
Grid size	$20 \times 20 \times 20$	$40 \times 40 \times 20$	$60 \times 60 \times 20$
Molecules	2 million	8 million	18 million
Processors	1	4	9
r	0.1	0.1	0.1
Fr	3.5	3.5	3.5
Kn	0.01	0.01	0.01
Ra	10,159	10,159	10,159

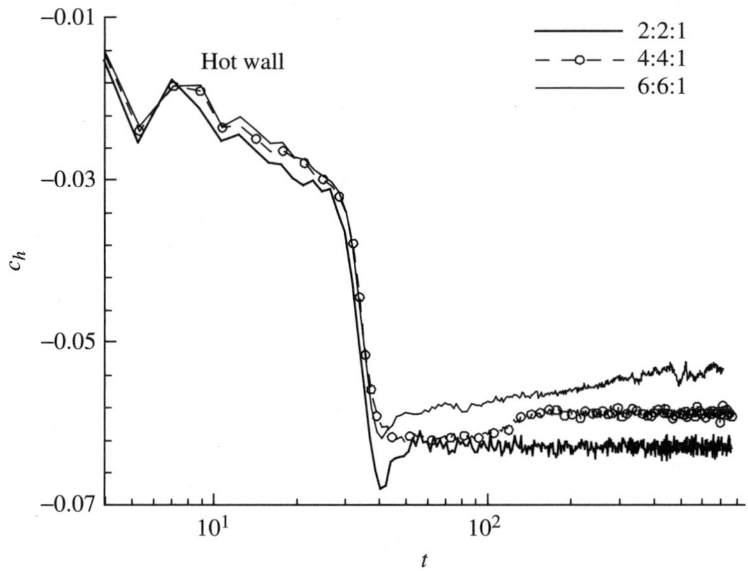

Figure 11.3.10 Evolution of heat transfer coefficient on the hot wall in 3D cases, with aspect ratio of 2:2:1, 4:4:1, and 6:6:1.

The evolutions of the Nusselt number on both the hot and cold walls are presented in Fig. 11.3.11. In light of the similarity between the development of the c_h in the first two stages shown in Fig. 11.3.10 between the 2D and the 3D cases, it is not too surprising to find that the Nusselt numbers for the three 3D cases also develop in parallel among themselves as did the 2D cases. In the third stage, the Nusselt numbers on the walls converge and develop steady averages of 3.34 and 3.14 for the 2:2:1 and the 4:4:1 cases, respectively. For the largest domain case, the Nusselt number appears to begin to level off toward the end of the simulations.

Figure 11.3.12 shows the pressure coefficient history on the upper cold wall. Despite the small variations for the 6:6:1 case, the average values in the third stage for all the three simulations stays constant at about 0.3. The difference in the wall pressure development among the three 3D cases is far less apparent compared to the wall heat transfer.

Flow structures in the third stage. The surface heat transfer and the surface pressure data shown above indicate that, during the first two stages of the simulations, the flow structures in the 3D simulations develop in a similar manner to those in the 2D simulations. As noted, inertia effects dominate in the first stage and the heat transfer is achieved mainly by heat conduction. In the second stage, buoyancy effects become more

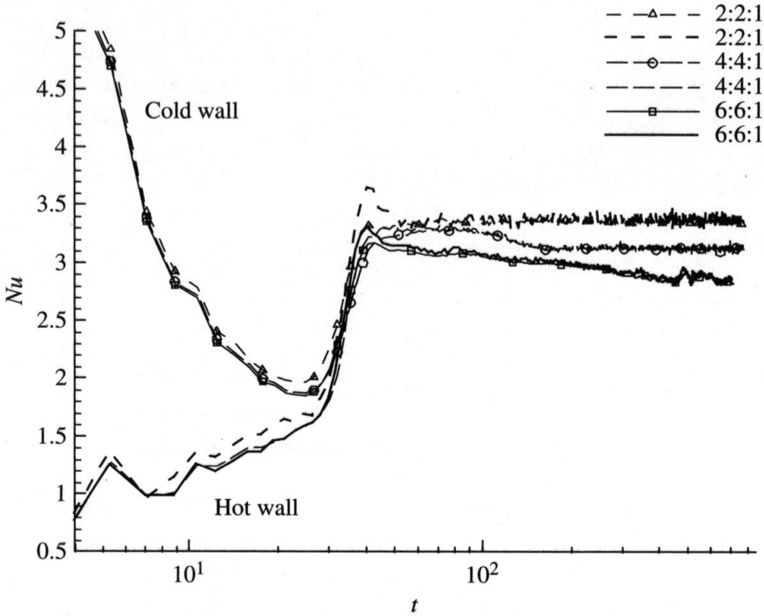

Figure 11.3.11 Evolution of Nusselt numbers on both hot and cold wall in 3D cases, with aspect ratio 2:2:1, 4:4:1, and 6:6:1.

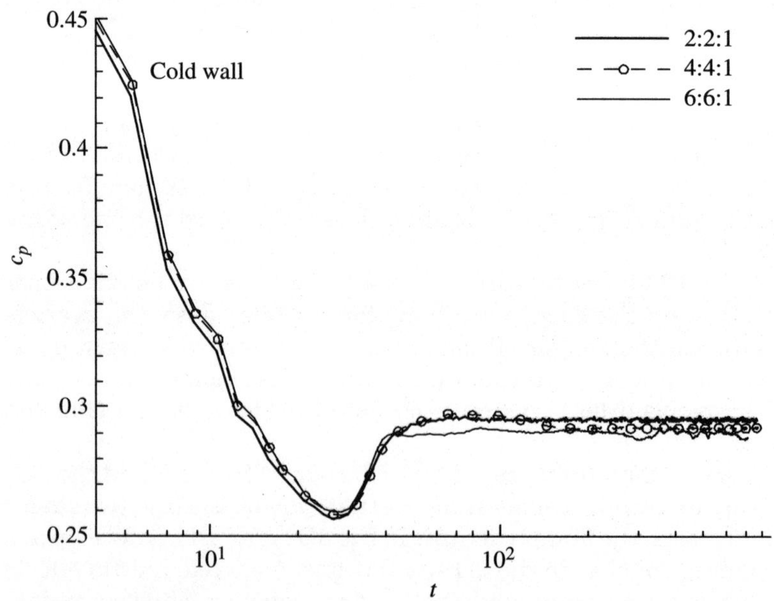

Figure 11.3.12 Evolution of pressure coefficient on the cold wall for the 3D cases.

significant and drive the formation of vortical flow structures. Initially more apparent in the region near the top wall, the vortical structures quickly grow in size and strength and eventually extend over the entire channel height. There are, however, significant differences between the third stages of their development, as indicated by the surface data. In the following, the flow evolution in the third stage for each of the three 3D cases are examined in details. As will be seen, the flow structures in the three 3D cases are fundamentally different from each other and, for the 4:4:1 and the 6:6:1 cases, from those in the corresponding 2D flows.

2:2:1. For this smallest domain case, the average surface heat transfer and pressure are constant, indicating that the flow has become stationary. Figure 11.3.13 shows the pathlines and vorticity contours of the steady flow at time $t = 762.16$. The flow field is sampled after time $t = 88.62$. In Fig. 11.3.13*a*, the pathlines are plotted on the planes at $z = 20$, 100, and 180 with contours representing the vorticity in the z-direction. Figure 11.3.13*b* shows the flow vorticity contours on three orthogonal planes at $x = 20$, $y = 50$ and $z = 20$. Figures 11.3.13*a*,*b* show a flow pattern of two parallel Rayleigh-Bénard convection rolls aligned with the z-direction. In fact, the flow is homogeneous in the z-direction.

Figure 11.3.14 presents the velocity profiles of u and v at $y = 20$ in the current 3D case and also the corresponding 2D simulation results shown in Fig. 11.3.9*a*.

Other than an apparent difference in phase in the periodic x-direction, the velocity component distributions are very similar. Therefore, for the 2:2:1 case, the cross-sectional flow in the 3D simulation produces essentially the same results as that in the corresponding 2D simulation. It should be noted that periodic boundary conditions are applied in the x- and the z-directions. The orientation of the stationary conduction rolls may be affected by many factors, such as the initial sampling and the subsequent nonlinear effects in the flows. In fact, as will be seen in the following case with AR $= 4$, the vortical structures can also align in the x-direction.

4:4:1. Figures 11.3.15 and 11.3.16 shows the pathlines and vertical velocity contours of the instantaneous flow field at time $t = 53.17$ and $t = 762.16$, respectively. $t = 53.17$ corresponds roughly to the beginning of the third stage and at $t = 762.16$ the wall heat transfer indicates that the flow is stationary. At $t = 53.17$, the in-plane pathlines show two pairs of vortex rolls of roughly equal size that are aligned with the x-direction and the velocity contours show that the rollers are slightly skewed. Figure 11.3.16 shows that at the later time of $t = 762.16$, two counter rotating vortex rolls can still be identified, but their widths vary significantly along their axes. It then becomes interesting to examine

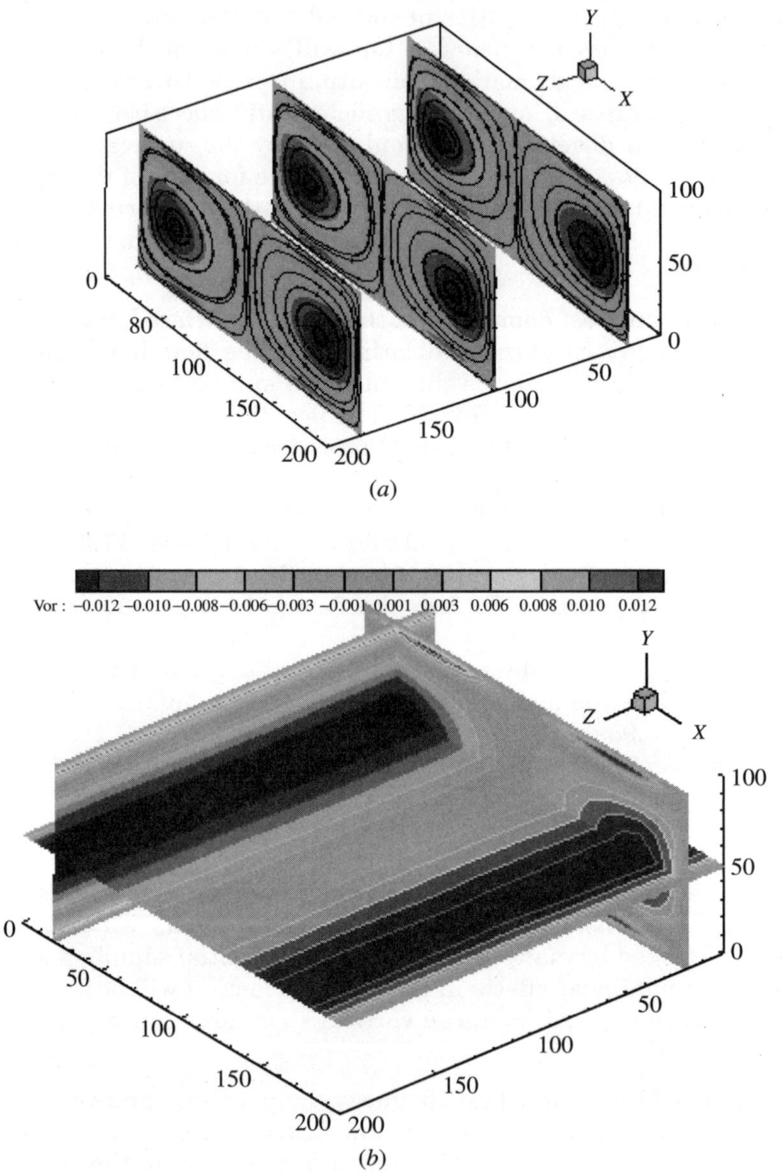

Figure 11.3.13 Path-lines and vorticity contours of the long-time-averaged flow field for the 2:2:1 case at $t = 762.16$: (a) On three z-planes at $z = 20$, 100 and 180; (b) On three orthogonal planes at $y = 50$, $x = 20$, and $z = 20$.

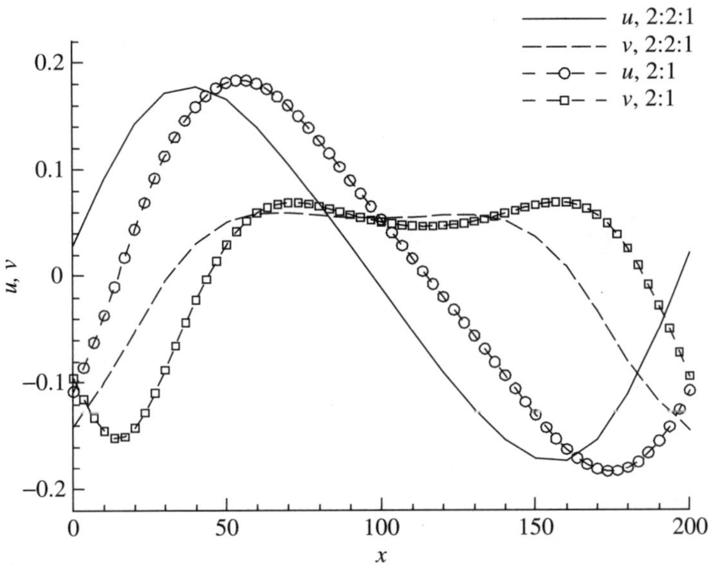

Figure 11.3.14 Velocity field comparisons between the 3D and the 2D simulations at $y = 20$ with the aspect ratio of 2:1.

how the vortical flow structures evolve in this stage of the flow development where the flow eventually becomes stationary, according to the surface data.

The evolution can be viewed with the series of snapshots presented in Figs. 11.3.17*a–f*. The figures show, for a time sequence that begins at $t = 53.17$ and ends at $t = 762.16$, the evolution of the instantaneous vorticity contours in the plane through $y = 50$. The times for Figs. 11.3.17*a–f* are 53.17, 88.62, 124.07, 159.52, 372.22, and 762.16, respectively. Therefore, the flows described by Figs. 11.3.17*a*,*b* have decreasing wall heat transfer and those by Figs. 11.3.17*c–f* have constant wall heat transfer. The interface boundaries between two counter rotating vortex rolls skew slightly in Figs. 11.3.17*a*,*b* and the vortex contours are wrinkled, especially compared to that in Fig. 11.3.13*b* for the 2:2:1 case. These features suggest that the vortex pairs experience some levels of perturbation that modulate and roughen their appearances during this period. The perturbation eventually results in the wavy formation of the vortex rolls at later times, as shown in Figs. 11.3.17*c–f*. The lateral widths of the vortex rolls vary in a sinusoidal manner along their axes. The interface boundaries between the "pinched" counter rotating vortex rolls, however, are not wavy, as they were earlier (Figs. 11.3.17*a*,*b*), but are quite straight. Moreover, a closer examination of the flow development in this period shows that the pinched vortex rolls are not

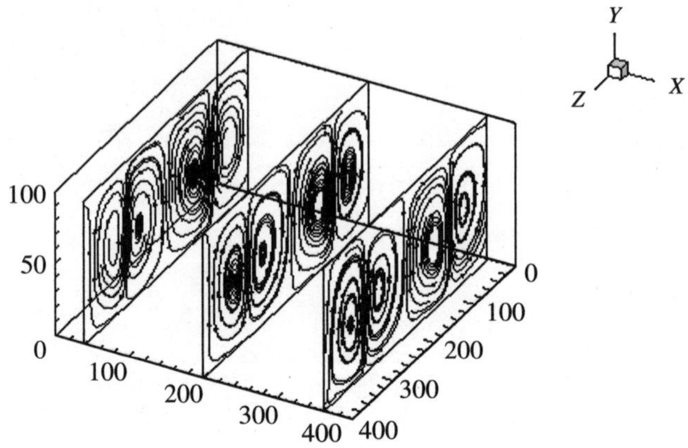

(a)

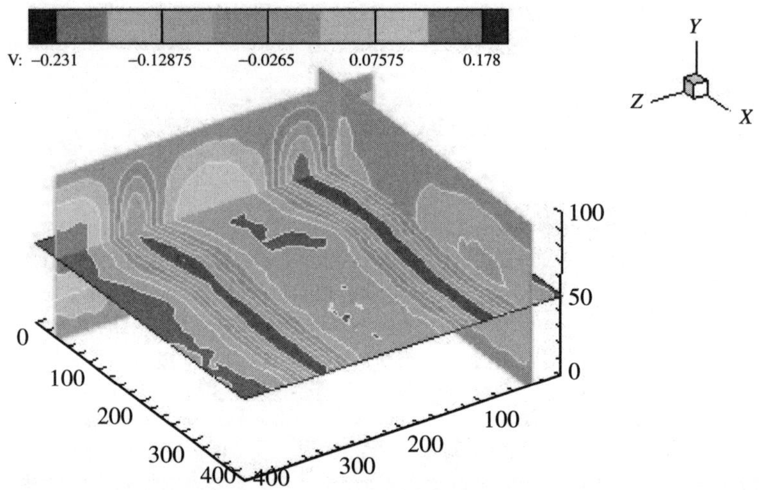

(b)

Figure 11.3.15 Path-lines and vertical velocity contours of the instantaneous flow field for the 4:4:1 case at $t = 53.17$: (a) Path-lines on three x-planes at $x = 40$, 200, and 360; (b) Vertical velocity field on three orthogonal planes at $y = 50$, $x = 40$, and $z = 40$.

steady. The vortex rolls are found to move in the positive z-direction as the simulation progresses.

Figure 11.3.18 shows the distribution of the spanwise velocity component in the spanwise direction on the walls and at $h/4$ intervals between the walls at $t = 53.17$ and 762.16, respectively. One can define an interface between two vortex rolls as the vertical plane where the

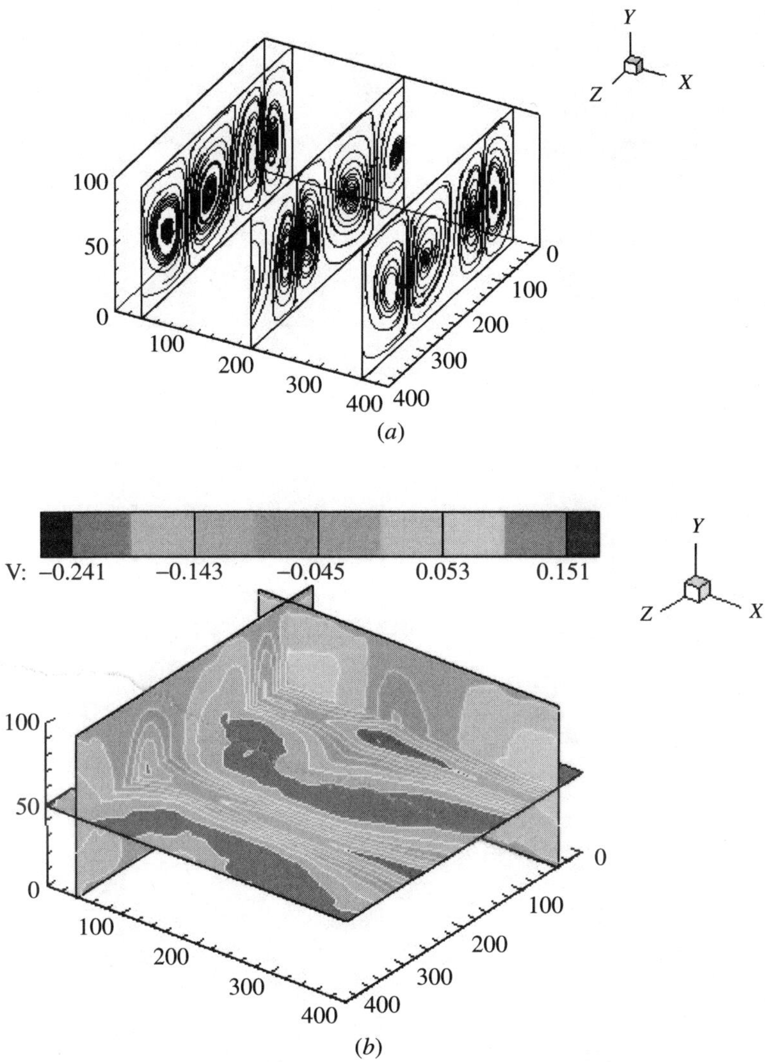

Figure 11.3.16 Path-lines and vertical velocity contours of the instantaneous 4:4:1 flow field case at $t = 762.16$: (*a*) Path-lines on three x-planes at $x = 40$, 200, and 360; (*b*) Vertical velocity field on three orthogonal planes at $y = 50$, $x = 40$, and $z = 40$.

spanwise velocity component is zero ($w = 0$). Figure 11.3.18*a* shows that the z-coordinates of the interface (denoted by P) at the different heights across the channel are the same, indicating that the interface P is in an upright position, stretching between and perpendicular to the walls, at $t = 53.17$. In fact, this interface P remains perpendicular to

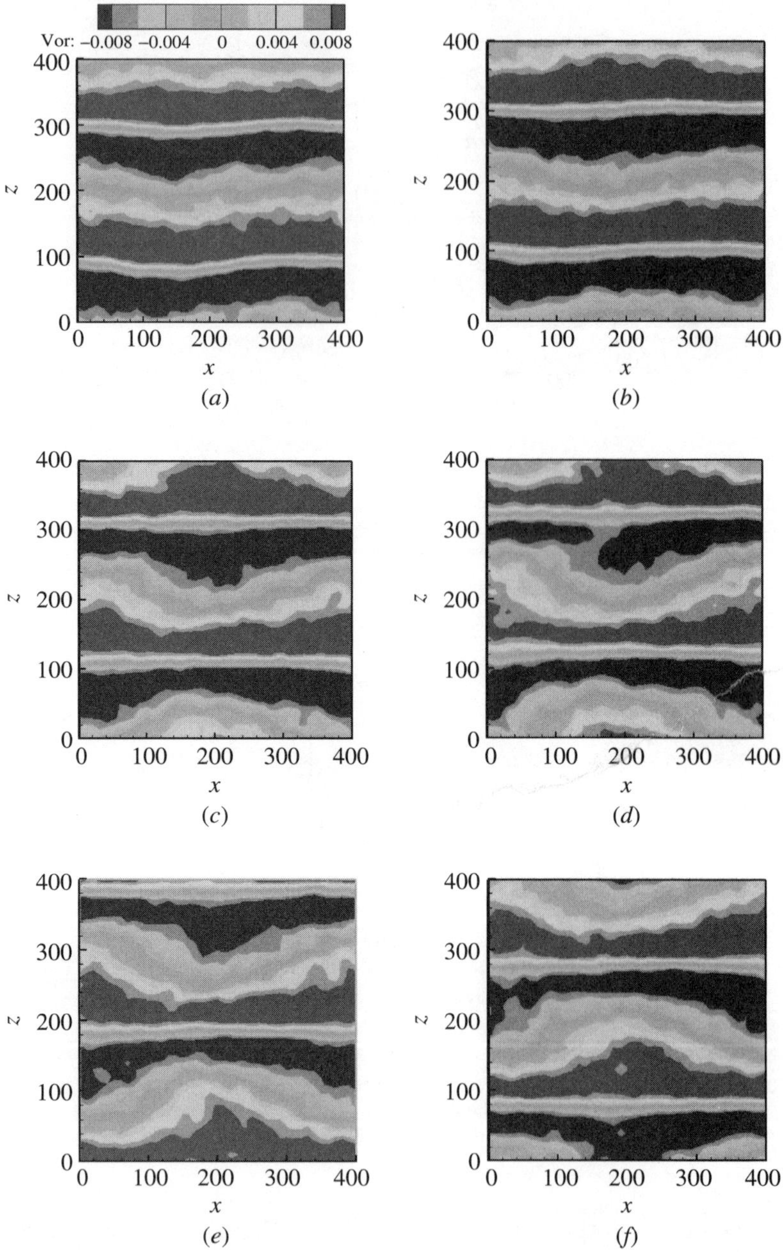

Figure 11.3.17 Evolutions of vorticity contours for the instantaneous 4:4:1 flow field case on the mid-plane of $y = 50$: (*a*) $t = 53.17$; (*b*) $t = 88.62$; (*c*) $t = 124.07$; (*d*) $t = 159.52$; (*e*) $t = 372.22$; (*f*) $t = 762.16$.

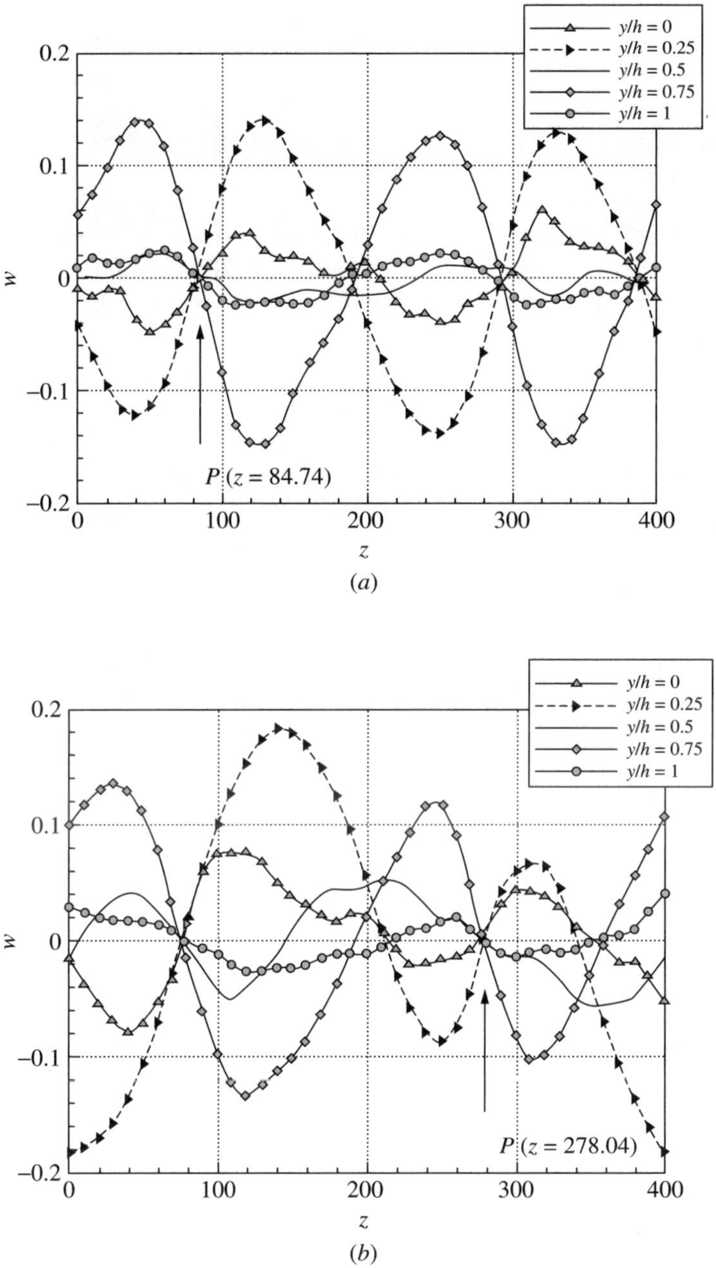

Figure 11.3.18 w velocity distributions along the z-direction for 4:4:1 at $x = 200$, $y/h = 0, 0.25, 0.5, 0.75$, and 1: (a) $t = 53.17$; (b) $t = 762.17$.

the walls at the later time of $t = 762.16$ (Fig. 11.3.18*b*). Moreover, the interface has obviously moved from $z = 84.74$ to $z = 278.04$ during this time period.

To track the translational movement of the flow, the locations of the interface P at different times are recorded and shown in Fig. 11.3.19. A linear curve fit shows a constant speed of 0.267. Therefore, the vertical flow structures in the present 3D, 4:4:1 case move in the z-direction at a constant speed of 0.267. Such an advective movement of the vortex rolls in the three-dimensional, unsteady MRBC problem, to the best of our knowledge, has not been previously reported in the open literature.

6:6:1. Figures 11.3.20 and 11.3.21 show the pathlines and vertical velocity contours of the instantaneous flow field at time $t = 53.17$ and $t = 762.16$, respectively, which are the same times as those for Figs. 11.3.15 and 11.3.16 for the 4:4:1 case. The flow patterns are more complex for the large domain case. At $t = 53.17$, the pathlines plot shows that there are six roller structures at $x = 60$ and 300. At $x = 540$, there are seven rollers that extend between the walls and two of smaller size near the walls at about $z = 200$. The vertical velocity contours of Fig. 11.3.20*b* on the plane midway between the walls shows the emergence of a new vortex roll in between the $x = 300$ and the $x = 540$ plane, as is evident by the "Y" branching in the positive vertical velocity contours. At $t = 762.16$, Fig. 11.3.21*a* indicates that there are four vortex rolls in the flow. As can be seen from Fig. 11.3.21*a*, the patterns of the vortex rolls are complex. It is apparent that the flow structures in the

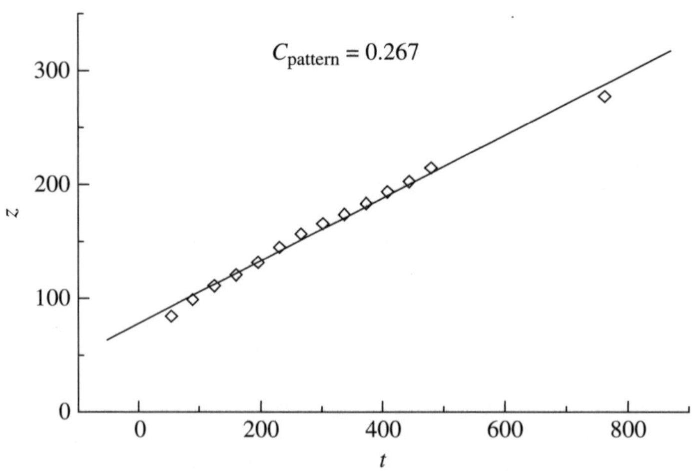

Figure 11.3.19 Locations of the interface P for the 4:4:1 case.

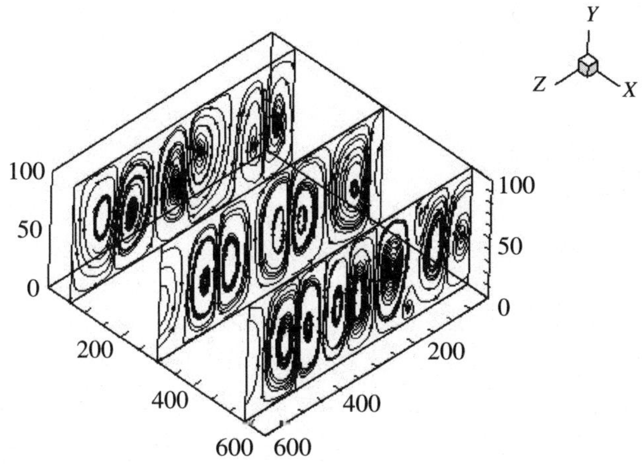

(*a*)

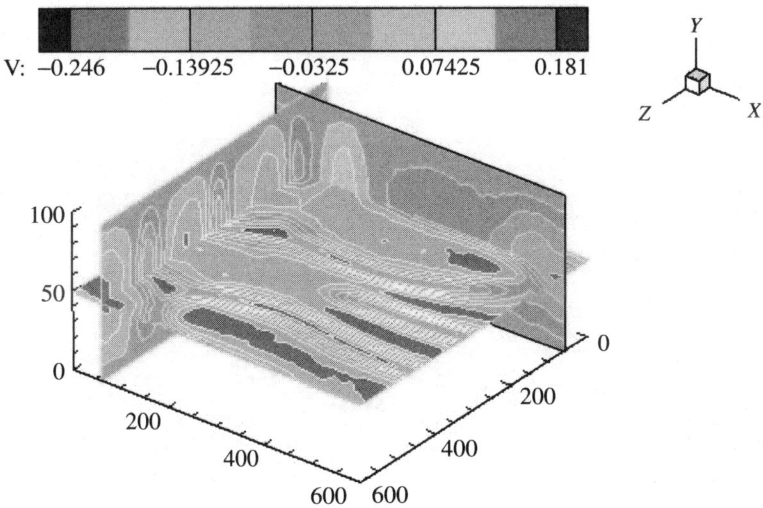

(*b*)

Figure 11.3.20 Path-lines and vertical velocity contours of the instantaneous flow field for 6:6:1 at $t = 53.17$: (*a*) Path-lines on three z-planes at $x = 60$, 300, and 540; (*b*) Vertical velocity field on three orthogonal planes at $y = 50$, $x = 60$, and $z = 60$.

current 6:6:1 case change significantly in this stage, with continuous generation, merging, and dissipation of vortex rolls.

Figures 11.3.22*a–j* show, in a sequence of time that begins at $t = 53.17$ and ends at $t = 762.16$, the evolution of the instantaneous vorticity contours in the plane of $y = 50$. The times for Figs. 11.3.22*a–j* are

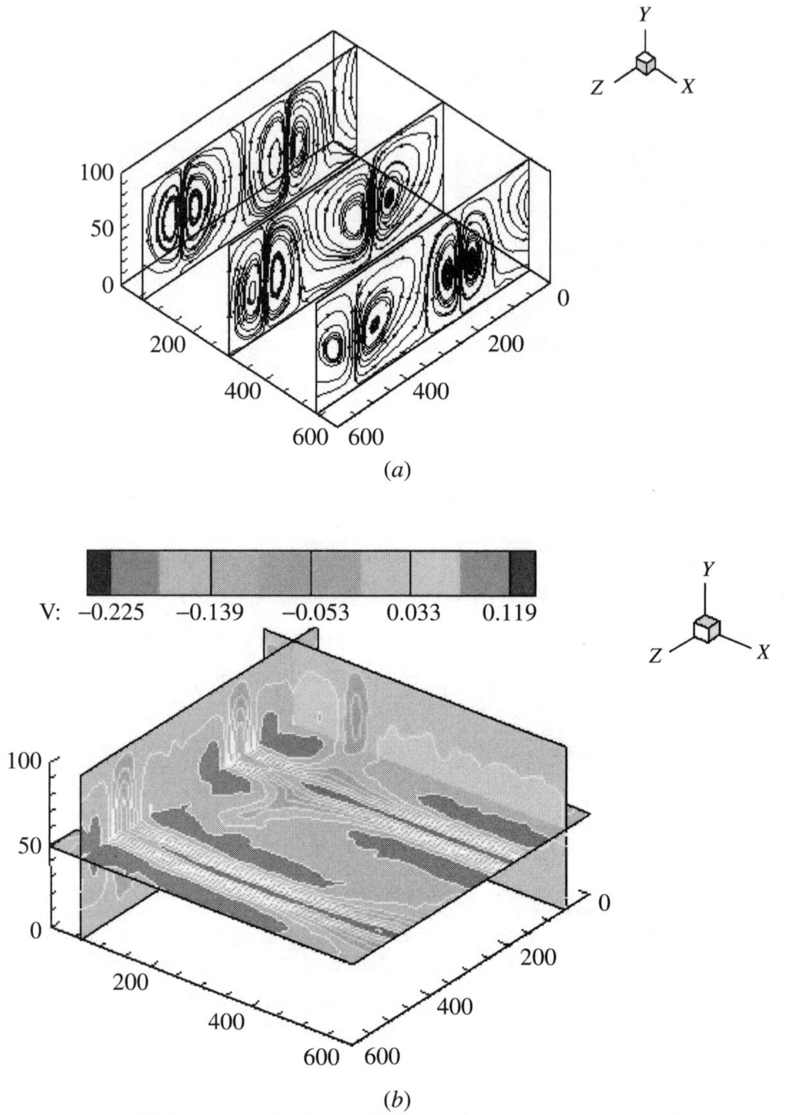

Figure 11.3.21 Path-lines and vertical velocity contours of the instantaneous flow field for 6:6:1 at $t = 762.16$: (a) Path-lines on the three x-planes at $x = 60$, 300, and 540; (b) vertical velocity field on the three orthogonal planes at $y = 50$, $x = 60$, and $z = 60$.

53.17, 88.62, 124.07, 159.52, 194.97, 301.32, 336.77, 443.11, 478.56, and 762.16, respectively. As was described earlier, the vortex rolls continues to split and merge initially (Figs. 11.3.22a,b). At later times, no more splitting can be found and the merging of existing vortex rolls continues

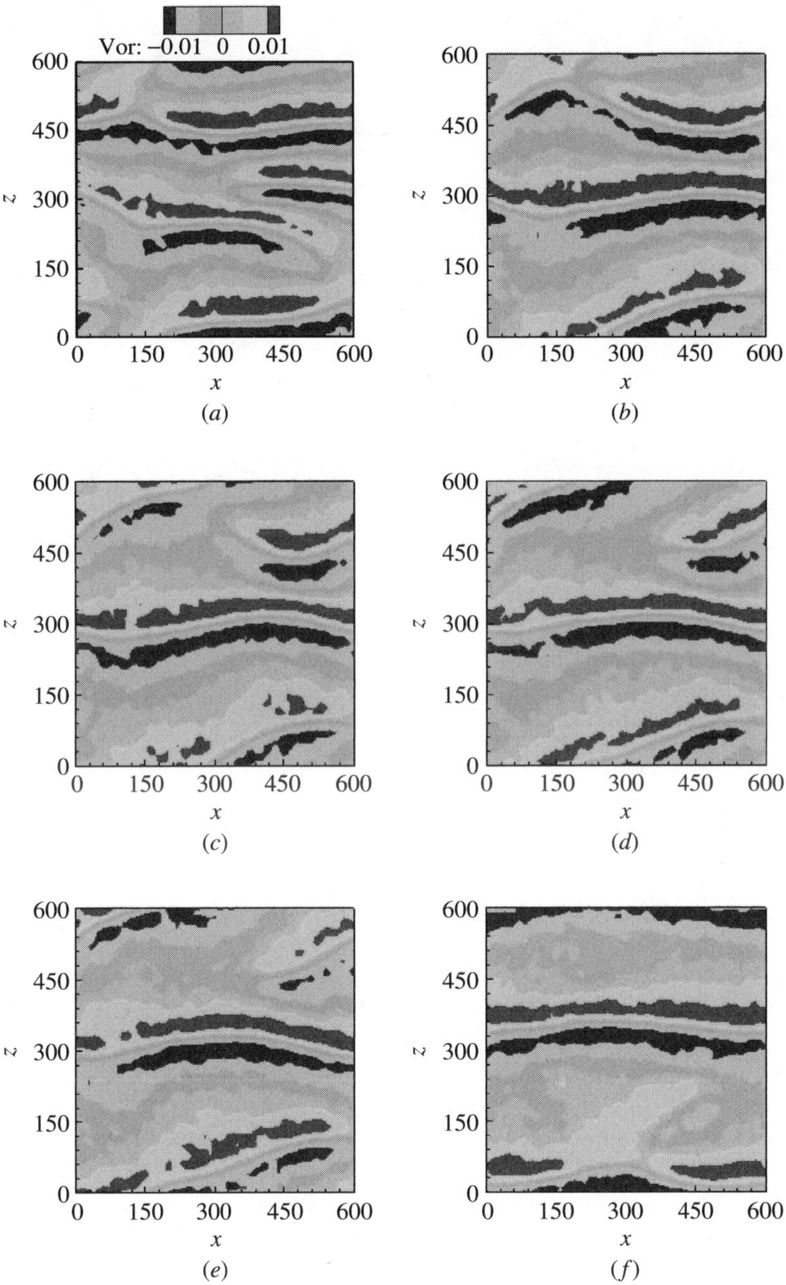

Figure 11.3.22 Evolutions of vorticity contours for the instantaneous flow field for 6:6:1 on the mid-plane of $y = 50$: (*a*) $t = 53.17$; (*b*) $t = 88.62$; (*c*) $t = 124.07$; (*d*) $t = 159.52$; (*e*) $t = 194.97$; (*f*) $t = 301.32$; (*g*) $t = 336.77$; (*h*) $t = 443.11$; (*i*) $t = 478.56$; (*j*) $t = 762.16$.

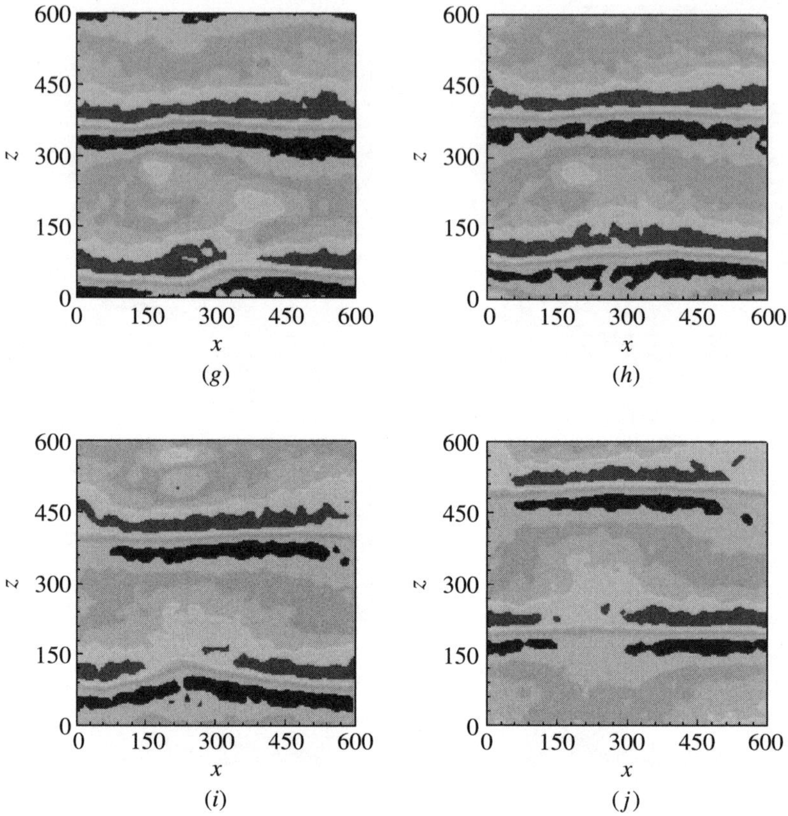

Figure 11.3.22 *(Continued)*

to about $t = 336.77$ (Fig. 11.3.22*g*) when there are four distorted vortex rolls in the flow. The four vortex rolls continues to evolve and at $t = 762.16$, the interface between two nearby counter rotating vortex cores become rather straight. Figure 11.3.23 shows the variation of the location of the lower interface with time after $t = 336.77$. The figure shows that the flow pattern moves at a speed of 0.332 in the positive z-direction.

Two and three-dimensional MRBC flows have been simulated numerically by using the DSMC method in an unsteady time-dependent manner with different aspect ratios for a high Rayleigh number of 10,159 with the Knudsen number of 0.01. The flow fields were initialized homogeneously based on the temperature of the bottom, hot wall. Evolutions of the convection flow patterns were studied, and their corresponding pressure and heat transfer on the wall were calculated in terms of the averaged pressure coefficient, the heat transfer rate c_h and the Nusselt number.

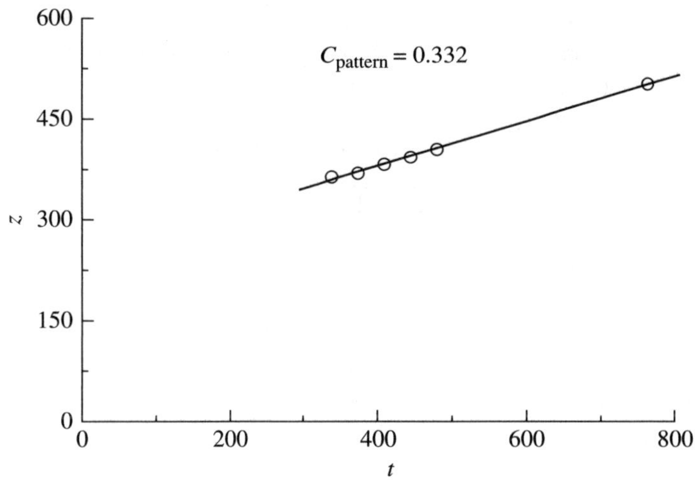

Figure 11.3.23 Locations of the moving pattern along the z-direction for 6:6:1.

In the 2D cases simulated, the flows are fully developed to a stationary pattern of vortex cells. There are one, two, and four pairs of vortices in the case with aspect ratio of 2:1, 4:1, and 6:1, respectively. The aspect ratio does not have significant effects on the wall pressure and the wall heat transfer. But it affects the final flow patterns and the process of their formation.

For the 3D cases with aspect ratio of 2:2:1, the cross-sectional view of the stationary flow is essentially the same as that in the 2D case with aspect ratio of 2:1. The 3D cases with aspect ratios of 4:4:1 and 6:6:1, however, produce unsteady, periodic flows, while the corresponding 2D cases (4:1 and 6:1) are steady.

The unsteady DSMC method successfully captures three stages of the pattern formation of the 2D and the 3D microRayleigh-Bénard problems. The inertia effect dominates the first stage with conduction being the main heat transfer mechanism. Cellular vortex rolls appear and continue to develop in the second stage where convective heat transfer becomes important. In the first two stages, evolutions of the flow structures, and their corresponding wall heat transfer and wall pressure for the 2D and 3D cases appear very similar. In the third stage, the flow structures in the three 3D cases are fundamentally different from each other and, for the 4:4:1 and the 6:6:1 cases, from those in the corresponding 2D flows.

For the 4:4:1 case, the flow develops into a pattern of two pairs of pinched vortex rolls that translate parallel to the plates at nearly a constant speed. The lateral widths of the vortex rolls vary in a sinusoidal manner along their axes. The interface boundaries between the "pinched" counter rotating vortex rolls, however, are quite straight. In the 6:6:1 case, similar flow patterns in translational movement can also be easily identified. The advective nature of the vortex rolls in the 3D, unsteady MRBC flows simulated is a characteristic of the final states of the MRBC problem that has not been reported at macroscales.

References

Ahlers, G., and Behringer, R.P., Evolution of turbulence from the Rayleigh-Bénard instability, *Phys., Rev. Letters*, Vol. 40, pp. 712–716, 1978.

Behringer, R.P., and Ahlers, G., Heat transport and temporal evolution of fluid flow near the Rayleigh-Bénard instability in cylindrical containers, *J. Fluid Mech.*, Vol. 125, pp. 219–258, 1982.

Bird, G.A., Direct simulation of high-vorticity gas flows, *Phys. Fluids*, Vol. 30, pp. 364–366, 1987.

Bird, G.A., Private communication, 2001.

Bird, G.A., DSMC study of three-dimensional forced chaotic flows, *Proceedings of the 23rd International Symposium on Rarefied Gas Dynamics*, Whistler, Canada, July 2002.

Bodenschatz, E., Pesch, W., and Ahlers, G., Recent developments in Rayleigh-Bénard convection, *Annu. Rev. Fluid Mech.*, Vol. 32, pp. 709–778, 2000.

Chapman, S., and Cowling, T.G., *The Mathematical Theory of Non-Uniform Gases*, Cambridge University Press, 3rd ed., 1970.

Fang, Y.C., and Liou, W.W., Computations of the flow and heat transfer in microdevices using DSMC with implicit boundary conditions, *J. Heat Transfer*, Vol. 124, pp. 338–345, 2002.

Fang, Y.C., Liou, W.W., and Bird, G.A., Three-dimensional simulations of Rayleigh-Benard convection by DSMC, 37th AIAA Thermalphysics Conference, Portland, OR, June 28–July 1, AIAA 2004-2671, 2004.

Godrèche, C., and Manneville, P., *Hydrodynamics and Nonlinear Instabilities*, Cambridge University Press, 1998c.

He, X., Chen, S., and Doolen, G.D., A novel thermal model for the lattice Boltzmann method in incompressible limit, *J. Computational Phys.*, Vol. 146, pp. 282–300, 1998.

Hixon, R., Shih, S.-H., Mankbadi, R.R., and Scott, J.R., Time domain solution of the airfoil gust problem using a high-order compact scheme, AIAA Paper 98-3241, 1998.

Kessler, R., Nonlinear transition in three-dimensional convection, *J. Fluid Mech.*, Vol. 174, pp. 357–379, 1987.

Landau, L.D., and Lifshitz, E.M., *Statistical Physics*, Pergamon Press, Reading, MA, 1958.

Lele, S.K., Compact finite difference schemes with spectral-like resolution, *J. Computational Phys.*, Vol. 103, pp. 16–42, 1992.

Liou, W.W., and Fang, Y.C., Forced Couette flow simulations using DSMC, *Phys. Fluids*, Vol. 16, pp. 4211–4220, 2004.

Liou, W.W., Fang, Y.C., and Bird, G.A., Three-dimensional vortical simulations using direct simulation Monte Carlo Method, AIAA paper 2003-1288, 2003*a*.

Liou, W.W., Fang, Y.C., and Bird, G.A., DSMC simulations of forced chaotic flows, AIAA paper 2003-3595, 2003*b*.

Liou, W.W., Liu, F., Fang, Y.C., and Bird, G.A., Navier-Stokes and DSMC simulations of forced chaotic microflows, AIAA paper 2003-3583, 2003*c*.

Malek Mansour, M., A simple model of hydrodynamic instability, *Microscopic Simulations of Complex Flows*, Ed. Michel Mareschal, Plenum Press, New York and London, pp. 189–198, 1990.

Nanbu, K., Watanabe, Y., and Igarashi, S., Conservation of angular momentum in the direct simulation Monte Carlo method, *J. Phys. Soc. of Japan*, Vol. 57, pp. 2877–2880, 1988.

Ozal, H., and Hara, T., Numerical analysis for oscillatory natural convection of low Prandtl number fluid heated from blow, *Num. Heat Trans., Part A*, Vol. 27, pp. 307–318, 1995.
Rayleigh, L., On convective currents in a horizontal layer of fluid when the higher temperature is on the under side, *Philos. Mag.*, Vol. 32, pp. 529–546, 1916.
Riechelmann, D., and Nanbu, K., Monte Carlo direct simulation of the Taylor instability in rarefied gas, *Phys. Fluids A*, Vol. 5, pp. 2585–2587, 1993.
Shan, X., Simulation of Rayleigh-Bénard convection using a lattice Boltzmann method, *Phys. Rev. E*, Vol. 55, pp. 2780–2788, 1997.
Sone, Y., Aoki, K., and Sugimoto, H., The Bénard problem for a rarefied gas: Formation of steady flow patterns and stability of array of rolls, *Phys. Fluids*, Vol. 9, pp. 3898–3914, 1997.
Stefanov, S., and Cercignani, C., Monte Carlo simulation of the Taylor-Couette flow of a rarefied gas, *J. Fluid Mech.*, Vol. 256, pp. 199–211, 1993.
Stefanov, S., Roussinov, V., and Cercignani, C., Rayleigh-Benard flow of a rarefied gas and its attractors. I. Convection regime, *Phys. Fluids*, Vol. 14, pp. 2255–2269, 2002*a*.
Stefanov, S., Roussinov, V., and Cercignani, C., Rayleigh-Benard flow of a rarefied gas and its attractors. II. Chaotic and periodic convective regimes, *Phys. Fluids*, Vol. 14, pp. 2270–2288, 2002*b*.
Sugimoto, H., Sone, Y., Aoki, K., and Motohashi, H., The Bénard problem of rarefied gas dynamics, *Proceedings of the 19th Rarefied Gas Dynamics Symposium*, Oxford University Press, 1995.
Tolstykh, A., and Lipavskii, M.V., On performance of methods with third- and fifth-order compact upwind differencing, *J. Computational Phys.*, Vol. 140, pp. 205–232, 1998.
Watanabe, T., Flow pattern and heat transfer rate in Rayleigh-Bénard convection, *Phys. Fluids*, Vol. 16, pp. 972–978, 2004.
Williamson, J.H., Low-storage Runge-Kutta schemes, *J. Computational Phys.*, Vol. 35, pp. 48–56, 1980.
Wu, M., Ahlers, G., and Cannell, D.S. Thermally induced fluctuations below the onset of Rayleigh-Bénard convection, *Phys. Rev. Lett.*, Vol. 75, pp. 1743–1746, 1995.

Index

ABOUT THE AUTHORS

WILLIAM W. LIOU has worked as a research associate at the Institute for Computational Mechanics in Propulsion at NASA Glenn Research Center for a six year period. Dr. Liou's research activities at NASA involved studying the laminar and turbulent flow physics. Since 1997, he has been working as an assistant, associate, and full professor at Western Michigan University. Dr. Liou's research activities in the past several years have been in the analytical and computational studies of fluid flows using continuum as well as molecular approaches. Dr. Liou received his B.S. degree from National Cheng-Kung University of Taiwan, and M.S. and Ph.D. from Penn State University in 1990.

YICHUAN FANG received his M.S. degree from the University of Science and Technology of China in 1998 and Ph.D. degree in mechanical engineering from Western Michigan University in 2003. Dr. Fang worked at Western Michigan University in 2003 to 2004. Near the completion of this manuscript, Dr. Fang joined Aerospace Engineering at Georgia Institute of Technology as a research fellow. Dr. Fang began his research work in 1998 in microfluid dynamics. His work includes theoretical and computational modeling by using both molecular and continuum methods.